VOLUME THREE

ADVANCES IN STEM CELLS AND THEIR NICHES

Epidermal Stem Cell Niche

VOLUME THREE

ADVANCES IN STEM CELLS AND THEIR NICHES

Epidermal Stem Cell Niche

Edited by

MIRNA PEREZ-MORENO
Department of Biology,
University of Copenhagen,
Copenhagen, Denmark

ELSEVIER

ACADEMIC PRESS
An imprint of Elsevier

Academic Press is an imprint of Elsevier
50 Hampshire Street, 5th Floor, Cambridge, MA 02139, United States
525 B Street, Suite 1650, San Diego, CA 92101, United States
The Boulevard, Langford Lane, Kidlington, Oxford OX5 1GB, United Kingdom
125 London Wall, London EC2Y 5AS, United Kingdom

First edition 2019

ISBN: 978-0-12-818446-2
ISSN: 2468-5097

For information on all Academic Press publications
visit our website at https://www.elsevier.com/books-and-journals

Publisher: Zoe Kruze
Acquisition Editor: Ashlie M. Jackman
Editorial Project Manager: Joanna Collett
Production Project Manager: James Selvam
Cover Designer: Miles Hitchen

Typeset by SPi Global, India

Contents

Contributors

Gopakumar Changarathil
Life Science Center for Survival Dynamics, Tsukuba Advanced Research Alliance (TARA); Graduate School of Comprehensive Human Sciences, University of Tsukuba, Tsukuba, Japan

Carlos Clavel
Hair & Pigmentation Development, Skin Research Institute of Singapore (SRIS); Lee Kong Chian School of Medicine, Nanyang Technological University, Singapore, Singapore

John T. Connelly
Centre for Cell Biology and Cutaneous Research, Barts and the London School of Medicine and Dentistry, Queen Mary University of London, London, United Kingdom

Agnes Forsthuber
Skin & Endothelium Research Division, Department of Dermatology, Medical University of Vienna, Vienna, Austria

Sophie Frech
Skin & Endothelium Research Division, Department of Dermatology, Medical University of Vienna, Vienna, Austria

Ana Korosec
Skin & Endothelium Research Division, Department of Dermatology, Medical University of Vienna, Vienna, Austria

Paola Kuri
Department of Dermatology, Institute for Regenerative Medicine, University of Pennsylvania Perelman School of Medicine, Philadelphia, PA, United States

Lu Q. Le
Department of Dermatology; Integrative Molecular and Biomedical Sciences Program Hamon Center for Regenerative Science and Medicine; Simmons Comprehensive Cancer Center, University of Texas Southwestern Medical Center, Dallas, TX, United States

Chung-Ping Liao
Department of Dermatology, University of Texas Southwestern Medical Center, Dallas, TX, United States

Beate M. Lichtenberger
Skin & Endothelium Research Division, Department of Dermatology, Medical University of Vienna, Vienna, Austria

Jamien Lim
Hair & Pigmentation Development, Skin Research Institute of Singapore (SRIS); Lee Kong Chian School of Medicine, Nanyang Technological University, Singapore, Singapore

Krystle Joy Ng
Hair & Pigmentation Development, Skin Research Institute of Singapore (SRIS); Lee Kong Chian School of Medicine, Nanyang Technological University, Singapore, Singapore

Yen Xuan Ngo
Life Science Center for Survival Dynamics, Tsukuba Advanced Research Alliance (TARA); Ph.D. Program in Human Biology, School of Integrative and Global Majors, University of Tsukuba, Tsukuba, Japan

Lalhaba Oinam
Life Science Center for Survival Dynamics, Tsukuba Advanced Research Alliance (TARA); Ph.D. Program in Human Biology, School of Integrative and Global Majors, University of Tsukuba, Tsukuba, Japan

Gabriella Rice
Department of Dermatology, Institute for Regenerative Medicine, University of Pennsylvania Perelman School of Medicine, Philadelphia, PA, United States

Panteleimon Rompolas
Department of Dermatology, Institute for Regenerative Medicine, University of Pennsylvania Perelman School of Medicine, Philadelphia, PA, United States

Aiko Sada
Life Science Center for Survival Dynamics, Tsukuba Advanced Research Alliance (TARA), University of Tsukuba, Tsukuba, Japan

Edem Tchegnon
Department of Dermatology; Integrative Molecular and Biomedical Sciences Program, University of Texas Southwestern Medical Center, Dallas, TX, United States

Etienne C.E. Wang
National Skin Centre; Skin Research Institute of Singapore, Singapore

Hiromi Yanagisawa
Life Science Center for Survival Dynamics, Tsukuba Advanced Research Alliance (TARA); Faculty of Medicine, University of Tsukuba, Tsukuba, Japan

Preface

In the past years, new knowledge has provided a remarkable insight about the contributions of the milieu in which epidermal stem cells reside to the regenerative powers of the skin. The novel emerging interactions of epidermal stem cells with their neighboring cells, as well as with cells in the microenvironment including dermal cells, melanocytes, vasculature, and immune cells, are adding structural and functional complexities to the already multifaceted stem cells located in the epidermis and its derived hair follicle appendages.

In this third volume of the Book Series: *Advances in Stem Cells and their Niches*, experts in the skin research field cover most of the recent conceptual and experimental findings in the epidermal stem cell niche arena. All the chapters in this book include comprehensive information and illustrations, providing the necessary framework to introduce the readers to the fascinating world of the epidermal stem niche. These concepts expand further covering the latest developments in the field, involving state-of-the-art methodologies, new ideas, and clinical implications. Besides, all the chapters raised pressing questions for further investigation within the field.

This book will be of interest to research scientists and clinicians, providing them with an updated view of these research fields, and enlarging concepts that may be complementary to their areas. It will also stimulate students to learn and engage on the exciting epidermal stem cell niche field, and it will appeal to a wide circle of readers that are interested in increasing their perception of the intricacies of the epidermal stem cell niche in skin regeneration and its implications in disease.

I thank Dominique Bonnet, Serial Editor of the Book Series: *Advances in Stem Cells and their Niches*, for the invitation to edit this volume. I would also like to thank Fiona Pattison, Publisher; a special thank you goes to Joanna Collet, Editorial Project Manager, and the rest of the production group for the efficient and smooth production of the volume.

Finally, I am grateful to the authors of the chapters for their invaluable contributions to this volume, and for moving the epidermal stem cell niche field forward.

Mirna Perez-Moreno
Department of Biology
University of Copenhagen
Copenhagen, Denmark

CHAPTER ONE

Biophysical regulation of epidermal fate and function

John T. Connelly[1]
Centre for Cell Biology and Cutaneous Research, Barts and the London School of Medicine and Dentistry, Queen Mary University of London, London, United Kingdom
[1]Corresponding author: e-mail address: j.connelly@qmul.ac.uk

Contents

1. Introduction

Mechanical and biophysical cues are powerful regulators of cell and tissue functions in organs throughout the body. This concept of "mechanobiology" or the biological response to mechanics has existed since as early as the 19th century when Julius Wolff proposed that bone adapts to altered mechanical loading (Wolff, 1986), and it has since become well established that mechanical signals control a diverse range of biological processes, such tissue development, homeostasis, and repair. Over the last 20 years there has been an even greater interest in understanding the precise molecular mechanisms by which cells sense and respond to their mechanical environment (Iskratsch, Wolfenson, & Sheetz, 2014). Through the use of sophisticated bioengineering tools and technologies, alongside advances in

Advances in Stem Cells and their Niches, Volume 3
ISSN 2468-5097
https://doi.org/10.1016/bs.asn.2019.04.001

imaging and molecular biology methods, many of the key signaling molecules, structural proteins, and downstream transcription programs involved in mechanotransduction have now been identified. Still, there are major unanswered questions in the field of mechanobiology, including the integration of mechanotransduction pathways with other signaling networks, identification of cell and tissue specific mechano-sensing mechanisms, and the translation of this knowledge into improved diagnosis and treatment of degenerative diseases.

The epidermis of the skin is one such tissue where biomechanical signals play an essential role in the maintenance of normal homeostasis and repair. As a physical barrier from the external environment, the epidermis is continually subjected to a wide range of mechanical forces and capable of adapting to changes in its mechanical environment. At the cellular level, epidermal keratinocytes sense mechanical forces and transduce these into downstream signaling pathways through a complex network of adhesion receptors, cytoskeletal proteins, and enzymes. These signaling pathways in turn control fundamental cellular functions and fate decisions and ultimately, how the tissue responds to mechanical stress. This chapter will outline the mechanotransduction signaling pathways that coordinate epidermal homeostasis and discuss how they may contribute to the pathogenesis of specific skin diseases. In addition, it will highlight key questions and on-going areas of investigation within the field.

2. Mechano-sensing across multiple length scales

The epidermis is a stratified squamous epithelium that covers the entire surface of the body and provides a protective barrier from chemicals, pathogens, and ultraviolet irradiation. The tissue undergoes continuous turnover through a process of self-renewal of progenitors in the basal layer and a tightly controlled program of terminal differentiation and shedding in the upper layers (Blanpain & Fuchs, 2009). These proliferative and differentiation processes rely on an intricate network of adhesion receptors and cytoskeletal proteins, where the specific expression pattern and organization varies through the different layers of the tissue. In the basal layer, integrin receptors anchor the keratinocytes to the extracellular matrix (ECM) of the basement membrane, while upregulation of cadherins and tight junctions promotes strong cell-cell adhesion and barrier formation within the suprabasal and granular layers. Integrins and cadherins also integrate with different cytoskeletal components, which support the mechanical and structural properties of the cell. Thus, the distinct patterns of receptor expression and cytoskeletal organization at different stages of differentiation underly the varied mechano-sensing functions of the epidermis (Fig. 1A). Here, we will examine the different mechanisms by which keratinocytes sense mechanical forces transmitted through cell-ECM and cell-cell adhesions and the downstream impact on cell behavior.

2.1 Cell-ECM adhesions

Cell adhesion to the extracellular matrix (ECM) anchors cells within tissues and provides a mechanical linkage between the surrounding environment and the cell's own cytoskeleton. These points of contact are key mediators of mechano-sensing and signal transduction. The integrin family of transmembrane receptors are the primary cell-ECM adhesion receptors, and they form α/β heterodimers, with large extracellular domains and short cytoplasmic tails. There are 18 types of α subunits and 8 β subunits, which together can form 24 different receptor pairs for binding ECM proteins with sequence specificity (see cited reviews for in depth analysis of integrin structure and function) (Giancotti & Ruoslahti, 1999; Hynes, 2002). Upon ligand binding, integrins cluster together and recruit scaffolding proteins, such as vinculin and talin, to form large macromolecular complexes known as focal adhesions (Giancotti & Ruoslahti, 1999; Hynes, 2002). These adhesion complexes link to the F-actin cytoskeleton and recruit signaling molecules including focal adhesion kinase (FAK) (Tomakidi, Schulz, Proksch, Weber, & Steinberg, 2014) and integrin-linked kinase (ILK) (Wickström, Lange, Montanez, & Fässler, 2010). The focal adhesion is a major site for mechanotransduction, whereby pulling forces on the integrins induce conformational changes in talin and vinculin to facilitate focal adhesion maturation and recruitment of additional signaling molecules (Carisey et al., 2013; del Rio et al., 2009; Grashoff et al., 2010; Humphries et al., 2007; Yao et al., 2016). Myosin-dependent contractility and tension within the F-actin cytoskeleton also provides resistance to extracellular forces transmitted through focal adhesions and mediates adaptation to mechanical stress (Choquet, Felsenfeld, & Sheetz, 1997; Galbraith, Yamada, & Sheetz, 2002).

Within the epidermis, integrins anchor the basal layer of keratinocytes to the basement membrane and play an important role in tissue homeostasis. Under normal homeostatic conditions basal keratinocytes express the $\alpha3\beta1$ and $\alpha6\beta4$ receptors, which bind laminin 332 within the basement membrane,

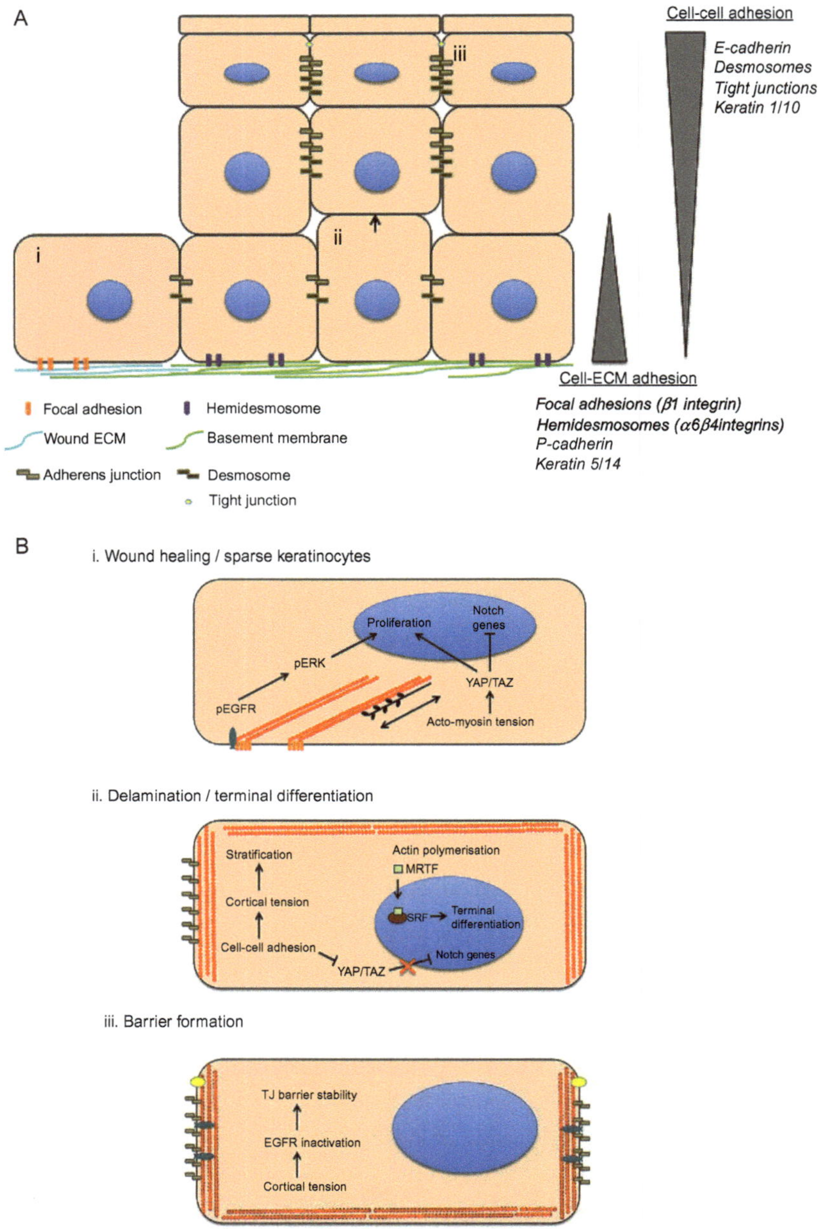

Fig. 1 Biomechanical regulation of keratinocyte function during different stages of homeostasis and repair. (A) Overview of adhesions, ECM, and keratin expression during (i) wound healing, (ii) commitment to differentiation and delamination, and (iii) terminal differentiation and barrier formation. (B) (i) In wound healing or sparse cell culture conditions focal adhesion assembly regulates EGFR and YAP/TAZ signaling. (ii) During early stages of terminal differentiation and commitment cell-cell adhesion facilitates delamination, while SRF and YAP/TAZ transcriptionally regulate differentiation gene expression programs. (iii) In the granular layer cell-cell adhesions regulate tight junction and barrier formation via modulation of EGFR signaling.

and the collagen receptor α2β1 (Adams & Watt, 1991; Geuijen & Sonnenberg, 2002). While the α3β1 and α2β1 receptors can form focal adhesions as described above, the α6β4 receptors form stable adhesions known as hemidesmosomes, which connect to the keratin cytoskeleton via the cytolinker plectin (de Pereda, Lillo, & Sonnenberg, 2009; Dowling, Yu, & Fuchs, 1996; Koster, Geerts, Favre, Borradori, & Sonnenberg, 2003). Integrin-mediated adhesion to the basement membrane is a primary determinant of epidermal cell fate and function. Early studies of human keratinocytes in suspension culture were the first to establish that loss of adhesion rapidly induces terminal differentiation and that this process is specifically mediated by ECM ligation of the β1 subunit (Adams & Watt, 1989; Jones & Watt, 1993; Watt, Kubler, Hotchin, Nicholson, & Adams, 1993). More recently, our own studies using micro-patterned ECM surfaces demonstrated that the strength of this response depends on the degree of adhesion, whereby differentiation increases with decreasing adhesive area (Connelly et al., 2010). Furthermore, high levels of β1 or α6 integrin expression marks a self-renewing population epidermal stem cells with increased capacity for clonal growth, and rapid adhesion to ECM distinguishes these stem cells from keratinocytes committed to terminal differentiation (Jones & Watt, 1993; Kaur & Li, 2000; Li, Simmons, & Kaur, 1998). Thus, integrins control epidermal homeostasis through the regulation of individual fate decisions of basal keratinocytes.

In addition to the influences of cell-ECM adhesions at the micro-scale, nano-scale organization of ECM molecules and integrin receptors contribute to the regulation of keratinocyte behavior. As integrin receptors require a minimum ligand spacing of approximately 70 nm in order to cluster and form stable focal adhesions, the nano-scale positioning and presentation of ECM molecules can significantly impact cell adhesion and signal transduction (Cavalcanti-Adam et al., 2006, 2007; Maheshwari, Brown, Lauffenburger, Wells, & Griffith, 2000). Studies using nanopatterned substrates have shown that as the size of the ECM nanopattern decreases from 3000 to 100 nm, keratinocytes display increased terminal differentiation alongside reduced cell spreading and focal adhesion assembly (Gautrot et al., 2014). Similar responses have also been observed on fiber-shaped nanopatterns, and the sensitivity to these nanoscale cues depends on differential expression of β1 and β3 integrins (Di Cio, Bøggild, Connelly, Sutherland, & Gautrot, 2017). How ECM molecules are tethered to the surrounding tissue or support structures can also impact how cells sense and respond to nanoscale cues. Without sufficient anchoring, ECM proteins such as collagen or fibronectin cannot provide enough mechanical resistance for keratinocytes to stabilize their adhesions, which in turn inhibits cell spreading and stimulates terminal differentiation (Trappmann et al., 2011). Taken together, these studies demonstrate that both the micro- and nano-scale presentation of adhesive cues from the ECM can modulate epidermal fate decisions yet affect differentiation through a common cell shape-dependent mechanism.

2.2 Rigidity and dynamic stretch

The transmission and transduction of forces through focal adhesion scaffolding proteins and the F-actin cytoskeleton allows cells to sense mechanical cues such as tissue rigidity or deformational strains. Over the past 20 years, numerous studies have shown that the elasticity of the ECM governs not only cell adhesion and migration, but also the growth, survival, and differentiation of various adult and progenitor cells (Guilak et al., 2009; Iskratsch et al., 2014). Likewise, human keratinocytes undergo efficient terminal differentiation when cultured on soft polyacrylamide gels (0.5 kPa); however, this response is dominated by changes in ECM tethering due to altered gel pore size, rather than bulk rigidity sensing (Trappmann et al., 2011). Interestingly, keratinocytes also have the ability to proliferate on liquid materials provided there is a sufficiently rigid surfactant/ECM complex at the interface to support adhesion assembly (Kong, Megone, et al., 2018; Kong, Peng, Di Cio, Novak, & Gautrot, 2018). These studies therefore suggest that nanoscale interfacial mechanics of the ECM are critical for the regulation of keratinocyte adhesion and in certain instances override the bulk material properties. Nevertheless, work in our own lab using non-porous elastomers to modulate substrate stiffness indicates that bulk material elasticity regulates the rate of keratinocyte proliferation independently of effects on terminal differentiation, and the response is mediated by activation of epidermal growth factor receptor (EGFR) signaling at focal adhesions (Kenny et al., 2018). Similar effects on EGFR signaling have also been observed on polyacrylamide gels (Wickert, Pomerenke, Mitchell, Masters, & Kreeger, 2016), suggesting that the proliferative response of keratinocytes to altered matrix stiffness is material independent. Through this body of work, the current model of keratinocyte mechano-sensing suggests that micro- and nano-scale ECM topology primarily control cell fate and terminal differentiation, while bulk ECM rigidity modulates proliferation.

Similar to tissue stiffness, dynamic stresses and strains also influence cell and tissue functions. Stretching of keratinocyte monolayers stimulates focal adhesion assembly (Knies, Bernd, Kaufmann, Bereiter-Hahn, & Kippenberger, 2006) and promotes proliferation and migration through activation of several major signaling axes (Kippenberger et al., 2000, 2005; Takei et al., 1997). Consistent with the response to increased matrix stiffness, mechanical strain upregulates EGFR signaling via increased interactions with focal adhesions and leads to downstream activation of ERK (Knies et al., 2006) and Akt (Kippenberger et al., 2005). Other studies have also reported induction of protein kinase C signaling (Takei et al., 1997) and release of intracellular calcium ions (Tsutsumi et al., 2009) in stretched keratinocytes. More recently, the proliferative response to mechanical strain has also been demonstrated in 3D organotypic skin models. Strained tissues displayed increased epidermal thickness, greater basement membrane deposition, and greater overall mechanical strength compared to unstrained tissues (Powell, McFarland, Butler, Supp, & Boyce, 2010; Tokuyama, Nagai, Takahashi, Kimata, & Naruse, 2015).

While much work has focused on focal adhesions in sensing forces transmitted from the ECM, a few studies have noted a potential role for hemidesmosomes in mechano-sensing. For example, mechanical strain promotes recruitment of the α6β4 integrins to the cell periphery in an Arf6 and caveolin-1 dependent manner (Osmani et al., 2018), and mechanical forces assist hemidesmosome maturation through p21-activated kinase (PAK) signaling (Zhang, Landmann, et al., 2011). Cross-talk and integration of focal adhesion and hemidesmosome mechanotransduction pathways in response to altered tissue rigidity and strain will certainly be an interesting and important area of future investigation.

2.3 Cell-cell adhesion

Strong cell-cell adhesions are essential for maintaining the integrity and barrier function of epithelia, and cell-cell junctions act as important sites for mechanotransduction and signaling. Within the epidermis, the two primary mechanical adhesions include adherens junctions and desmosomes. In adherens junctions, the classical calcium-dependent cadherin receptors, including epithelial (E)-, placental (P)-, neural (N)-, and vascular endothelial (VE)-cadherins, bind homotypically to receptors on adjacent cells (Saito, Tucker, Kohlhorst, Niessen, & Kowalczyk, 2012). On the intracellular side, they connect to the F-actin cytoskeleton via the adaptor proteins α-catenin and β-catenin (Saito et al., 2012). Adherens junctions facilitate the assembly of a dense cortical F-actin network, and interactions with myosin II generate intracellular tensile forces that resist external pulling forces transmitted across the junction (Borghi et al., 2012; Charras & Yap, 2018; Leerberg et al., 2014). F-actin remodeling at adherens junctions is mediated by the small Rho-GTPase RhoA (Arnold, Stephenson, & Miller, 2017), Arp2/3 (Verma et al., 2012) and formins, such as Dia1 (Carramusa, Ballestrem, Zilberman, & Bershadsky, 2007), to further stabilize and reinforce junctional mechanics. Tension-induced conformational changes in α-catenin are central to the recruitment of proteins such as vinculin and Ena/VASP proteins and may also play a direct role in further signal transduction (Huveneers et al., 2012; le Duc et al., 2010; Leerberg et al., 2014).

Functionally, force transmission through adherens junction controls cell division orientations (den Elzen, Buttery, Maddugoda, Ren, & Yap, 2009; Gloerich, Bianchini, Siemers, Cohen, & Nelson, 2017; Hart et al., 2017) and collective migration of simple epithelia (Borghi, Lowndes, Maruthamuthu, Gardel, & Nelson, 2010; Friedl & Mayor, 2017; Ladoux & Mège, 2017; Ng, Besser, Danuser, & Brugge, 2012; Seddiki et al., 2018), but in the epidermis junctional mechanics play a more complex role in the three-dimensional organization of a stratified epithelium. Basal keratinocytes express both E-cadherin and P-cadherin, but following commitment to differentiation in the suprabasal layer, there is a downregulation of P-cadherin and upregulation E-cadherin (Samuelov, Sprecher, & Paus, 2015; Tinkle, Lechler, Pasolli, & Fuchs, 2004), which coincides with increased F-actin at the junctions and higher cortical stiffness (Kao, Connelly, & Barber, 2016; Miroshnikova et al., 2018). While P-cadherin can partially compensate for a loss of E-cadherin in the basal layer (Jensen, Telegan, Lavker, & Wheelock, 1997; Tinkle et al., 2004; Young et al., 2003), conditional deletion of their common F-actin linkage, α-catenin, results in epidermal hyperproliferation (Vasioukhin, Bauer, Degenstein, Wise, & Fuchs, 2001) (Vasioukhin), involving activation of the transcriptional co-activator Yes-associated protein (YAP) (Schlegelmilch et al., 2011). Junctional mechanics also help maintain tissue homeostasis by coupling cell division in the basal layer to terminal differentiation. Loss of cortical tension in keratinocytes adjacent to proliferating cells, followed by suprabasal adhesion assembly and strengthening, promotes delamination from the basal layer and induction of the terminal differentiation process (Miroshnikova et al., 2018). Further increases in the mechanical strength of adherens junctions in the suprabasal and granular layers also tune tight junction assembly through negative regulation of EGFR (Rübsam et al., 2017). Finally in the context of wound healing, intercellular tension at adherens junctions helps maintain cohesion during collective keratinocyte migration (Vedula et al., 2014), and junction assembly supports a switch from cell-ECM to cell-cell force transmission and sensing (Mertz et al., 2013; Noethel et al., 2018).

Like adherens junctions, desmosomes provide important mechanical structure and resilience for the epidermis, and are essential for the tissue's barrier function. However, far less is known about their direct role in mechano-sensing and mechanotransduction. Desmosomes consist of the desmosomal cadherins desmoglein 1–4 or desmocolin 1–3, which integrate with the keratin cytoskeleton through the intracellular adaptors desmoplakin 1 and 2 (Saito et al., 2012). Consistent with the central role of keratins in determining keratinocyte mechanics (Ramms et al., 2013), the disruption of desmosomes through depletion of desmoplakins reduces cell-cell adhesion forces and overall cellular stiffness (Broussard et al., 2017). However, desmoglein 1 appears to specifically decrease tension at adherens junctions via interaction with its binding partners Tctex and cortactin and facilitates keratinocyte delamination from the basal layer (Nekrasova et al., 2018). These findings have provided some initial insight into the roles of desmosomes in mechano-sensing, and raise interesting questions about how other desmosomal signaling pathways (Dubash et al., 2013; Harmon et al., 2013) directly respond to forces and transduce mechanical signals.

3. Mechanisms of mechanotransduction and downstream pathways

3.1 Cytoskeletal machinery

The cytoskeleton provides structural support and mechanical integrity to nearly all cell types, and through interactions with motor proteins, it can generate contractile and protrusive forces to drive cell motility. As the individual components (microfilaments, intermediate filaments, and microtubules) interact with adhesion receptors and signaling molecules, the cytoskeleton also plays a central role in sensing and transducing a wide range of mechanical stimuli. The network of actin microfilaments (F-actin) is one of the most studied and important cytoskeletal elements involved in mechanotransduction. It interacts with the myosin motor proteins to generate intracellular tension and contractile forces, which as noted above, provides mechanical resistance to externally applied forces and facilitates the adaptation and maturation of focal adhesions and adherens junctions. The organization and mechanical properties of the F-actin cytoskeleton can also tune global cell mechanics ("the tensegrity model") and mechano-sensing (Ingber, 2003a, 2003b). For example, the polymerization state and contractility of the F-actin cytoskeleton regulates transcriptional activity of serum-response factor (SRF) (Miralles, Posern, Zaromytidou, & Treisman, 2003) and YAP (Dupont et al., 2011) (see further discussion below). In addition, monomeric globular (G)-actin itself can translocate to the nucleus and regulate processes such as chromatin remodeling, gene transcription, and DNA damage responses (Belin, Lee, & Mullins, 2015; Bohnsack, Stuven, Kuhn, Cordes, & Gorlich, 2006; Philimonenko et al., 2004; Song et al., 2007). In keratinocytes, nuclear actin negatively regulates the expression of cytoskeletal and adhesive genes involved in cell motility through inhibition of SRF transcriptional activity (Sharili, Kenny, Vartiainen, & Connelly, 2016). At the same time, mechanical strain promotes depletion of nuclear actin and epigenetic repression of epidermal lineage genes, thereby playing an important role in terminal differentiation and tissue homeostasis (Le et al., 2016).

While the actin cytoskeleton has been the focus of many mechanotransduction studies, substantially less is known about the role of intermediate filaments and microtubules. Intermediate filaments (IFs) include a broad group of cytoskeletal proteins that contribute to cell strength and resilience. In keratinocytes, the keratins (IF Types I and II) are the major determinants of cell stiffness (Ramms et al., 2013; Seltmann, Fritsch, Käs, & Magin, 2013) and are essential for the mechanical integrity and barrier function of the epidermis (Bär et al., 2014; Kumar et al., 2015). Despite their importance in cell and tissue mechanics, only a few studies have linked these properties to cellular mechano-sensing functions. In mouse keratinocytes, depletion of all type I and II keratins leads to destabilization of hemidesmosomes (Seltmann, Cheng, Wiche, Eriksson, & Magin, 2015) and desmosomes (Bär et al., 2014), disruption of p38 mitogen-activated protein kinase and PCK signaling, and increased mechanical fragility in dispase fragmentation assays (Vielmuth et al., 2018). In addition, deletion of the cytolinker plectin, which mediates keratin bundle organization and linkage to hemidesmosomes, alters the mechanical behavior and signaling in keratinocytes. Plectin deficient keratinocytes display increased MAP kinase, PKC delta, and c-Src kinase activity, alongside increased migratory behavior (Osmanagic-Myers et al., 2006). Moreover, plectin and keratins play an important role in protection of the nucleus from mechanical deformation

(Almeida et al., 2015; Lee & Coulombe, 2009). Similar roles for other types of intermediate filaments, such as vimentin, desmin and glial fibrillary acidic protein, have been observed in multiple cell types (Burgstaller, Gregor, Winter, & Wiche, 2010; De Pascalis et al., 2018; Keeling, Flores, Dodhy, Murray, & Gavara, 2017; Schröder et al., 2002). Finally, nuclear lamins (type V IFs) possess a unique function in nuclear mechanotransduction as they are mechanically coupled to the cytoskeleton and coordinate chromatin remodeling and DNA damage responses (Lammerding et al., 2004; Swift et al., 2013) (see further discussion below).

Cytoplasmic microtubules are a highly dynamic component of the cytoskeleton, and in addition to supporting overall cell structure, they are essential for intracellular trafficking and mitosis. In vitro studies and computation modeling suggest that microtubules are more rigid and resistant to bending and compressive stresses than microfilaments or intermediate filaments (Hawkins, Mirigian, Selcuk Yasar, & Ross, 2010). While a few studies have demonstrated a mechanical function for cytoplasmic microtubules in processes such as centrosome positioning (Wu et al., 2011), branching morphogenesis (Myers, Applegate, Danuser, Fischer, & Waterman, 2011), and directional cell migration (Adams et al., 2016; Cao et al., 2015), axonemal microtubules in primary cilia have a more well defined role in mechanotransduction. Primary cilia are stable microtubule structures that protrude from the membrane and act as signaling hubs for the Hedgehog pathway. In chondrocytes, the primary cilium mediates calcium and hedgehog signaling in response to compressive strain (Thompson, Chapple, & Knight, 2014; Wann et al., 2012), while in osteoblasts and MSCs, it controls the mechanical response to fluid shear stress (Chen, Hoey, Chua, Bellon, & Jacobs, 2016; Temiyasathit & Jacobs, 2010). In the epidermis, cytoplasmic microtubules are crucial for stratification and terminal differentiation through the control of symmetric and asymmetric division orientations in the basal layer (Lechler & Fuchs, 2005). During terminal differentiation, microtubules also switch from a radial orientation from the centrosome to becoming localized at cell-cell junctions (Lechler & Fuchs, 2007). This process involves the desmosomal protein desmoplakin and suggests an important cross-talk with the keratin cytoskeleton (Lechler & Fuchs, 2007). Likewise, microtubules interact with focal adhesions through the adapter ACF7, which helps to target growing microtubules and protein trafficking to these mechano-sensing adhesions (Wu, Kodama, & Fuchs, 2008, p. 7). Finally, the primary cilium of epidermal cells regulates skin morphogenesis and differentiation through several major signaling pathways, such as p63, Notch, and Hedgehog (Croyle et al., 2011; Ezratty et al., 2011), but whether it also has direct mechano-sensing function in the skin has yet to determined.

3.2 Mechano-sensitive transcriptional regulators

Defined gene expression programs specify cell type and function, and commitment and differentiation along particular lineages requires controlled activation or repression of these programs. Thus, in order for mechanical and biophysical cues to direct cell differentiation, forces must be transduced into specific transcription responses. While adhesion receptors and the cytoskeleton sense and adapt to a wide range of mechanical cues, these signals must ultimately be transmitted to the nucleus in order to impact cell fate decisions. Key mechano-sensitive transcriptional mediators identified over the past 10 years, include serum response factor (SRF), yes-associated protein (YAP), and the nuclear lamina, which each regulate epidermal differentiation and tissue homeostasis.

The SRF transcription factor is expressed in many different cell types and activates a large set of target genes, such as immediate-early and muscle related genes, all of which contain the serum response element in their promoter regions. SRF selectivity and transcriptional activity are controlled by interactions with its co-activators the myocardin related transcription factors (MRTFs) or ternary complex factors (TCFs) (Selvaraj & Prywes, 2004). MRTF availability depends on its interaction with monomeric G-actin, which can sequester MRTF and block SRF activation (Miralles et al., 2003; Sotiropoulos, Gineitis, Copeland, & Treisman, 1999). Thus, biophysical cues that promote F-actin polymerization can activate SRF target gene expression through the release of MRTF. SRF is required for terminal differentiation in the epidermis (Connelly et al., 2010; Dubash et al., 2013; Koegel et al., 2009; Luxenburg, Pasolli, Williams, & Fuchs, 2011), and it regulates inflammatory (Koegel et al., 2009) and migratory responses of keratinocytes (Sharili et al., 2016). Loss of SRF in the mouse epidermis causes hyperproliferation associated with increased Stat3 activation (Koegel et al., 2009), as well as defects in cortical F-actin organization, which leads to dysregulated spindle orientation in basal cells and impaired differentiation (Luxenburg et al., 2011). In addition, SRF depletion inhibits normal hair follicle development in mice (Lin et al., 2013). Our own studies using micro-patterned substrates to control cell shape have further shown that SRF mediates human keratinocyte differentiation in response to altered biophysical cues via changes in actin polymerization and MRTF interactions (Connelly et al., 2010). Similarly, the guanine exchange factor Bcr and Rho-GTPase RhoA have are important upstream regulators of MRTF/SRF activity (Dubash et al., 2013). Additional mechanotransduction activities of SRF have been observed in other cell types (Busche, Descot, Julien, Genth, & Posern, 2008; Ho, Jaalouk, Vartiainen, & Lammerding, 2013; Iyer, Pulford, Mogilner, & Shivashankar, 2012), suggesting that the MRTF-SRF pathway is a conserved mechanism for mechanically regulated transcriptional responses.

YAP and its homologue, transcriptional co-activator with PDZ binding motif (TAZ), are highly conserved regulators of cell growth and organ size, and over the last several years, they have emerged as central mediators of many transcriptional responses to mechanical force. YAP/TAZ are transcriptional co-activators that cooperate with TEAD transcriptional factors to promote expression of a wide range of growth and differentiation genes (Totaro, Panciera, & Piccolo, 2018; Zhao et al., 2008). Nuclear localization of YAP/TAZ is negatively regulated by the canonical Hippo pathway components Mst1/2 and Lats1/2, which phosphorylate YAP/TAZ and target it for cytoplasmic retention or degradation (Totaro et al., 2018). RhoA-dependent cytoskeletal tension also controls the nuclear localization of YAP/TAZ, whereby F-actin stress-fiber assembly induced by cell spreading, increased substrate rigidity, or mechanical strain promotes nuclear accumulation of YAP/TAZ (Aragona et al., 2013; Dupont et al., 2011). Recent evidence suggests that biomechanical regulation of YAP/TAZ activity can be both independent and dependent on the Hippo pathway and involves signaling through focal adhesions and the Ras GTPase RAP2(Aragona et al., 2013; Chakraborty et al., 2017; Meng et al., 2018; Walko et al., 2017). Moreover, mechanical forces can influence YAP nuclear entry via direct deformation of the nuclear pore complex and changes in nuclear membrane permeability (Elosegui-Artola et al., 2017). For in depth analysis of YAP/TAZ signaling, please see the recent review by Totaro et al. (2018).

In the epidermis, YAP/TAZ localizes to the nucleus in basal keratinocytes and is required for the proliferative capacity of undifferentiated progenitors (Beverdam et al., 2013; Elbediwy et al., 2016; Walko et al., 2017; Zhang, Pasolli, & Fuchs, 2011). Signals from integrin-mediated adhesion to the ECM, Src, and cytoskeletal tension promote YAP translocation to the nucleus (Elbediwy et al., 2016; Walko et al., 2017), where interactions with TEAD proteins and WBP2 modulate the expression of cell cycle regulators, Hox genes, and components of the Notch pathway

(Liu, Zhao, Lin, & Wang, 2015; Totaro et al., 2017; Walko et al., 2017). Indeed, YAP/TAZ mechano-sensing contributes to the shape-induced differentiation of keratinocytes on micro-patterned substrates (Totaro et al., 2017), and inter-cellular tension regulates nuclear YAP localization on substrates mimicking the topography of the dermal-epidermal junction (Walko et al., 2017). Interestingly, the biophysical regulation of YAP/TAZ and proliferation in the basal layer appears to be independent of the Hippo pathway, and although Lats1 is upregulated in the suprabasal layers, its function in the epidermis is not completely clear (Elbediwy et al., 2016; Walko et al., 2017). Beyond tissue homeostasis, YAP/TAZ signaling also regulates cutaneous wound repair and hair follicle morphogenesis (Elbediwy et al., 2016; Lee, Ran Byun, Furutani-Seiki, Hong, & Jung, 2013; Zhang, Pasolli, & Fuchs, 2011). Taken together, these studies demonstrate that YAP/TAZ mechano-sensing plays an important role in epidermal homeostasis and the tissue's regenerative capacity.

3.3 Force sensing within the nucleus

In addition to the mechanical activation of transcriptional regulators, the nucleus itself can act as a direct force sensing element within the cell. The nuclear lamina, which consists of a network of A and B type lamins, provides structural support and integrity for the nucleus (Lammerding et al., 2004). The nucleus interacts with each major cytoskeletal element through the linker of nucleoskeleton and cytoskeleton (LINC) complex, which comprises nesprin proteins at the outer nuclear membrane and SUN1 and 2 on the inner nuclear membrane (Crisp et al., 2006). Nesprins 1 and 2 bind F-actin filaments, while nesprin 3 interacts with intermediate filaments via plectin, and nesprin 4 interacts with the microtubule motor protein kinesin. SUN1/2 bind nesprins within the inter-nuclear membrane space and physically connects to the lamin meshwork inside the inner nuclear membrane. Thus, the LINC complex mechanically integrates the nucleus with the cell's cytoskeleton and facilitates direct transmission of forces to the nucleus.

A number of recent studies have shown that mechanical and biophysical cues can deform the nucleus and induce dramatic changes in nuclear architecture, chromatin remodeling, and gene expression. For example, endothelial cell spreading-induced cytoskeletal tension promotes nuclear elongation and global chromatin condensation (Versaevel, Grevesse, & Gabriele, 2012), while direct mechanical stimulation of MSCs can stimulate similar changes in chromatin condensation, which is mediated by ATP release (Heo et al., 2016, 2015). In addition, nuclear mechanics influence longer range chromatin remodeling, such as telomere dynamics and chromosome positioning (Makhija, Jokhun, & Shivashankar, 2016; Wang, Nagarajan, Uhler, & Shivashankar, 2017), and altered chromatin states reciprocally affect nuclear stiffness and stability (Heo et al., 2015). Many of these responses are mediated by the nuclear lamina through its interactions with chromatin, nuclear pore complexes, and the DNA repair machinery. In particular, expression of A type lamins adapts to changing mechanical environments (Swift et al., 2013) and specifies nuclear stiffness (Lammerding et al., 2006) and multiple mechanobiological responses, such as SRF and retinoic acid signaling (Ho et al., 2013; Ivanovska et al., 2017; Swift et al., 2013). Recently, the nuclear lamina has been shown to interact with DNA damage machinery and facilitate repair of mechanically induced DNA damage (Denais et al., 2016; Xia et al., 2018).

Similar nuclear mechanosensing mechanisms have also been observed within the epidermis and are important for the control of tissue homeostasis. Most notably, the application of dynamic strain to human keratinocytes induces the accumulation of emerin, actin, and myosin at the outer nuclear membrane. This response in turn causes an upregulation of polycomb-mediated gene silencing and suppression of terminal differentiation (Le et al., 2016). In contrast, limited keratinocyte adhesion and spreading leads to reduced nuclear volume and downregulation histone acetylation, which is required for terminal differentiation (Connelly, Mishra, Gautrot, & Watt, 2011). The mechanical regulation of nuclear size and shape involves a complex interaction between different cytoskeletal elements (Almeida et al., 2015; Lee, Kim, Chung, Leahy, & Coulombe, 2012). While acto-myosin contractility is a key driver of nuclear deformation, the keratin network acts to mechanically restrain and buffer the nucleus from excess deformation (Almeida et al., 2015). Moreover, the nuclear-cytoskeletal linkage is essential for maintaining correct nuclear shape (Luke et al., 2008) and positioning (Stewart et al., 2015) within epidermal keratinocytes. While these recent findings clearly implicate the nucleus in keratinocyte mechano-sensing, much work is still required to understand the integrated functions of nuclear mechanotransduction and epigenetic gene regulation in epidermal mechanobiology.

4. Mechanobiology in skin disease

As described throughout this review, epidermal keratinocytes employ a wide range of adhesion receptors, cytoskeletal elements, and transcrsiptional regulators to coordinate appropriate biomechanical functions at different stages of growth and differentiation (Fig. 1). Therefore, defects in these regulatory processes can disrupt normal mechano-sensing functions and contribute to a variety of skin diseases. Examples include, blistering skin diseases, keratoderma, scarring, and skin cancer. In this section, we will highlight some recent findings and insights into the role of biomechanics in skin disease pathogenesis.

The epidermolysis bullosa (EB) family of blistering skin diseases are caused by mutations in the genes for structural proteins involved in anchorage of the epidermis to the basement membrane. The recessive dystrophic form (RDEB) involves a deficiency in type VII collagen anchoring fibers within the basement membrane zone, while the junctional form (JEB) is caused by mutations in the basement membrane proteins themselves, including laminin 332 and type XVII collagen. In contrast, epidermolysis bullosa simplex (EBS) results from mutations in the basal keratins (K5 or K14) or the keratin cytolinker plectin. While each form of EB causes epidermal blistering and fragility, the EBS subtype may have important implications for cellular mechano-sensing, given the significance of the keratin cytoskeleton in cell biomechanics. Keratinocytes with K14 associated with severe generalized EBS display rapid keratin fragmentation and aggregate formation in response to mechanical stress or osmotic shock. These responses correlate with desmosome instability, increased motility, increased stress-associated Jun kinase (Jnk) signaling, and an overall reduction in cell stiffness (D'Alessandro, Russell, Morley, Davies, & Lane, 2002; Homberg et al., 2015; Morley et al., 2003; Russell, Andrews, James, & Lane, 2004). Similar effects on migration and stress responses have been observed in plectin deficient mouse keratinocytes, which mimic the EBS with muscular dystrophy form of the disease (Osmanagic-Myers et al., 2006).

Analogous to keratin mutations, desmoplakin mutations in cells derived from palmoplantar keratoderma also display reduced cell stiffness and disrupted junction assembly (Puzzi et al., 2018), suggesting a potential role for impaired mechano-sensing at desmosomes in the keratoderma family of skin diseases (Brooke, Nitoiu, & Kelsell, 2012). One form of palmoplantar keratoderma known as tylosis with oesophageal cancer (TOC) is caused by gain of function mutations in the inactive rhomboid protease iRHOM2 and has recently been implicated in defective stress responses (Brooke et al., 2014; Maruthappu et al., 2017). In TOC keratinocytes, mutant iRHOM2 upregulates and interacts with the stress-associated keratin 16, which is normally restricted to wounded skin, palms, and soles of the feet (Maruthappu et al., 2017). As iRHOM2 also regulates inflammatory responses and p63 signaling (Arcidiacono et al., 2018), cross-talk between mechanical stress, the keratin cytoskeleton, and iRHOM2 may influence these major downstream pathways. Together, these findings suggest that the keratin cytoskeleton is a central component of the mechano-biological response to blistering and keratoderma skin diseases.

In contrast to the epidermal fragility caused by different forms of EB, scar formation and chronic fibrosis are characterized by overall tissue stiffening and a loss of elasticity. Recent studies in our laboratory revealed that the elastic modulus of the dermis increases 10- to 50-fold in extreme forms of scarring, such as keloids, and this tissue stiffening stimulates epidermal hyperproliferation through activation of the EGFR signaling pathway (Kenny et al., 2018). In many types of cancer, tissue biomechanics can similarly drive disease progression, whereby increased stromal stiffness promotes primary tumor growth, as well as invasion and metastasis (Levental et al., 2009; Paszek et al., 2005; Samuel et al., 2011). These changes in tissue biomechanics are a result of paracrine activation of the "cancer associated fibroblasts" (CAFs), which assemble an ECM with altered composition, structure, and mechanics (Miles & Sikes, 2014; Ng, Pourreyron, et al., 2012; Pearce et al., 2018). Stiffening of the stromal ECM in cutaneous squamous cell carcinoma associated with RDEB (RDEB-cSCC) results from increased collagen deposition and cross-linking, derived from TGF-β activated CAFs (Mittapalli et al., 2016). These biomechanical changes in the stromal ECM in turn promote RDEB-SCC invasion through focal adhesion kinase and Akt signaling (Mittapalli et al., 2016). Consistent with the response of normal keratinocytes to stiffened keloid scars, head and neck SCC keratinocytes upregulate EGFR signaling, which promotes actomyosin contractility and increased invasive behavior (Grasset et al., 2018). These shared mechanobiology responses to tissue stiffening suggest that ECM biomechanics is a major contributing factor to the response of the epithelium to dermal remodeling in scarring and cancer. Greater understanding of the downstream signaling and molecular mechanisms of mechano-sensing in these disease states may therefore be important for the development of new therapies targeting SCC growth and metastasis.

5. Conclusions and future perspectives

As described throughout this chapter, epidermal keratinocytes sense mechanical forces and biophysical cues through a range of different mechanisms (Fig. 1). Cell-ECM interactions primarily mediate rigidity and topography sensing and regulate proliferation and migration, both of which are important in the context of wound healing. In addition, integrin-mediated adhesion controls commitment to terminal differentiation in basal keratinocytes. By contrast, dynamic changes in cell-cell adhesion forces facilitate stratification, terminal differentiation, and barrier formation. These mechano-sensing mechanisms each signal to several major pathways, including EGFR, SRF, and YAP/TAZ. Furthermore, the nucleoskeleton is a direct force sensing element within the cell and regulates overall nuclear architecture, chromatin remodeling, and gene expression.

Mechano-sensing also contributes to the progression of several major skin diseases. Defects in the keratin cytoskeleton and keratin-interacting proteins lead to impaired mechanical function in EB and keratoderma, but additional influences on cell signaling suggest further defects in mechanotransduction processes in these diseases. Finally, stromal stiffening associated with scarring and tumor formation promotes keratinocyte hyperproliferation and SCC invasion, respectively.

While a number of the fundamental responses and molecular mechanisms by which keratinocytes sense their mechanical environment have now been mapped out, there are still several important questions for the field. First, how do these signals cross-talk with each other and integrate with other biochemical cues and intrinsic gene expression programs? There are a few notable examples of interplay between different cytoskeletal components and growth factor signaling, but much more work is needed in order to develop a systems-level understanding of skin mechanobiology. Another important question is whether mechano-sensing plays a role in a wider range of skin diseases, including common conditions, such as psoriasis and eczema. The occurrence of psoriatic plaques in areas of high stress, like the elbows, suggests that there may indeed be a biomechanical component.

Finally, a major unanswered question is whether knowledge of skin mechanobiology can be translated into novel therapeutic approaches for treating skin diseases or promoting skin regeneration. Can these mechanotransduction pathways be targeted and can modulation of keratinocyte mechano-sensing be harnessed to improve skin health? Advances in translational mechanobiology will most likely require further mechanistic understanding of the contribution of mechano-sensing in skin disease, but also careful development of targeting and delivery strategies (e.g., small molecules, biologics, biomaterials). Determining which pathways to target, as well as how and when to apply certain therapies will be key considerations. While these are all extremely challenging questions, the discovery of new treatments for common and rare skin conditions would most certainly have a significant beneficial impact on a large number of patients.

References

Adams, J. C., & Watt, F. M. (1989). Fibronectin inhibits the terminal differentiation of human keratinocytes. *Nature*, *340*(6231), 307–309.

Adams, J. C., & Watt, F. M. (1991). Expression of beta 1, beta 3, beta 4, and beta 5 integrins by human epidermal keratinocytes and non-differentiating keratinocytes. *The Journal of Cell Biology*, *115*(3), 829–841.

Adams, G., Zhou, J., Wang, W., Wu, H., Quan, J., Liu, Y., et al. (2016). The microtubule plus end tracking protein TIP150 interacts with Cortactin to steer directional cell migration. *The Journal of Biological Chemistry*, *291*(39), 20692–20706. https://doi.org/10.1074/jbc.M116.732719.

Almeida, F. V., Walko, G., McMillan, J. R., McGrath, J. A., Wiche, G., Barber, A. H., et al. (2015). The cytolinker plectin regulates nuclear mechanotransduction in keratinocytes. *Journal of Cell Science*, *128*(24), 4475–4486. https://doi.org/10.1242/jcs.173435.

Aragona, M., Panciera, T., Manfrin, A., Giulitti, S., Michielin, F., Elvassore, N., et al. (2013). A mechanical checkpoint controls multicellular growth through YAP/TAZ regulation by actin-processing factors. *Cell*, *154*(5), 1047–1059. https://doi.org/10.1016/j.cell.2013.07.042.

Arcidiacono, P., Webb, C. M., Brooke, M. A., Zhou, H., Delaney, P. J., Ng, K.-E., et al. (2018). p63 is a key regulator of iRHOM2 signalling in the keratinocyte stress response. *Nature Communications*, *9*(1), 1021. https://doi.org/10.1038/s41467-018-03470-y.

Arnold, T. R., Stephenson, R. E., & Miller, A. L. (2017). Rho GTPases and actomyosin: Partners in regulating epithelial cell-cell junction structure and function. *Experimental Cell Research*, *358*(1), 20–30. https://doi.org/10.1016/j.yexcr.2017.03.053.

Bär, J., Kumar, V., Roth, W., Schwarz, N., Richter, M., Leube, R. E., et al. (2014). Skin fragility and impaired desmosomal adhesion in mice lacking all keratins. *The Journal of Investigative Dermatology*, *134*(4), 1012–1022. https://doi.org/10.1038/jid.2013.416.

Belin, B. J., Lee, T., & Mullins, R. D. (2015). DNA damage induces nuclear actin filament assembly by Formin-2 and Spire-1/2 that promotes efficient DNA repair. *eLife*, *4*, e07735. https://doi.org/10.7554/eLife.07735.

Beverdam, A., Claxton, C., Zhang, X., James, G., Harvey, K. F., & Key, B. (2013). Yap controls stem/progenitor cell proliferation in the mouse postnatal epidermis. *The Journal of Investigative Dermatology*, *133*(6), 1497–1505. https://doi.org/10.1038/jid.2012.430.

Blanpain, C., & Fuchs, E. (2009). Epidermal homeostasis: A balancing act of stem cells in the skin. *Nature Reviews. Molecular Cell Biology*, *10*(3), 207–217. https://doi.org/10.1038/nrm2636.

Bohnsack, M. T., Stuven, T., Kuhn, C., Cordes, V. C., & Gorlich, D. (2006). A selective block of nuclear actin export stabilizes the giant nuclei of Xenopus oocytes. *Nature Cell Biology*, *8*(3), 257–263. https://doi.org/10.1038/ncb1357.

Borghi, N., Lowndes, M., Maruthamuthu, V., Gardel, M. L., & Nelson, W. J. (2010). Regulation of cell motile behavior by crosstalk between cadherin- and integrin-mediated adhesions. *Proceedings of the National Academy of Sciences of the United States of America*, *107*(30), 13324–13329. https://doi.org/10.1073/pnas.1002662107.

Borghi, N., Sorokina, M., Shcherbakova, O. G., Weis, W. I., Pruitt, B. L., Nelson, W. J., et al. (2012). E-cadherin is under constitutive actomyosin-generated tension that is increased at cell–cell contacts upon externally applied stretch. *Proceedings of the National Academy of Sciences of the United States of America*, *109*(31), 12568–12573. https://doi.org/10.1073/pnas.1204390109.

Brooke, M. A., Etheridge, S. L., Kaplan, N., Simpson, C., O'Toole, E. A., Ishida-Yamamoto, A., et al. (2014). iRHOM2-dependent regulation of ADAM17 in cutaneous disease and epidermal barrier function. *Human Molecular Genetics*, *23*(15), 4064–4076. https://doi.org/10.1093/hmg/ddu120.

Brooke, M. A., Nitoiu, D., & Kelsell, D. P. (2012). Cell-cell connectivity: Desmosomes and disease. *The Journal of Pathology*, *226*(2), 158–171. https://doi.org/10.1002/path.3027.

Broussard, J. A., Yang, R., Huang, C., Nathamgari, S. S. P., Beese, A. M., Godsel, L. M., et al. (2017). The desmoplakin-intermediate filament linkage regulates cell mechanics. *Molecular Biology of the Cell*, *28*(23), 3156–3164. https://doi.org/10.1091/mbc.E16-07-0520.

Burgstaller, G., Gregor, M., Winter, L., & Wiche, G. (2010). Keeping the vimentin network under control: Cell-matrix adhesion-associated plectin 1f affects cell shape and polarity of fibroblasts. *Molecular Biology of the Cell*, *21*(19), 3362–3375. https://doi.org/10.1091/mbc.E10-02-0094.

Busche, S., Descot, A., Julien, S., Genth, H., & Posern, G. (2008). Epithelial cell-cell contacts regulate SRF-mediated transcription via Rac-actin-MAL signalling. *Journal of Cell Science*, *121*(Pt. 7), 1025–1035.

Cao, D., Su, Z., Wang, W., Wu, H., Liu, X., Akram, S., et al. (2015). Signaling scaffold protein IQGAP1 interacts with microtubule plus-end tracking protein SKAP and links dynamic microtubule plus-end to steer cell migration. *The Journal of Biological Chemistry*, *290*(39), 23766–23780. https://doi.org/10.1074/jbc.M115.673517.

Carisey, A., Tsang, R., Greiner, A. M., Nijenhuis, N., Heath, N., Nazgiewicz, A., et al. (2013). Vinculin regulates the recruitment and release of core focal adhesion proteins in a force-dependent manner. *Current Biology: CB*, *23*(4), 271–281. https://doi.org/10.1016/j.cub.2013.01.009.

Carramusa, L., Ballestrem, C., Zilberman, Y., & Bershadsky, A. D. (2007). Mammalian diaphanous-related formin Dia1 controls the organization of E-cadherin-mediated cell-cell junctions. *Journal of Cell Science*, *120*(Pt. 21), 3870–3882. https://doi.org/10.1242/jcs.014365.

Cavalcanti-Adam, E. A., Micoulet, A., Blümmel, J., Auernheimer, J., Kessler, H., & Spatz, J. P. (2006). Lateral spacing of integrin ligands influences cell spreading and focal adhesion assembly. *European Journal of Cell Biology*, *85*(3–4), 219–224. https://doi.org/10.1016/j.ejcb.2005.09.011.

Cavalcanti-Adam, E. A., Volberg, T., Micoulet, A., Kessler, H., Geiger, B., & Spatz, J. P. (2007). Cell spreading and focal adhesion dynamics are regulated by spacing of integrin ligands. *Biophysical Journal*, *92*(8), 2964–2974. https://doi.org/10.1529/biophysj.106.089730.

Chakraborty, S., Njah, K., Pobbati, A. V., Lim, Y. B., Raju, A., Lakshmanan, M., et al. (2017). Agrin as a mechanotransduction signal regulating YAP through the hippo pathway. *Cell Reports*, *18*(10), 2464–2479. https://doi.org/10.1016/j.celrep.2017.02.041.

Charras, G., & Yap, A. S. (2018). Tensile forces and mechanotransduction at cell-cell junctions. *Current Biology: CB*, *28*(8), R445–R457. https://doi.org/10.1016/j.cub.2018.02.003.

Chen, J. C., Hoey, D. A., Chua, M., Bellon, R., & Jacobs, C. R. (2016). Mechanical signals promote osteogenic fate through a primary cilia-mediated mechanism. *FASEB Journal: Official Publication of the Federation of American Societies for Experimental Biology*, *30*(4), 1504–1511. https://doi.org/10.1096/fj.15-276402.
Choquet, D., Felsenfeld, D. P., & Sheetz, M. P. (1997). Extracellular matrix rigidity causes strengthening of integrin-cytoskeleton linkages. *Cell*, *88*(1), 39–48.
Connelly, J. T., Gautrot, J. E., Trappmann, B., Tan, D. W., Donati, G., Huck, W. T., et al. (2010). Actin and serum response factor transduce physical cues from the microenvironment to regulate epidermal stem cell fate decisions. *Nature Cell Biology*, *12*(7), 711–718. https://doi.org/10.1038/ncb2074.
Connelly, J. T., Mishra, A., Gautrot, J. E., & Watt, F. M. (2011). Shape-induced terminal differentiation of human epidermal stem cells requires p38 and is regulated by histone acetylation. *PLoS One*, *6*(11), e27259. https://doi.org/10.1371/journal.pone.0027259.
Crisp, M., Liu, Q., Roux, K., Rattner, J. B., Shanahan, C., Burke, B., et al. (2006). Coupling of the nucleus and cytoplasm: Role of the LINC complex. *The Journal of Cell Biology*, *172*(1), 41–53. https://doi.org/10.1083/jcb.200509124.
Croyle, M. J., Lehman, J. M., O'Connor, A. K., Wong, S. Y., Malarkey, E. B., Iribarne, D., et al. (2011). Role of epidermal primary cilia in the homeostasis of skin and hair follicles. *Development (Cambridge, England)*, *138*(9), 1675–1685. https://doi.org/10.1242/dev.060210.
D'Alessandro, M., Russell, D., Morley, S. M., Davies, A. M., & Lane, E. B. (2002). Keratin mutations of epidermolysis bullosa simplex alter the kinetics of stress response to osmotic shock. *Journal of Cell Science*, *115*(Pt. 22), 4341–4351.
De Pascalis, C., Pérez-González, C., Seetharaman, S., Boëda, B., Vianay, B., Burute, M., et al. (2018). Intermediate filaments control collective migration by restricting traction forces and sustaining cell-cell contacts. *The Journal of Cell Biology*, *217*(9), 3031–3044. https://doi.org/10.1083/jcb.201801162.
de Pereda, J. M., Lillo, M. P., & Sonnenberg, A. (2009). Structural basis of the interaction between integrin alpha6beta4 and plectin at the hemidesmosomes. *The EMBO Journal*, *28*(8), 1180–1190.
del Rio, A., Perez-Jimenez, R., Liu, R., Roca-Cusachs, P., Fernandez, J. M., & Sheetz, M. P. (2009). Stretching single talin rod molecules activates vinculin binding. *Science (New York, N.Y.)*, *323*(5914), 638–641. https://doi.org/10.1126/science.1162912.
den Elzen, N., Buttery, C. V., Maddugoda, M. P., Ren, G., & Yap, A. S. (2009). Cadherin adhesion receptors orient the mitotic spindle during symmetric cell division in mammalian epithelia. *Molecular Biology of the Cell*, *20*(16), 3740–3750. https://doi.org/10.1091/mbc.e09-01-0023.
Denais, C. M., Gilbert, R. M., Isermann, P., McGregor, A. L., te Lindert, M., Weigelin, B., et al. (2016). Nuclear envelope rupture and repair during cancer cell migration. *Science (New York, N.Y.)*, *352*(6283), 353–358. https://doi.org/10.1126/science.aad7297.
Di Cio, S., Bøggild, T. M. L., Connelly, J., Sutherland, D. S., & Gautrot, J. E. (2017). Differential integrin expression regulates cell sensing of the matrix nanoscale geometry. *Acta Biomaterialia*, *50*, 280–292. https://doi.org/10.1016/j.actbio.2016.11.069.
Dowling, J., Yu, Q. C., & Fuchs, E. (1996). Beta4 integrin is required for hemidesmosome formation, cell adhesion and cell survival. *The Journal of Cell Biology*, *134*(2), 559–572. Retrieved from http://jcb.rupress.org/content/134/2/559.abstract.
Dubash, A. D., Koetsier, J. L., Amargo, E. V., Najor, N. A., Harmon, R. M., & Green, K. J. (2013). The GEF Bcr activates RhoA/MAL signaling to promote keratinocyte differentiation via desmoglein-1. *The Journal of Cell Biology*, *202*(4), 653–666. https://doi.org/10.1083/jcb.201304133.

Dupont, S., Morsut, L., Aragona, M., Enzo, E., Giulitti, S., Cordenonsi, M., et al. (2011). Role of YAP/TAZ in mechanotransduction. *Nature*, *474*(7350), 179–183. https://doi.org/10.1038/nature10137.

Elbediwy, A., Vincent-Mistiaen, Z. I., Spencer-Dene, B., Stone, R. K., Boeing, S., Wculek, S. K., et al. (2016). Integrin signalling regulates YAP and TAZ to control skin homeostasis. *Development (Cambridge, England)*, *143*(10), 1674–1687. https://doi.org/10.1242/dev.133728.

Elosegui-Artola, A., Andreu, I., Beedle, A. E. M., Lezamiz, A., Uroz, M., Kosmalska, A. J., et al. (2017). Force triggers YAP nuclear entry by regulating transport across nuclear pores. *Cell*, *171*(6), 1397–1410.e14. https://doi.org/10.1016/j.cell.2017.10.008.

Ezratty, E. J., Stokes, N., Chai, S., Shah, A. S., Williams, S. E., & Fuchs, E. (2011). A role for the primary cilium in notch signaling and epidermal differentiation during skin development. *Cell*, *145*(7), 1129–1141. https://doi.org/10.1016/j.cell.2011.05.030.

Friedl, P., & Mayor, R. (2017). Tuning collective cell migration by cell-cell junction regulation. *Cold Spring Harbor Perspectives in Biology*, *9*(4), a029199. https://doi.org/10.1101/cshperspect.a029199.

Galbraith, C. G., Yamada, K. M., & Sheetz, M. P. (2002). The relationship between force and focal complex development. *The Journal of Cell Biology*, *159*(4), 695–705. https://doi.org/10.1083/jcb.200204153.

Gautrot, J. E., Malmström, J., Sundh, M., Margadant, C., Sonnenberg, A., & Sutherland, D. S. (2014). The nanoscale geometrical maturation of focal adhesions controls stem cell differentiation and mechanotransduction. *Nano Letters*, *14*(7), 3945–3952. https://doi.org/10.1021/nl501248y.

Geuijen, C. A. W., & Sonnenberg, A. (2002). Dynamics of the alpha6beta4 integrin in keratinocytes. *Molecular Biology of the Cell*, *13*(11), 3845–3858. https://doi.org/10.1091/mbc.02-01-0601.

Giancotti, F. G., & Ruoslahti, E. (1999). Integrin signaling. *Science (New York, N.Y.)*, *285*(5430), 1028–1032.

Gloerich, M., Bianchini, J. M., Siemers, K. A., Cohen, D. J., & Nelson, W. J. (2017). Cell division orientation is coupled to cell-cell adhesion by the E-cadherin/LGN complex. *Nature Communications*, *8*, 13996. https://doi.org/10.1038/ncomms13996.

Grashoff, C., Hoffman, B. D., Brenner, M. D., Zhou, R., Parsons, M., Yang, M. T., et al. (2010). Measuring mechanical tension across vinculin reveals regulation of focal adhesion dynamics. *Nature*, *466*(7303), 263–266. https://doi.org/10.1038/nature09198.

Grasset, E. M., Bertero, T., Bozec, A., Friard, J., Bourget, I., Pisano, S., et al. (2018). Matrix stiffening and EGFR cooperate to promote the collective invasion of cancer cells. *Cancer Research*, *78*(18), 5229–5242. https://doi.org/10.1158/0008-5472.CAN-18-0601.

Guilak, F., Cohen, D. M., Estes, B. T., Gimble, J. M., Liedtke, W., & Chen, C. S. (2009). Control of stem cell fate by physical interactions with the extracellular matrix. *Cell Stem Cell*, *5*(1), 17–26. https://doi.org/10.1016/j.stem.2009.06.016.

Harmon, R. M., Simpson, C. L., Johnson, J. L., Koetsier, J. L., Dubash, A. D., Najor, N. A., et al. (2013). Desmoglein-1/Erbin interaction suppresses ERK activation to support epidermal differentiation. *The Journal of Clinical Investigation*, *123*(4), 1556–1570. https://doi.org/10.1172/JCI65220.

Hart, K. C., Tan, J., Siemers, K. A., Sim, J. Y., Pruitt, B. L., Nelson, W. J., et al. (2017). E-cadherin and LGN align epithelial cell divisions with tissue tension independently of cell shape. *Proceedings of the National Academy of Sciences of the United States of America*, *114*(29), E5845–E5853. https://doi.org/10.1073/pnas.1701703114.

Hawkins, T., Mirigian, M., Selcuk Yasar, M., & Ross, J. L. (2010). Mechanics of microtubules. *Journal of Biomechanics*, *43*(1), 23–30. https://doi.org/10.1016/j.jbiomech.2009.09.005.

Heo, S.-J., Han, W. M., Szczesny, S. E., Cosgrove, B. D., Elliott, D. M., Lee, D. A., et al. (2016). Mechanically induced chromatin condensation requires cellular contractility in mesenchymal stem cells. *Biophysical Journal, 111*(4), 864–874. https://doi.org/10.1016/j.bpj.2016.07.006.

Heo, S.-J., Thorpe, S. D., Driscoll, T. P., Duncan, R. L., Lee, D. A., & Mauck, R. L. (2015). Biophysical regulation of chromatin architecture instills a mechanical memory in mesenchymal stem cells. *Scientific Reports, 5,* 16895. https://doi.org/10.1038/srep16895.

Ho, C. Y., Jaalouk, D. E., Vartiainen, M. K., & Lammerding, J. (2013). Lamin A/C and emerin regulate MKL1-SRF activity by modulating actin dynamics. *Nature, 497*(7450), 507–511. https://doi.org/10.1038/nature12105.

Homberg, M., Ramms, L., Schwarz, N., Dreissen, G., Leube, R. E., Merkel, R., et al. (2015). Distinct impact of two keratin mutations causing epidermolysis bullosa simplex on keratinocyte adhesion and stiffness. *The Journal of Investigative Dermatology, 135*(10), 2437–2445. https://doi.org/10.1038/jid.2015.184.

Humphries, J. D., Wang, P., Streuli, C., Geiger, B., Humphries, M. J., & Ballestrem, C. (2007). Vinculin controls focal adhesion formation by direct interactions with talin and actin. *The Journal of Cell Biology, 179*(5), 1043–1057. https://doi.org/10.1083/jcb.200703036.

Huveneers, S., Oldenburg, J., Spanjaard, E., van der Krogt, G., Grigoriev, I., Akhmanova, A., et al. (2012). Vinculin associates with endothelial VE-cadherin junctions to control force-dependent remodeling. *The Journal of Cell Biology, 196*(5), 641–652. https://doi.org/10.1083/jcb.201108120.

Hynes, R. O. (2002). Integrins: Bidirectional, allosteric signaling machines. *Cell, 110*(6), 673–687.

Ingber, D. E. (2003a). Tensegrity I. Cell structure and hierarchical systems biology. *Journal of Cell Science, 116*(Pt. 7), 1157–1173.

Ingber, D. E. (2003b). Tensegrity II. How structural networks influence cellular information processing networks. *Journal of Cell Science, 116*(Pt. 8), 1397–1408.

Iskratsch, T., Wolfenson, H., & Sheetz, M. P. (2014). Appreciating force and shape—The rise of mechanotransduction in cell biology. *Nature Reviews. Molecular Cell Biology, 15*(12), 825–833. https://doi.org/10.1038/nrm3903.

Ivanovska, I. L., Swift, J., Spinler, K., Dingal, D., Cho, S., & Discher, D. E. (2017). Cross-linked matrix rigidity and soluble retinoids synergize in nuclear lamina regulation of stem cell differentiation. *Molecular Biology of the Cell, 28*(14), 2010–2022. https://doi.org/10.1091/mbc.E17-01-0010.

Iyer, K. V., Pulford, S., Mogilner, A., & Shivashankar, G. V. (2012). Mechanical activation of cells induces chromatin remodeling preceding MKL nuclear transport. *Biophysical Journal, 103*(7), 1416–1428. https://doi.org/10.1016/j.bpj.2012.08.041.

Jensen, P. J., Telegan, B., Lavker, R. M., & Wheelock, M. J. (1997). E-cadherin and P-cadherin have partially redundant roles in human epidermal stratification. *Cell and Tissue Research, 288*(2), 307–316.

Jones, P. H., & Watt, F. M. (1993). Separation of human epidermal stem cells from transit amplifying cells on the basis of differences in integrin function and expression. *Cell, 73*(4), 713–724.

Kao, A. P., Connelly, J. T., & Barber, A. H. (2016). 3D nanomechanical evaluations of dermal structures in skin. *Journal of the Mechanical Behavior of Biomedical Materials, 57,* 14–23. https://doi.org/10.1016/j.jmbbm.2015.11.017.

Kaur, P., & Li, A. (2000). Adhesive properties of human basal epidermal cells: An analysis of keratinocyte stem cells, transit amplifying cells, and postmitotic differentiating cells. *Journal of Investigative Dermatology, 114*(3), 413–420. Retrieved from https://doi.org/10.1046/j.1523-1747.2000.00884.x.

Keeling, M. C., Flores, L. R., Dodhy, A. H., Murray, E. R., & Gavara, N. (2017). Actomyosin and vimentin cytoskeletal networks regulate nuclear shape, mechanics and chromatin organization. *Scientific Reports*, 7(1), 5219. https://doi.org/10.1038/s41598-017-05467-x.

Kenny, F. N., Drymoussi, Z., Delaine-Smith, R., Kao, A. P., Laly, A. C., Knight, M. M., et al. (2018). Tissue stiffening promotes keratinocyte proliferation via activation of epidermal growth factor signaling. *Journal of Cell Science*, *131*(10), https://doi.org/10.1242/jcs.215780.

Kippenberger, S., Bernd, A., Loitsch, S., Guschel, M., Muller, J., Bereiter-Hahn, J., et al. (2000). Signaling of mechanical stretch in human keratinocytes via MAP kinases. *The Journal of Investigative Dermatology*, *114*(3), 408–412.

Kippenberger, S., Loitsch, S., Guschel, M., Müller, J., Knies, Y., Kaufmann, R., et al. (2005). Mechanical stretch stimulates protein kinase B/Akt phosphorylation in epidermal cells via angiotensin II type 1 receptor and epidermal growth factor receptor. *The Journal of Biological Chemistry*, *280*(4), 3060–3067. https://doi.org/10.1074/jbc.M409590200.

Knies, Y., Bernd, A., Kaufmann, R., Bereiter-Hahn, J., & Kippenberger, S. (2006). Mechanical stretch induces clustering of beta1-integrins and facilitates adhesion. *Experimental Dermatology*, *15*(5), 347–355. https://doi.org/10.1111/j.0906-6705.2006.00422.x.

Koegel, H., von Tobel, L., Schafer, M., Alberti, S., Kremmer, E., Mauch, C., et al. (2009). Loss of serum response factor in keratinocytes results in hyperproliferative skin disease in mice. *The Journal of Clinical Investigation*, *119*(4), 899–910.

Kong, D., Megone, W., Nguyen, K. D. Q., Di Cio, S., Ramstedt, M., & Gautrot, J. E. (2018). Protein nanosheet mechanics controls cell adhesion and expansion on low-viscosity liquids. *Nano Letters*, *18*(3), 1946–1951. https://doi.org/10.1021/acs.nanolett.7b05339.

Kong, D., Peng, L., Di Cio, S., Novak, P., & Gautrot, J. E. (2018). Stem cell expansion and fate decision on liquid substrates are regulated by self-assembled nanosheets. *ACS Nano*, *12*(9), 9206–9213. https://doi.org/10.1021/acsnano.8b03865.

Koster, J., Geerts, D., Favre, B., Borradori, L., & Sonnenberg, A. (2003). Analysis of the interactions between BP180, BP230, plectin and the integrin alpha6beta4 important for hemidesmosome assembly. *Journal of Cell Science*, *116*(Pt. 2), 387–399.

Kumar, V., Bouameur, J.-E., Bär, J., Rice, R. H., Hornig-Do, H.-T., Roop, D. R., et al. (2015). A keratin scaffold regulates epidermal barrier formation, mitochondrial lipid composition, and activity. *The Journal of Cell Biology*, *211*(5), 1057–1075. https://doi.org/10.1083/jcb.201404147.

Ladoux, B., & Mège, R.-M. (2017). Mechanobiology of collective cell behaviours. *Nature Reviews. Molecular Cell Biology*, *18*(12), 743–757. https://doi.org/10.1038/nrm.2017.98.

Lammerding, J., Fong, L. G., Ji, J. Y., Reue, K., Stewart, C. L., Young, S. G., et al. (2006). Lamins A and C but not lamin B1 regulate nuclear mechanics. *The Journal of Biological Chemistry*, *281*(35), 25768–25780. https://doi.org/10.1074/jbc.M513511200.

Lammerding, J., Schulze, P. C., Takahashi, T., Kozlov, S., Sullivan, T., Kamm, R. D., et al. (2004). Lamin A/C deficiency causes defective nuclear mechanics and mechanotransduction. *The Journal of Clinical Investigation*, *113*(3), 370–378. https://doi.org/10.1172/JCI19670.

le Duc, Q., Shi, Q., Blonk, I., Sonnenberg, A., Wang, N., Leckband, D., et al. (2010). Vinculin potentiates E-cadherin mechanosensing and is recruited to actin-anchored sites within adherens junctions in a myosin II-dependent manner. *The Journal of Cell Biology*, *189*(7), 1107–1115. https://doi.org/10.1083/jcb.201001149.

Le, H. Q., Ghatak, S., Yeung, C.-Y. C., Tellkamp, F., Günschmann, C., Dieterich, C., et al. (2016). Mechanical regulation of transcription controls polycomb-mediated gene silencing during lineage commitment. *Nature Cell Biology*, *18*(8), 864–875. https://doi.org/10.1038/ncb3387.

Lechler, T., & Fuchs, E. (2005). Asymmetric cell divisions promote stratification and differentiation of mammalian skin. *Nature, 437*(7056), 275–280. https://doi.org/10.1038/nature03922.

Lechler, T., & Fuchs, E. (2007). Desmoplakin: An unexpected regulator of microtubule organization in the epidermis. *The Journal of Cell Biology, 176*(2), 147–154. https://doi.org/10.1083/jcb.200609109.

Lee, C.-H., & Coulombe, P. A. (2009). Self-organization of keratin intermediate filaments into cross-linked networks. *The Journal of Cell Biology, 186*(3), 409–421. https://doi.org/10.1083/jcb.200810196.

Lee, C.-H., Kim, M.-S., Chung, B. M., Leahy, D. J., & Coulombe, P. A. (2012). Structural basis for heteromeric assembly and perinuclear organization of keratin filaments. *Nature Structural & Molecular Biology, 19*(7), 707–715. https://doi.org/10.1038/nsmb.2330.

Lee, M.-J., Ran Byun, M., Furutani-Seiki, M., Hong, J.-H., & Jung, H.-S. (2013). YAP and TAZ regulate skin wound healing. *The Journal of Investigative Dermatology*. https://doi.org/10.1038/jid.2013.339.

Leerberg, J. M., Gomez, G. A., Verma, S., Moussa, E. J., Wu, S. K., Priya, R., et al. (2014). Tension-sensitive actin assembly supports contractility at the epithelial zonula adherens. *Current Biology: CB, 24*(15), 1689–1699. https://doi.org/10.1016/j.cub.2014.06.028.

Levental, K. R., Yu, H., Kass, L., Lakins, J. N., Egeblad, M., Erler, J. T., et al. (2009). Matrix crosslinking forces tumor progression by enhancing integrin signaling. *Cell, 139*(5), 891–906. Retrieved from http://www.sciencedirect.com/science/article/pii/S0092867409013531.

Li, A., Simmons, P. J., & Kaur, P. (1998). Identification and isolation of candidate human keratinocyte stem cells based on cell surface phenotype. *Proceedings of the National Academy of Sciences of the United States of America, 95*(7), 3902–3907.

Lin, C., Hindes, A., Burns, C. J., Koppel, A. C., Kiss, A., Yin, Y., et al. (2013). Serum response factor controls transcriptional network regulating epidermal function and hair follicle morphogenesis. *The Journal of Investigative Dermatology, 133*(3), 608–617. https://doi.org/10.1038/jid.2012.378.

Liu, M., Zhao, S., Lin, Q., & Wang, X.-P. (2015). YAP regulates the expression of Hoxa1 and Hoxc13 in mouse and human oral and skin epithelial tissues. *Molecular and Cellular Biology, 35*(8), 1449–1461. https://doi.org/10.1128/MCB.00765-14.

Luke, Y., Zaim, H., Karakesisoglou, I., Jaeger, V. M., Sellin, L., Lu, W., et al. (2008). Nesprin-2 Giant (NUANCE) maintains nuclear envelope architecture and composition in skin. *Journal of Cell Science, 121*(Pt. 11), 1887–1898. https://doi.org/10.1242/jcs.019075.

Luxenburg, C., Pasolli, H. A., Williams, S. E., & Fuchs, E. (2011). Developmental roles for Srf, cortical cytoskeleton and cell shape in epidermal spindle orientation. *Nature Cell Biology, 13*(3), 203–214. https://doi.org/10.1038/ncb2163.

Maheshwari, G., Brown, G., Lauffenburger, D. A., Wells, A., & Griffith, L. G. (2000). Cell adhesion and motility depend on nanoscale RGD clustering. *Journal of Cell Science, 113*(Pt. 10), 1677–1686.

Makhija, E., Jokhun, D. S., & Shivashankar, G. V. (2016). Nuclear deformability and telomere dynamics are regulated by cell geometric constraints. *Proceedings of the National Academy of Sciences of the United States of America, 113*(1), E32–E40. https://doi.org/10.1073/pnas.1513189113.

Maruthappu, T., Chikh, A., Fell, B., Delaney, P. J., Brooke, M. A., Levet, C., et al. (2017). Rhomboid family member 2 regulates cytoskeletal stress-associated Keratin 16. *Nature Communications, 8*, 14174. https://doi.org/10.1038/ncomms14174.

Meng, Z., Qiu, Y., Lin, K. C., Kumar, A., Placone, J. K., Fang, C., et al. (2018). RAP2 mediates mechanoresponses of the hippo pathway. *Nature, 560*(7720), 655–660. https://doi.org/10.1038/s41586-018-0444-0.

Mertz, A. F., Che, Y., Banerjee, S., Goldstein, J. M., Rosowski, K. A., Revilla, S. F., et al. (2013). Cadherin-based intercellular adhesions organize epithelial cell-matrix traction forces. *Proceedings of the National Academy of Sciences of the United States of America, 110*(3), 842–847. https://doi.org/10.1073/pnas.1217279110.

Miles, F. L., & Sikes, R. A. (2014). Insidious changes in stromal matrix fuel cancer progression. *Molecular Cancer Research: MCR, 12*(3), 297–312. https://doi.org/10.1158/1541-7786.MCR-13-0535.

Miralles, F., Posern, G., Zaromytidou, A. I., & Treisman, R. (2003). Actin dynamics control SRF activity by regulation of its coactivator MAL. *Cell, 113*(3), 329–342.

Miroshnikova, Y. A., Le, H. Q., Schneider, D., Thalheim, T., Rübsam, M., Bremicker, N., et al. (2018). Adhesion forces and cortical tension couple cell proliferation and differentiation to drive epidermal stratification. *Nature Cell Biology, 20*(1), 69–80. https://doi.org/10.1038/s41556-017-0005-z.

Mittapalli, V. R., Madl, J., Löffek, S., Kiritsi, D., Kern, J. S., Römer, W., et al. (2016). Injury-driven stiffening of the dermis expedites skin carcinoma progression. *Cancer Research, 76*(4), 940–951. https://doi.org/10.1158/0008-5472.CAN-15-1348.

Morley, S. M., D'Alessandro, M., Sexton, C., Rugg, E. L., Navsaria, H., Shemanko, C. S., et al. (2003). Generation and characterization of epidermolysis bullosa simplex cell lines: Scratch assays show faster migration with disruptive keratin mutations. *The British Journal of Dermatology, 149*(1), 46–58.

Myers, K. A., Applegate, K. T., Danuser, G., Fischer, R. S., & Waterman, C. M. (2011). Distinct ECM mechanosensing pathways regulate microtubule dynamics to control endothelial cell branching morphogenesis. *The Journal of Cell Biology, 192*(2), 321–334. https://doi.org/10.1083/jcb.201006009.

Nekrasova, O., Harmon, R. M., Broussard, J. A., Koetsier, J. L., Godsel, L. M., Fitz, G. N., et al. (2018). Desmosomal cadherin association with Tctex-1 and cortactin-Arp2/3 drives perijunctional actin polymerization to promote keratinocyte delamination. *Nature Communications, 9*(1), 1053. https://doi.org/10.1038/s41467-018-03414-6.

Ng, M. R., Besser, A., Danuser, G., & Brugge, J. S. (2012). Substrate stiffness regulates cadherin-dependent collective migration through myosin-II contractility. *The Journal of Cell Biology, 199*(3), 545–563. https://doi.org/10.1083/jcb.201207148.

Ng, Y.-Z., Pourreyron, C., Salas-Alanis, J. C., Dayal, J. H. S., Cepeda-Valdes, R., Yan, W., et al. (2012). Fibroblast-derived dermal matrix drives development of aggressive cutaneous squamous cell carcinoma in patients with recessive dystrophic epidermolysis bullosa. *Cancer Research, 72*(14), 3522–3534. https://doi.org/10.1158/0008-5472.CAN-11-2996.

Noethel, B., Ramms, L., Dreissen, G., Hoffmann, M., Springer, R., Rübsam, M., et al. (2018). Transition of responsive mechanosensitive elements from focal adhesions to adherens junctions on epithelial differentiation. *Molecular Biology of the Cell, 29*(19), 2317–2325. https://doi.org/10.1091/mbc.E17-06-0387.

Osmanagic-Myers, S., Gregor, M., Walko, G., Burgstaller, G., Reipert, S., & Wiche, G. (2006). Plectin-controlled keratin cytoarchitecture affects MAP kinases involved in cellular stress response and migration. *The Journal of Cell Biology, 174*(4), 557–568. https://doi.org/10.1083/jcb.200605172.

Osmani, N., Pontabry, J., Comelles, J., Fekonja, N., Goetz, J. G., Riveline, D., et al. (2018). An Arf6- and caveolae-dependent pathway links hemidesmosome remodeling and mechanoresponse. *Molecular Biology of the Cell, 29*(4), 435–451. https://doi.org/10.1091/mbc.E17-06-0356.

Paszek, M. J., Zahir, N., Johnson, K. R., Lakins, J. N., Rozenberg, G. I., Gefen, A., et al. (2005). Tensional homeostasis and the malignant phenotype. *Cancer Cell, 8*(3), 241–254. https://doi.org/10.1016/j.ccr.2005.08.010.

Pearce, O. M. T., Delaine-Smith, R. M., Maniati, E., Nichols, S., Wang, J., Böhm, S., et al. (2018). Deconstruction of a metastatic tumor microenvironment reveals a common matrix response in human cancers. *Cancer Discovery*, *8*(3), 304–319. https://doi.org/10.1158/2159-8290.CD-17-0284.

Philimonenko, V. V., Zhao, J., Iben, S., Dingova, H., Kysela, K., Kahle, M., et al. (2004). Nuclear actin and myosin I are required for RNA polymerase I transcription. *Nature Cell Biology*, *6*(12), 1165–1172. https://doi.org/10.1038/ncb1190.

Powell, H. M., McFarland, K. L., Butler, D. L., Supp, D. M., & Boyce, S. T. (2010). Uniaxial strain regulates morphogenesis, gene expression, and tissue strength in engineered skin. *Tissue Engineering. Part A*, *16*(3), 1083–1092. https://doi.org/10.1089/ten.TEA.2009.0542.

Puzzi, L., Borin, D., Martinelli, V., Mestroni, L., Kelsell, D. P., & Sbaizero, O. (2018). Cellular biomechanics impairment in keratinocytes is associated with a C-terminal truncated desmoplakin: An atomic force microscopy investigation. *Micron (Oxford, England: 1993)*, *106*, 27–33. https://doi.org/10.1016/j.micron.2017.12.005.

Ramms, L., Fabris, G., Windoffer, R., Schwarz, N., Springer, R., Zhou, C., et al. (2013). Keratins as the main component for the mechanical integrity of keratinocytes. *Proceedings of the National Academy of Sciences of the United States of America*, *110*(46), 18513–18518. https://doi.org/10.1073/pnas.1313491110.

Rübsam, M., Mertz, A. F., Kubo, A., Marg, S., Jüngst, C., Goranci-Buzhala, G., et al. (2017). E-cadherin integrates mechanotransduction and EGFR signaling to control junctional tissue polarization and tight junction positioning. *Nature Communications*, *8*(1), 1250. https://doi.org/10.1038/s41467-017-01170-7.

Russell, D., Andrews, P. D., James, J., & Lane, E. B. (2004). Mechanical stress induces profound remodelling of keratin filaments and cell junctions in epidermolysis bullosa simplex keratinocytes. *Journal of Cell Science*, *117*(Pt. 22), 5233–5243. https://doi.org/10.1242/jcs.01407.

Saito, M., Tucker, D. K., Kohlhorst, D., Niessen, C. M., & Kowalczyk, A. P. (2012). Classical and desmosomal cadherins at a glance. *Journal of Cell Science*, *125*(Pt. 11), 2547–2552. https://doi.org/10.1242/jcs.066654.

Samuel, M. S., Lopez, J. I., McGhee, E. J., Croft, D. R., Strachan, D., Timpson, P., et al. (2011). Actomyosin-mediated cellular tension drives increased tissue stiffness and β-catenin activation to induce epidermal hyperplasia and tumor growth. *Cancer Cell*, *19*(6), 776–791. https://doi.org/10.1016/j.ccr.2011.05.008.

Samuelov, L., Sprecher, E., & Paus, R. (2015). The role of P-cadherin in skin biology and skin pathology: Lessons from the hair follicle. *Cell and Tissue Research*, *360*(3), 761–771. https://doi.org/10.1007/s00441-015-2114-y.

Schlegelmilch, K., Mohseni, M., Kirak, O., Pruszak, J., Rodriguez, J. R., Zhou, D., et al. (2011). Yap1 acts downstream of α-catenin to control epidermal proliferation. *Cell*, *144*(5), 782–795. https://doi.org/10.1016/j.cell.2011.02.031.

Schröder, R., Kunz, W. S., Rouan, F., Pfendner, E., Tolksdorf, K., Kappes-Horn, K., et al. (2002). Disorganization of the desmin cytoskeleton and mitochondrial dysfunction in plectin-related epidermolysis bullosa simplex with muscular dystrophy. *Journal of Neuropathology and Experimental Neurology*, *61*(6), 520–530.

Seddiki, R., Narayana, G. H. N. S., Strale, P.-O., Balcioglu, H. E., Peyret, G., Yao, M., et al. (2018). Force-dependent binding of vinculin to α-catenin regulates cell-cell contact stability and collective cell behavior. *Molecular Biology of the Cell*, *29*(4), 380–388. https://doi.org/10.1091/mbc.E17-04-0231.

Seltmann, K., Cheng, F., Wiche, G., Eriksson, J. E., & Magin, T. M. (2015). Keratins stabilize Hemidesmosomes through regulation of β4-integrin turnover. *The Journal of Investigative Dermatology*, *135*(6), 1609–1620. https://doi.org/10.1038/jid.2015.46.

Seltmann, K., Fritsch, A. W., Käs, J. A., & Magin, T. M. (2013). Keratins significantly contribute to cell stiffness and impact invasive behavior. *Proceedings of the National Academy of Sciences of the United States of America*, *110*(46), 18507–18512. https://doi.org/10.1073/pnas.1310493110.

Selvaraj, A., & Prywes, R. (2004). Expression profiling of serum inducible genes identifies a subset of SRF target genes that are MKL dependent. *BMC Molecular Biology*, *5*, 13.

Sharili, A. S., Kenny, F. N., Vartiainen, M. K., & Connelly, J. T. (2016). Nuclear actin modulates cell motility via transcriptional regulation of adhesive and cytoskeletal genes. *Scientific Reports*, *6*, 33893. https://doi.org/10.1038/srep33893.

Song, Z., Wang, M., Wang, X., Pan, X., Liu, W., Hao, S., et al. (2007). Nuclear actin is involved in the regulation of CSF1 gene transcription in a chromatin required, BRG1 independent manner. *Journal of Cellular Biochemistry*, *102*(2), 403–411. https://doi.org/10.1002/jcb.21300.

Sotiropoulos, A., Gineitis, D., Copeland, J., & Treisman, R. (1999). Signal-regulated activation of serum response factor is mediated by changes in actin dynamics. *Cell*, *98*(2), 159–169.

Stewart, R. M., Zubek, A. E., Rosowski, K. A., Schreiner, S. M., Horsley, V., & King, M. C. (2015). Nuclear-cytoskeletal linkages facilitate cross talk between the nucleus and intercellular adhesions. *The Journal of Cell Biology*, *209*(3), 403–418. https://doi.org/10.1083/jcb.201502024.

Swift, J., Ivanovska, I. L., Buxboim, A., Harada, T., Dingal, P. C. D. P., Pinter, J., et al. (2013). Nuclear Lamin-A scales with tissue stiffness and enhances matrix-directed differentiation. *Science (New York, N.Y.)*, *341*(6149), 1240104. https://doi.org/10.1126/science.1240104.

Takei, T., Han, O., Ikeda, M., Male, P., Mills, I., & Sumpio, B. E. (1997). Cyclic strain stimulates isoform-specific PKC activation and translocation in cultured human keratinocytes. *Journal of Cellular Biochemistry*, *67*(3), 327–337.

Temiyasathit, S., & Jacobs, C. R. (2010). Osteocyte primary cilium and its role in bone mechanotransduction. *Annals of the New York Academy of Sciences*, *1192*, 422–428. https://doi.org/10.1111/j.1749-6632.2009.05243.x.

Thompson, C. L., Chapple, J. P., & Knight, M. M. (2014). Primary cilia disassembly down-regulates mechanosensitive hedgehog signalling: A feedback mechanism controlling ADAMTS-5 expression in chondrocytes. *Osteoarthritis and Cartilage*, *22*(3), 490–498. https://doi.org/10.1016/j.joca.2013.12.016.

Tinkle, C. L., Lechler, T., Pasolli, H. A., & Fuchs, E. (2004). Conditional targeting of E-cadherin in skin: Insights into hyperproliferative and degenerative responses. *Proceedings of the National Academy of Sciences of the United States of America*, *101*(2), 552–557. https://doi.org/10.1073/pnas.0307437100.

Tokuyama, E., Nagai, Y., Takahashi, K., Kimata, Y., & Naruse, K. (2015). Mechanical stretch on human skin equivalents increases the epidermal thickness and develops the basement membrane. *PLoS One*, *10*(11), https://doi.org/10.1371/journal.pone.0141989.

Tomakidi, P., Schulz, S., Proksch, S., Weber, W., & Steinberg, T. (2014). Focal adhesion kinase (FAK) perspectives in mechanobiology: Implications for cell behaviour. *Cell and Tissue Research*, *357*(3), 515–526. https://doi.org/10.1007/s00441-014-1945-2.

Totaro, A., Castellan, M., Battilana, G., Zanconato, F., Azzolin, L., Giulitti, S., et al. (2017). YAP/TAZ link cell mechanics to notch signalling to control epidermal stem cell fate. *Nature Communications*, *8*, 15206. https://doi.org/10.1038/ncomms15206.

Totaro, A., Panciera, T., & Piccolo, S. (2018). YAP/TAZ upstream signals and downstream responses. *Nature Cell Biology*, *20*(8), 888–899. https://doi.org/10.1038/s41556-018-0142-z.

Trappmann, B., Gautrot, J. E., Connelly, J. T., Strange, D. G., Li, Y., Oyen, M. L., et al. (2011). Extracellular-matrix tethering regulates stem-cell fate. *Nature Materials*, *11*(7), 642–649. https://doi.org/10.1038/nmat3339.

Tsutsumi, M., Inoue, K., Denda, S., Ikeyama, K., Goto, M., & Denda, M. (2009). Mechanical-stimulation-evoked calcium waves in proliferating and differentiated human keratinocytes. *Cell and Tissue Research*, *338*(1), 99–106. https://doi.org/10.1007/s00441-009-0848-0.

Vasioukhin, V., Bauer, C., Degenstein, L., Wise, B., & Fuchs, E. (2001). Hyperproliferation and defects in epithelial polarity upon conditional ablation of alpha-catenin in skin. *Cell*, *104*(4), 605–617.

Vedula, S. R. K., Hirata, H., Nai, M. H., Brugués, A., Toyama, Y., Trepat, X., et al. (2014). Epithelial bridges maintain tissue integrity during collective cell migration. *Nature Materials*, *13*(1), 87–96. https://doi.org/10.1038/nmat3814.

Verma, S., Han, S. P., Michael, M., Gomez, G. A., Yang, Z., Teasdale, R. D., et al. (2012). A WAVE2-Arp2/3 actin nucleator apparatus supports junctional tension at the epithelial zonula adherens. *Molecular Biology of the Cell*, *23*(23), 4601–4610. https://doi.org/10.1091/mbc.E12-08-0574.

Versaevel, M., Grevesse, T., & Gabriele, S. (2012). Spatial coordination between cell and nuclear shape within micropatterned endothelial cells. *Nature Communications*, *3*, 671. https://doi.org/10.1038/ncomms1668.

Vielmuth, F., Wanuske, M.-T., Radeva, M. Y., Hiermaier, M., Kugelmann, D., Walter, E., et al. (2018). Keratins regulate the adhesive properties of desmosomal cadherins through signaling. *The Journal of Investigative Dermatology*, *138*(1), 121–131. https://doi.org/10.1016/j.jid.2017.08.033.

Walko, G., Woodhouse, S., Pisco, A. O., Rognoni, E., Liakath-Ali, K., Lichtenberger, B. M., et al. (2017). A genome-wide screen identifies YAP/WBP2 interplay conferring growth advantage on human epidermal stem cells. *Nature Communications*, *8*, 14744. https://doi.org/10.1038/ncomms14744.

Wang, Y., Nagarajan, M., Uhler, C., & Shivashankar, G. V. (2017). Orientation and repositioning of chromosomes correlate with cell geometry-dependent gene expression. *Molecular Biology of the Cell*, *28*(14), 1997–2009. https://doi.org/10.1091/mbc.E16-12-0825.

Wann, A. K. T., Zuo, N., Haycraft, C. J., Jensen, C. G., Poole, C. A., McGlashan, S. R., et al. (2012). Primary cilia mediate mechanotransduction through control of ATP-induced Ca2+ signaling in compressed chondrocytes. *FASEB Journal: Official Publication of the Federation of American Societies for Experimental Biology*, *26*(4), 1663–1671. https://doi.org/10.1096/fj.11-193649.

Watt, F. M., Kubler, M. D., Hotchin, N. A., Nicholson, L. J., & Adams, J. C. (1993). Regulation of keratinocyte terminal differentiation by integrin-extracellular matrix interactions. *Journal of Cell Science*, *106*(Pt. 1), 175–182.

Wickert, L. E., Pomerenke, S., Mitchell, I., Masters, K. S., & Kreeger, P. K. (2016). Hierarchy of cellular decisions in collective behavior: Implications for wound healing. *Scientific Reports*, *6*, 20139. https://doi.org/10.1038/srep20139.

Wickström, S. A., Lange, A., Montanez, E., & Fässler, R. (2010). The ILK/PINCH/parvin complex: The kinase is dead, long live the pseudokinase! *The EMBO Journal*, *29*(2), 281–291. https://doi.org/10.1038/emboj.2009.376.

Wolff, J. (1986). *The law of bone remodelling*. Berlin Heidelberg: Springer-Verlag. Retrieved from www.springer.com/us/book/9783642710339.

Wu, X., Kodama, A., & Fuchs, E. (2008). ACF7 regulates cytoskeletal-focal adhesion dynamics and migration and has ATPase activity. *Cell*, *135*(1), 137–148. https://doi.org/10.1016/j.cell.2008.07.045.

Wu, J., Misra, G., Russell, R. J., Ladd, A. J. C., Lele, T. P., & Dickinson, R. B. (2011). Effects of dynein on microtubule mechanics and centrosome positioning. *Molecular Biology of the Cell, 22*(24), 4834–4841. https://doi.org/10.1091/mbc.E11-07-0611.

Xia, Y., Ivanovska, I. L., Zhu, K., Smith, L., Irianto, J., Pfeifer, C. R., et al. (2018). Nuclear rupture at sites of high curvature compromises retention of DNA repair factors. *The Journal of Cell Biology, 217*(11), 3796–3808. https://doi.org/10.1083/jcb.201711161.

Yao, M., Goult, B. T., Klapholz, B., Hu, X., Toseland, C. P., Guo, Y., et al. (2016). The mechanical response of Talin. *Nature Communications*, 7, 11966. https://doi.org/10.1038/ncomms11966.

Young, P., Boussadia, O., Halfter, H., Grose, R., Berger, P., Leone, D. P., et al. (2003). E-cadherin controls adherens junctions in the epidermis and the renewal of hair follicles. *The EMBO Journal, 22*(21), 5723–5733.

Zhang, H., Landmann, F., Zahreddine, H., Rodriguez, D., Koch, M., & Labouesse, M. (2011). A tension-induced mechanotransduction pathway promotes epithelial morphogenesis. *Nature, 471*(7336), 99–103. https://doi.org/10.1038/nature09765.

Zhang, H., Pasolli, H. A., & Fuchs, E. (2011). Yes-associated protein (YAP) transcriptional coactivator functions in balancing growth and differentiation in skin. *Proceedings of the National Academy of Sciences of the United States of America, 108*(6), 2270–2275. https://doi.org/10.1073/pnas.1019603108.

Zhao, B., Ye, X., Yu, J., Li, L., Li, W., Li, S., et al. (2008). TEAD mediates YAP-dependent gene induction and growth control. *Genes & Development, 22*(14), 1962–1971. https://doi.org/10.1101/gad.1664408.

CHAPTER TWO

Epidermal stem cell lineages

Lalhaba Oinam[a,b], Gopakumar Changarathil[a,c], Yen Xuan Ngo[a,b], Hiromi Yanagisawa[a,d], Aiko Sada[a,]*

[a]Life Science Center for Survival Dynamics, Tsukuba Advanced Research Alliance (TARA), University of Tsukuba, Tsukuba, Japan
[b]Ph.D. Program in Human Biology, School of Integrative and Global Majors, University of Tsukuba, Tsukuba, Japan
[c]Graduate School of Comprehensive Human Sciences, University of Tsukuba, Tsukuba, Japan
[d]Faculty of Medicine, University of Tsukuba, Tsukuba, Japan
*Corresponding author: e-mail address: aisada@tara.tsukuba.ac.jp

Contents

Advances in Stem Cells and their Niches, Volume 3
ISSN 2468-5097
https://doi.org/10.1016/bs.asn.2019.04.003

1. Introduction

Tissue homeostasis is maintained by resident stem cells, which can self-renew and differentiate, and function as a source for the cells that reconstitute the tissue. The epidermis, the outermost layer of skin, is a first barrier to protect our body from infection and dehydration. Different populations of stem cells reside at distinct compartments in the epidermis and participate in tissue homeostasis and injury repair. Due to the synchrony of the hair cycle and the hair cycle-dependent behavior of stem cells, the hair follicle serves as an excellent model for understanding the cellular and molecular basis of stem cell characters and regulations. Stem cells in the interfollicular epidermis (IFE) have been used for clinical purposes since 1980s, yet their basic knowledge lagged behind until recently due to the lack of molecular markers and tools to define "stem cell" populations. With recent advances in experimental techniques, we now have started to better understand features of stem cells in the IFE. In this review, we provide an overview of skin stem cells and highlight new insights into property and roles of stem cells and niches in the IFE.

2. Structure of skin and its stem cells

The skin comprises two compartments: (1) the epidermis, made up of keratinocytes, consisting of the IFE, skin appendages (hair follicles, sebaceous and sweat glands) and junctional zone (infundibulum, isthmus); and (2) the dermis, an underlying mesenchymal compartment mainly composed of fibroblasts and extracellular matrix (ECM) (Fig. 1). During homeostasis, the epidermis is highly-compartmentalized, and each epidermal compartment is renewed by lineage-specific stem cells within their own respective territory. Under injury or disease condition, they acquire plasticity and convert to different lineages (Adam et al., 2015; Ge et al., 2017; Hoeck et al., 2017; Ito et al., 2005; Jaks, Kasper, & Toftgard, 2010; Levy, Lindon, Harfe, & Morgan, 2005; Lu et al., 2012; Page, Lombard, Ng,

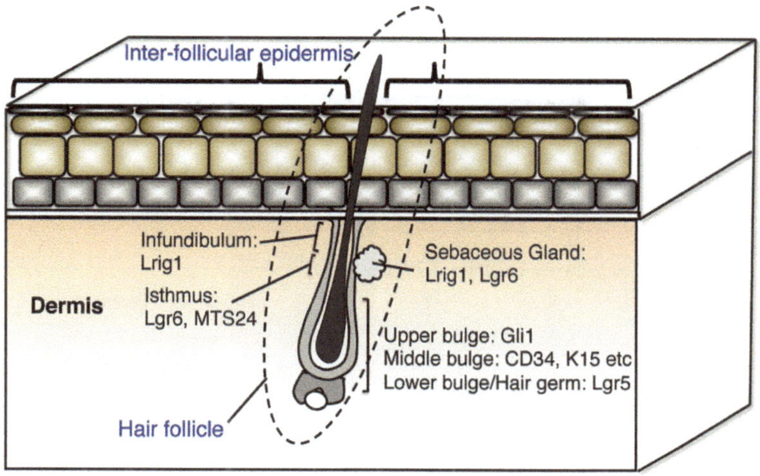

Fig. 1 Structure of mouse skin and its stem cells. Molecular markers are shown to define lineage-specific stem cell populations within each epidermal compartment.

Gottgens, & Jensen, 2013; Rognoni & Watt, 2018). This transient or permanent lineage conversion, named "lineage infidelity," is driven by activation of stress-responsive enhancers that override the lineage-specific enhancer (Ge et al., 2017).

2.1 The interfollicular epidermis and its stem cells

The IFE is a keratinized stratified squamous epithelium, covering the surface of our body. The IFE regenerates quickly and is maintained by stem cells located in the basal layer. Basal cells are in direct contact with the basement membrane and express basal marker proteins, e.g., α6/β4 integrins, Keratin5 (K5) and K14 (Fuchs, Coppock, Green, & Cleveland, 1981; Fuchs & Green, 1980; Watt & Jones, 1993). As they differentiate, they exit the cell cycle, cease expression of the basal markers and lose contact with the basement membrane. Differentiating basal cells move upwards to form the spinous, the granular and the terminally-differentiated cornified layers (Fig. 2A). The epidermal differentiation is accompanied by step-wise changes in structural protein expressions (Simpson, Patel, & Green, 2011): K1 and K10 are expressed during the early-stage differentiation (Fuchs & Green, 1980); Loricrin, Filaggrin and Involucrin are expressed during the late-stage differentiation (Candi, Schmidt, & Melino, 2005). The cells in the cornified layer are eventually shed out from the skin surface with turnover rates of 3–10 days (Potten, Saffhill, & Maibach, 1987; Rompolas et al., 2016; Sada et al., 2016),

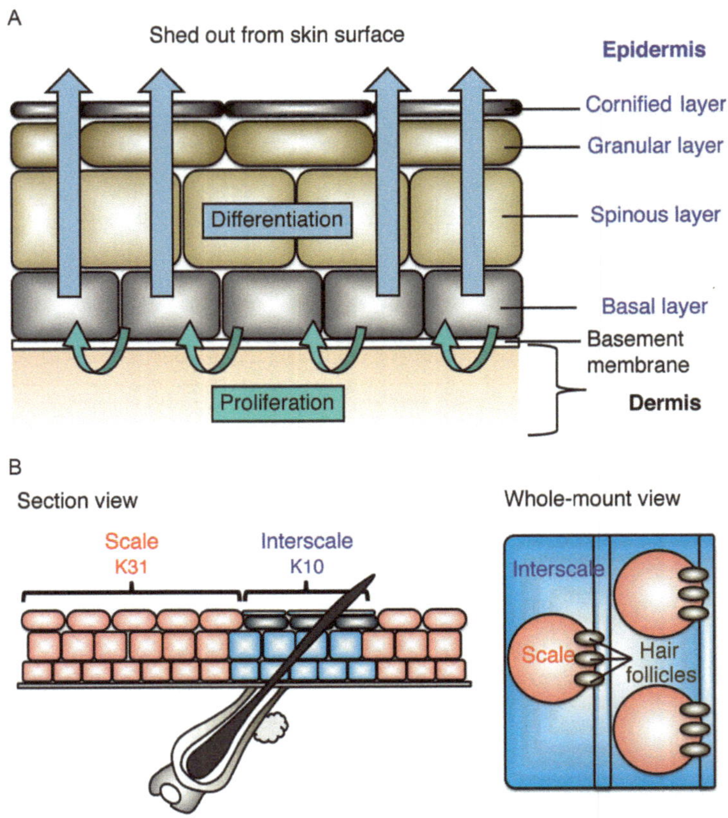

Fig. 2 Structure of the mouse interfollicular epidermis. (A) Schematic representation of mouse skin that undergoes a typical orthokeratotic keratinization. (B) Schematic representation shows mouse tail skin with sagittal section and whole-mount views. Interscale and scale regions are shown in blue and red, respectively.

thus, raising a need for continuous displacement of the lost cells (Mesa et al., 2018).

In mouse skin research, different body regions have been utilized. The divergence of skin appendages is dependent on the regional differences in mesenchymal niche (Lu, Polak, Keyes, & Fuchs, 2016; Yu et al., 2018). The dorsal and ventral skin are covered by high density of hair follicles, an ideal model for hair study; however, the IFE in adult mice is relatively thin containing only a few layers of cells. Similarly, the ear IFE commonly used for live imaging (Pineda et al., 2015), also consists of a thin layer of the IFE. In contrast, the paw and tail skin have thicker IFE than those in the dorsal or ear skin. The mouse paw skin is hairless and it harbors sweat glands in the footpad region (Lu & Fuchs, 2014). The tail IFE has characteristic structures, the interscale and scale, which were originally described in 1970s (Schweizer & Marks, 1977) and re-discovered by Gomez et al. more recently (Gomez et al., 2013) (Fig. 2B). The interscale IFE undergoes a typical orthokeratotic keratinization process accompanied by the loss of nucleus in the cornified layer. The scale forming IFE is specialized in the adult mouse tail, where the epidermis undergoes parakeratotic differentiation, resulting in the loss of granular layer and the retention of nucleus in the cells residing in the cornified layer (Gomez et al., 2013). The scale structures develop postnatally (Gomez et al., 2013). The interscale and scale IFE alternate and are regularly spaced within the tail skin, correlated with the location of triplets of hair follicles (Braun et al., 2003; Gomez et al., 2013). The scale also contains resident pigmented melanocytes (Schweizer & Marks, 1977). The interscale and scale IFE can be defined by their spatial distribution and by expressions of unique differentiation markers: K2, K10, Filaggrin and Involucrin in the interscale; K31 in the scale (Fig. 2B).

It has long been a question as to which cells act as stem cells in the IFE. Previously it was thought that epidermal homeostasis is maintained by rare, slow-cycling stem cells that give rise to short-lived, fast-dividing progeny called transit-amplifying (TA) cells. However, now it is becoming clear that there is a heterogeneity of stem/progenitor cells within the basal layer that behave differently in a context and region-specific manner. Later, we will discuss about the presence of such heterogeneous populations and their dynamic behaviors in the IFE.

2.2 Stem cells in hair follicles and other epidermal compartments

Hair follicle stem cells reside in a specialized structure, called the bulge, which was originally identified as a non-cycling, quiescent cell cluster in the resting hair follicles (Blanpain, Lowry, Geoghegan, Polak, & Fuchs, 2004; Cotsarelis, Sun, & Lavker, 1990; Tumbar et al., 2004). The bulge stem cells show several characteristic properties: infrequent cell divisions (Cotsarelis et al., 1990; Tumbar et al., 2004); unique abilities of self-renewal and multipotent differentiation (Blanpain et al., 2004); distinct molecular signature, including specific surface expression of α6-integrin and CD34 (Trempus et al., 2003), K19 and transcription factors Sox9, Lhx2 and Tcf3/4 (Blanpain et al., 2004; Morris et al., 2004; Tumbar et al., 2004) (Fig. 1). In addition, several promoters have been identified to mark distinct populations of bulge cells and confirmed their long-term contributions to the entire hair lineages by lineage tracing experiments (Jaks et al., 2010): Gli1 (Brownell, Guevara, Bai, Loomis, & Joyner, 2011), K15 (Liu, Lyle, Yang, & Cotsarelis, 2003; Morris et al., 2004) and Lgr5 (Jaks et al., 2008) preferentially marks upper-bulge, mid-bulge, lower-bulge/hair germ, respectively (Fig. 1).

Different stem cell populations have been identified and mapped in other epidermal compartments in skin (Fig. 1). In the upper junctional zone, the infundibulum and sebaceous glands are independently maintained and labeled by Lrig1 (Jensen et al., 2009; Page et al., 2013) and Lgr6 (Fullgrabe et al., 2015; Snippert, Haegebarth, et al., 2010). The lower junctional zone/isthmus is marked by Lgr6 (Fullgrabe et al., 2015; Snippert, Haegebarth, et al., 2010) and Mts24 (Nijhof et al., 2006). Recent study by Donati et al. suggested that Gata6 + terminally-differentiated population possesses remarkable plasticity to de-differentiate and become stem cells in response to injury (Donati et al., 2017) and it may also be the case for Blimp1 + differentiated population (Horsley et al., 2006; Kretzschmar et al., 2014; Montgomery et al., 2011).

In the glandular skin, such as food pads of paw skin in mice, at least three distinct, unipotent stem cell populations exist within sweat glands and its ducts (Lu & Fuchs, 2014; Lu et al., 2012). K14+ sweat duct basal cells can produce suprabasal duct cells during normal tissue turnover without contributing to the IFE or sweat glands. The sweat glands, consisting of an outer smooth muscle actin (SMA) + myoepithelial and an inner K8/K18/K19+ luminal cells, are largely quiescent during homeostasis but can repair the loss of their own lineages upon selective killing of gland cells (Lu et al., 2012).

2.3 Hair follicle stem cell dynamics

The hair follicle periodically dies out and regenerates by resident stem cells during hair cycles. The hair cycle is divided into three stages: hair growth (anagen), degeneration (catagen) and rest (telogen) (Alonso & Fuchs, 2006; Muller-Rover et al., 2001). In anagen, hair follicle consists of concentric multilayers of epidermal cells: the centrally located hair shaft, the inner root sheath (IRS), the companion layer, and the outer root sheath (ORS). At the lower part of the anagen follicle, there is a cluster of rapidly-dividing progenitor cells, called the matrix, surrounding a mesenchymal pocket of cells called the dermal papillae, a signaling center with fate instructive properties (Driskell, Clavel, Rendl, & Watt, 2011). During catagen, hair follicles regress from the lower part by apoptotic cell death or by upward movement and subsequent cell loss (Mesa et al., 2015). The telogen follicle is composed of: (1) K14/CD34+ quiescent stem cells in the outer bulge; (2) the hair germ, primed stem cells, located below the bulge; (3) a K6+ inner bulge of terminally-differentiated cells; and (4) the dermal papillae. The K6+ inner bulge acts as epithelial niche and sends growth inhibitory signals, Bmp6 and Fgf18, to outer bulge stem cells (Hsu, Pasolli, & Fuchs, 2011).

The cellular origin of K6+ inner bulge has been under debate. Possibilities include differentiating stem cells located in the lower ORS, cells located at the base of the bulb region, or terminally-differentiated cells in the anagen IRS (Hsu et al., 2011; Reichenbach et al., 2018; Sada, Jain, Wang, Leung, & Tumbar, 2017; Shinin & Panteleyev, 2017; Takeda et al., 2013). The discrepancy could be explained by less specific labeling of Cre, differences in Cre/reporter lines used, or induction of Cre in slightly different hair cycle stages; therefore, further studies will be needed.

How are hair follicle stem cells regulated during hair cycles? Here we highlight some latest studies on cellular fates of hair follicle stem cells and molecular aspects of stem cell regulation (also see reviews by Rompolas & Greco, 2014; Yi, 2017). In response to BMP growth inhibitory signals from the inner bulge (Hsu et al., 2011), stem cells in the telogen bulge express transcription factors, e.g., Nfatc1 (Horsley, Aliprantis, Polak, Glimcher, & Fuchs, 2008; Keyes et al., 2013) and Foxc1 (Lay, Kume, & Fuchs, 2016), to maintain their quiescence state. Upon activation of stem cells, "primed" stem cells appear in the hair germ region. This process is induced by activation of Wnt pathway, which compensates the BMP signals, in coordination with the "super-enhancer" activity, transcriptional and chromatin remodeling (Adam et al., 2018, 2015; Hsu, Li, & Fuchs, 2014; Lee et al., 2016; Lee & Tumbar, 2012; Lien et al., 2011; Rompolas et al., 2012; Veltri, Lang, & Lien, 2018). A transcription factor Runx1 starts to express at high levels in primed stem cells and acts as a key driver for stem cell activation (Adam et al., 2018; Lee et al., 2014; Osorio et al., 2008). The primed stem cells further proliferate and generate multipotent progenitors (Hsu et al., 2014; Zhang & Hsu, 2017). The multipotent progenitors continue to grow and become the matrix, which makes diverse lineage choices by combinatorial action of Lef1, a Wnt target, and other signaling molecules, e. g., Notch and BMP (Adam et al., 2018; Xin, Gonzalez,

Rompolas, & Greco, 2018; Yang, Adam, Ge, Hua, & Fuchs, 2017). In another fate, bulge cells self-renew symmetrically without undergoing differentiation to replenish their own stem cell pool during early anagen and return to quiescence by late anagen (Zhang, Cheong, Ciapurin, McDermitt, & Tumbar, 2009).

2.4 Non-epithelial cell types and ECM in skin

The skin dermis is divided into three sublayers (Rognoni & Watt, 2018; Sorrell & Caplan, 2004): (1) the papillary dermis, the uppermost part of the dermis, composed of papillary fibroblasts, thin and loosely-arranged collagen fibers; (2) the reticular dermis, located in the lower part of the dermis, consisting of reticular fibroblasts and mature collagen bundles; (3) the hypodermis or the dermal white adipose tissue, enriched in adipocytes and located underneath the reticular dermis. Specialized fibroblasts also play a role in skin, such as the dermal papillae (Driskell et al., 2011), the dermal sheath, the arrector pili muscle. These diverse fibroblast populations in the skin arise from at least two distinct lineages during development (Driskell et al., 2013; Philippeos et al., 2018; Rinkevich et al., 2015). The skin also accommodates melanocytes (Nishimura, 2011), immune cells (Ali et al., 2017; Heath & Carbone, 2013; Malissen, Tamoutounour, & Henri, 2014; Naik, Larsen, Cowley, & Fuchs, 2018), blood vessels, pericytes (Lawlor & Kaur, 2015), Merkel cells and nerves (Doucet, Woo, Ruiz, & Owens, 2013; Maksimovic et al., 2014).

The functional interactions between epidermal stem cells and non-epithelial cell types in skin have started to emerge (Fujiwara, Tsutsui, & Morita, 2018; Goldstein & Horsley, 2012; Rompolas & Greco, 2014). For example, adipocytes are one of the key factors in the regulation of hair cycle-dependent stem cell behavior (Festa et al., 2011). In parallel with the hair cycle, adipocytes also undergo cyclic proliferation and maturation and provide platelet-derived growth factor (PDGF) ligands to hair follicle stem cells to drive their activation in a hair cycle-dependent manner. The dermal-epidermal crosstalk within defined micro compartments or the "micro-niche" controls diverse fate choices of hair follicle stem cells (Yang et al., 2017). The arrector pili muscle and sensory nerves are specifically attached to the upper- and mid-bulge region, respectively, and form specialized microenvironments for their proper functions (Cheng et al., 2018; Fujiwara et al., 2011). To achieve this, hair follicle stem cells in each region express uniquely defined ECM and create a niche for muscle (Fujiwara et al., 2011) and nerves (Cheng et al., 2018). ECM also plays a vital role in sequestering various growth factors and cytokines, suggesting that the diverse composition of ECM may be playing an essential role in the stem cell compartmentalization (Chermnykh, Kalabusheva, & Vorotelyak, 2018; Watt & Fujiwara, 2011).

3. Experimental techniques in epidermal stem cell biology

Recent advancements in technology lead way to a better understanding of biology of skin stem cells and their niches. Here, we introduce the commonly used techniques in studying epidermal stem cells, mainly in mouse in vivo studies.

3.1 Label-retaining assay

Many mammalian tissues contain infrequently-dividing cells (so-called "slow-cycling" cells), which are identified as nucleotide or histone H2B-green fluorescent protein (GFP) label-retaining cells (LRCs) (Sada & Tumbar, 2013). In these experiments, when mice are given BrdU (bromodeoxyuridine), EDU (5-ethynyl-2′-deoxyuridine) or ^{3}H-Thymidine, these nucleotide analogs are incorporated in the newly-synthesized DNA of S-phase cells (= pulse). After the nucleotide analogs are withdrawn (= chase), over a period of time, cells dilute the label by divisions or to be shed out from the tissue by differentiation. In contrast, slower-cycling cells retain the label even after a long-term chase and can be detected as LRCs (Cotsarelis et al., 1990).

By adapting the same principle to mouse transgenic systems, Tumbar et al. successfully developed a histone H2B-GFP tet-off mouse system, allowing us to visualize and isolate slow-cycling cells in vivo (Tumbar et al., 2004) (Fig. 3). This system utilizes two transgenes: (1) GFP fused to histone H2B under the control of a tetracycline response element (TRE); (2) The tet repressor (tetR)-VP16 transgene expressed by a tissue- or a cell-type specific promoter such as K5, a promoter to mark skin epidermal cells (Diamond, Owolabi, Marco, Lam, & Glick, 2000) (Fig. 3A). In the double transgenic mice, the H2B-GFP is transcribed in the absence of tetracycline (doxycycline; doxy); therefore, all skin epidermal cells exhibit the GFP expression (= pulse). The doxy treatment inhibits the binding of tetR protein to the TRE promoter, causing the H2B-GFP transcription to be turned off (= chase). Since GFP is fused with histone H2B, GFP protein is maintained stably even after transcription is stopped. During doxy chase, cells dilute GFP protein by divisions and only cells dividing less frequently retain the

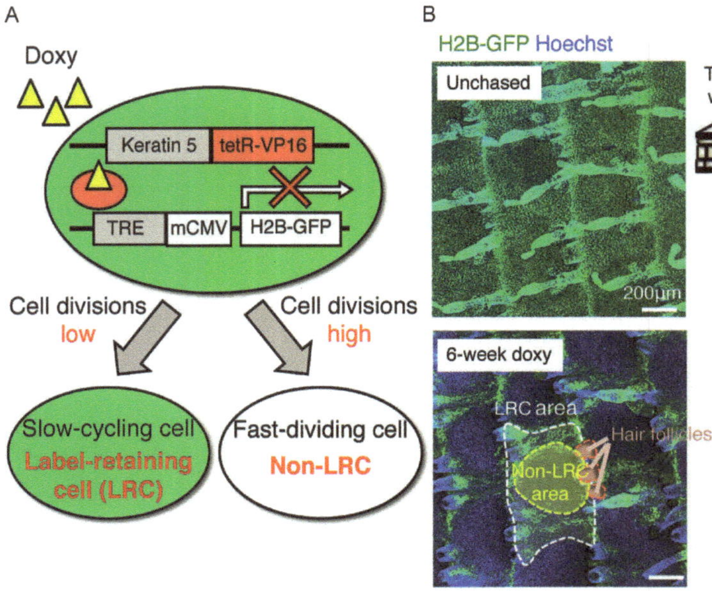

Fig. 3 H2B-GFP tet-off system to visualize slow-cycling cells in vivo. (A) Schematic representation of the H2B-GFP system to detect slow-cycling cells as label-retaining cell (LRC)s. Doxycycline (doxy) administration inhibits the transcription of H2B-GFP. Cells dilute H2B-GFP proteins after divisions, which enables quantification of the frequency of cell division during chase. TRE, tetracycline response element. (B) An example of H2B-GFP pulse-chase experiment in the tail epidermis (Sada et al., 2016). At 0-day chase, the K5 promoter drives GFP expression in epidermal cells. After 6-week-chase, LRC and non-LRC territories can be observed. Scale bars, 200 μm.

GFP label and therefore can be identified as LRCs (Fig. 3B). This system has several advantages over the nucleotide pulse-chase experiments: the pulse-labeling can be induced uniformly irrespective of proliferation ability in all K5+ epidermal cells; the proliferation history of cells can be quantitatively measured as GFP intensity by using flow cytometry; H2B-GFP LRCs can be sorted for molecular and functional characterization.

3.2 Cre-mediated cell fate tracking, clonal analysis and cell ablation

Since tissue stem cells are defined by two properties, i.e., the ability to self-renew and to differentiate, it has been challenging to precisely identify "stem cells" in intact tissues. To test such abilities, functional stem cell assays, such as cell transplantation or in vitro colony-forming assays, have been utilized in several mammalian tissues. However, these assays may alter the normal behavior/characters of stem cells due to cellular stress or changed microenvironments. Indeed, several examples have been reported that the stem cell potential can be broken down during isolation, culturing and transplantation of stem cells in skin (Blanpain et al., 2004; Lu et al., 2012) and other tissues (Nakagawa, Nabeshima, & Yoshida, 2007). In the last decade, an inducible Cre-mediated lineage tracing has been widely used for analyzing the fates of epidermal stem cells in vivo and provided us a better knowledge on how stem cells behave under steady state conditions and how they change their behaviors following injury (Alcolea & Jones, 2014).

In a lineage-tracing experiment, cells of interest can be labeled at one-time point (= pulse), and the progeny derived from the labeled cells can be analyzed at a later time point (= chase). This system uses an inducible CreER, a Cre recombinase fused with a mutant form of estrogen receptor (ER), expressed under the control of the specific promoter of gene of interest (Fig. 4A). The CreER expressing line is crossed with a reporter mouse line carrying fluorescent proteins (e.g., Tomato, GFP) or β-galactosidase inserted into a ubiquitously expressed locus such as Rosa26 where the expression is prevented by a stop cassette flanked by two loxP sites. After the Cre recombination by injection of Tamoxifen, an estrogen ligand, the stop sequence is removed and cells are labeled permanently. Since the reporter expression is integrated into daughter cells arising from the original cells, this system allows us to track cellular fate in vivo. Based on the theory that non-stem cells (short-lived progenitors or TA cells) divide and are lost from the tissue whereas stem cells with the ability to self-renew and differentiate can be found in the tissue even after repeated tissue turnover, this experiment enables us to distinguish stem cells versus non-stem cells (Fig. 4B). Multi-color reporters, called Confetti or Brainbow, can be combined with CreER lines, for the purpose of visualizing different clones in different colors (Livet et al., 2007; Roy, Neufeld, Livet, & Khosrotehrani, 2014). After Cre activation, each cell undergoes random recombination and expresses a distinct fluorescent protein. By using a Rosa26-Confetti or a "Skinbow" reporter, clonal dynamics of epidermal stem cells was studied during homeostasis and wound healing (Aragona et al., 2017; Roy et al., 2016).

When genetic labeling is introduced into single cells, their clonal fates can be traced and the quantitative data are obtained by counting the number of cells derived from original cells overtime. This assay, called clonal analysis, has been applied to study the kinetics of stem/progenitor cells in skin, testes, intestine and other tissues (Blanpain & Simons, 2013;

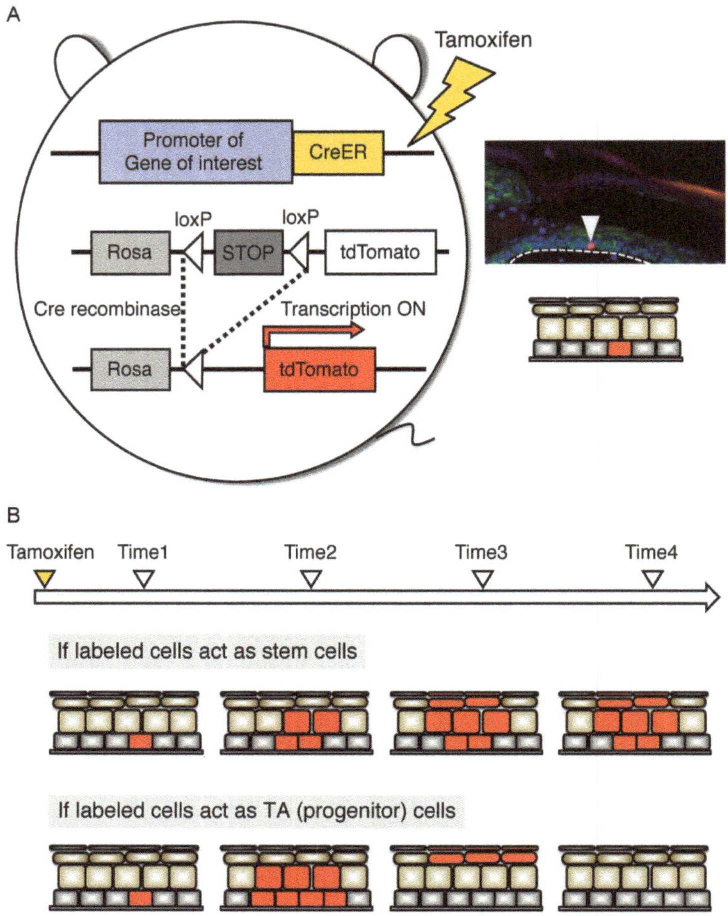

Fig. 4 Schematic representation of lineage tracing. (A) Schematic representation of Tamoxifen-inducible Cre-loxP system. Tamoxifen administration translocates CreER to the nucleus, where Cre recombination takes place. A picture shows an example of genetic-labeling of epidermal basal cells. (B) Expected outcome. If labeled cells act as stem cells, they self-renew and contribute to the entire basal and suprabasal layers after chase. If labeled cells act as short-lived progenitor or transit-amplifying (TA) cells, they proliferate and differentiate without self-renewal; therefore, their clones will be lost after long-term chase.

Rulands & Simons, 2016). By mathematical modeling, biological data, i.e., the clone size distributions, is fitted to a model and the stochastic rate parameters are inferred.

A cell killing experiment is a way to address the functional importance of cells of interest. The genetic ablation of cells can be achieved by using Cre mediated expression of diphtheria toxin fragment A (DTA) or a diphtheria toxin receptor (DTR) (Buch et al., 2005; Ivanova et al., 2005). Similar to the approach for lineage tracing, genetic cell ablation utilizes the DTA or DTR expression under a ubiquitously expressed locus like Rosa26, followed by stop sequence flanked by loxP sites. Under the expression of inducible CreER recombinase from a cell specific promoter, the loxP sites are recombined and DTA or DTR starts to express thus ablating the Cre-expressing cells. A selective cell ablation by using a laser has also been reported to target stem or niche cells at their native state (Pineda et al., 2015). Combining the laser with a high aperture objective lens, high energy of laser can be concentrated into a small area and selectively ablate that area.

3.3 Whole-mount staining, live imaging

Another important question in epidermal stem cell biology is how epidermal basal (stem) cells, epidermal appendages or other cell types are arranged in the three-dimensional structure of the skin. To address this point, the whole-mount staining of epidermal sheets or whole skin has been developed. For the preparation of whole-mount mouse epidermis, the intact epidermal sheet is separated from the dermis by peeling after incubating with EDTA (ethylenediaminetetraacetic acid) (Braun et al., 2003). The samples can be used for immunofluorescence staining with antibodies or detection of fluorescent proteins using a confocal microscope. The whole-mount staining techniques have been successfully used in visualization of the spatial organization along with clonal development of epidermal stem cells in adult skin, especially in the tail.

The procedure of whole-mount staining described above involves a step in which the epidermis is separated from the dermis. This is more challenging in hairy and thinner dorsal skin and often results in tissue damage and rupture of the intact hair follicles or other structures. It also overlooks the information of cell types underneath the epidermis. To overcome this problem, various tissue clearing methods have been developed and utilized in skin biology, including BBBA (benzyl benzoate: benzyl alcohol) (Chang & Nathans, 2013), 3DISCO (three-dimensional imaging of solvent cleared organs) (Erturk et al., 2012) and CUBIC (Clear, Unobstructed Brain Imaging cocktails and Computational analysis) (Liang et al., 2016; Susaki et al., 2014). These methods keep the tissue transparent and proteins intact, thereby allowing the detection of signal at a single cell resolution. There are some limitations in using antibodies or fluorescent proteins as the tissue clearing procedures sometimes diminish these signals.

The analyses of skin sections from the frozen- and paraffin-embedded samples or isolated cells are not always sufficient to provide the information of stem cell behavior in intact tissues. Skin being the outermost part of the mammalian body provides a direct access for observation that requires minimal, non-invasive preparation for imaging (Huang & Rompolas, 2017; Pineda et al., 2015; Rompolas et al., 2012). Live imaging with two-photon microscopy combined with genetic tools to mark stem cells has been successfully applied to study epidermal stem cell dynamics in their native state for long periods of time.

3.4 Gene profiling and functional analysis

To understand the molecular property of different epidermal compartments, transcriptome analysis of target cells has been employed in the skin field. This is achieved by FACS (fluorescence activated cell sorting)-based cell isolation of specific epidermal subpopulations by using cell surface markers or transgenic reporters (Soteriou et al., 2016). To prepare a single cell suspension, harvested mouse skin is treated with trypsin and scrapped to obtain cells from the epidermal layer. The isolated epidermal cells can be stained with suitable antibodies: α6-integrin and CD34 (Trempus et al., 2003) for isolation of the bulge stem cells; or α6-integrin and Sca1 (Jensen et al., 2008) for basal cells within the IFE. These isolated samples are utilized for various purposes such as cell culture, Western blotting, mRNA expression analysis including RT-PCR (reverse transcription polymerase chain reaction), microarray or RNA-sequencing. To study chromatin dynamics of stem cells, researchers have developed a method called "assay for transposase-accessible chromatin with high throughput sequencing" (ATAC-seq) (Buenrostro, Giresi, Zaba, Chang, & Greenleaf, 2013). This method utilizes a transposase sequence that could integrate into the open chromatin landscape, allowing for identification of areas with high transcriptional activity.

With the advent of single-cell sequencing, it enables us to analyze the unbiased cellular composition with high-throughput single-cell transcriptomics of cell populations within tissues, including the skin epidermis (Mok, Saxena, & Rendl, 2018). It also has an advantage in profiling stem cell dynamics during wound healing and cancer progression using a small number of pure wounds or tumor-induced cells for transcriptome analysis (Ge et al., 2017). The single cell RNA sequencing has been successfully applied to study the transcriptional organization of the mouse and human epidermis (Joost et al., 2016; Tan et al., 2013); and wound healing dynamics of Lgr5+ hair follicle and Lgr6+ IFE stem cell populations (Joost et al., 2018). It has also been applied to the normal and inflamed human epidermis to define their cellular states (Cheng et al., 2018).

For studying the role of identified genes, Beronja et al. established an elegant technique using an ultrasound guided, in utero delivery of siRNA (small interfering RNA)-containing lentivirus (Beronja, Livshits, Williams, & Fuchs, 2010). They determined a time point when the epidermis existed as a single layer ectoderm of the surface of developing embryo and injected the lentivirus containing the siRNA or Cre recombinase for selective transduction to the surface epithelium. RNAi (RNA interference) screening using pooled lentivirus library and high-throughput sequencing enabled comprehensive screening of genes in vitro (Chen et al., 2012) or in vivo (Beronja et al., 2013). The recently-developed genome-wide CRISPR/Cas9 library has also been combined (Sanjana, Shalem, & Zhang, 2014). Genome-wide knockout mouse screening using available mouse resources was carried out to identify novel skin phenotypes (Liakath-Ali et al., 2014). They innovated a method to allow large-scale phenotype screening by whole-mount staining and confocal imaging of the tail skin, and visualized and classified the phenotype within different epidermal compartments.

3.5 Keratinocyte culture

Even though the keratinocyte culture system has downside of culturing cells outside the original niche, it has a great advantage for studying molecular and cellular regulation of epidermal stem cells as well as for the use in clinical application. The development of culture system led to the first breakthrough for skin tissue engineering and research. In 1975, Rheinwald and Green managed to grow human primary keratinocytes on a feeder layer of irradiated 3T3 murine fibroblasts (Rheinwald & Green, 1975a, 1975b). After that, human keratinocyte culture had been improved to maintain their division capability and subculture in many passages. To assess the stem cell potential, the colony-forming assay was performed and the number of clones derived from the seeded stem or progenitor cells was examined (Barrandon & Green, 1987). Differentiation ability of keratinocytes can be assessed by inducing differentiation in culture by increasing the calcium concentration (Hennings et al., 1980). Protocol for isolation, culturing and transplantation of mouse skin epidermal keratinocytes had been developed and widely used (Lichti, Anders, & Yuspa, 2008).

4. Stem cell dynamics in the mouse interfollicular epidermis

With recent discoveries of molecular markers and technical advances, different populations of stem cells in the IFE have been reported and the field has greatly advanced over the past few years. Yet, there still seem to be some discrepancies on models of heterogeneity in IFE stem cells, depending on the markers, methods or regions of skin used. Here we summarize and discuss about key findings regarding cellular dynamics of IFE stem cells during normal homeostasis of mouse skin (also see review by Belokhvostova et al., 2018).

4.1 Invariant asymmetry versus population asymmetry

To achieve constant cellular turnover, there are at least three divisional strategies to balance the number of stem cells and differentiated progeny produced in a tissue (Morrison & Kimble, 2006; Sada & Tumbar, 2013) (Fig. 5). (1) Invariant asymmetric cell division (Fig. 5A): each and every stem

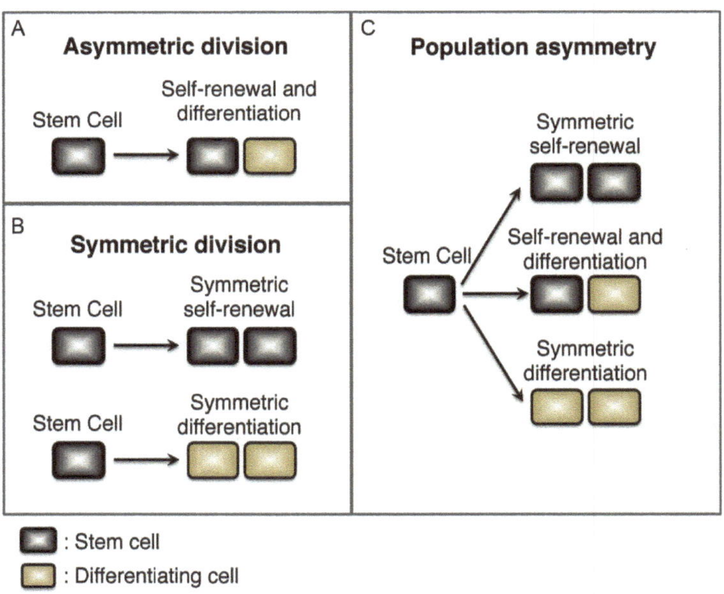

Fig. 5 Stem cell division strategies. (A) Invariant asymmetric division. (B) Invariant symmetric division. (C) Combination of asymmetric and symmetric divisions. If self-renewal and differentiation are balanced as a population, tissue homeostasis is achieved (population asymmetry).

cell generates one daughter stem cell and one differentiating daughter (e.g., germline stem cells in *Drosophila* (Betschinger & Knoblich, 2004) and the mouse developmental skin (Lechler & Fuchs, 2005)). (2) Symmetric cell division: each stem cell divides symmetrically to generate either two daughter stem cells or two differentiating daughters (Fig. 5B). In this case, if the probability of differentiation is matched by that of a self-renewal stem cell division, homeostasis can be maintained (e.g., hair follicle stem cells Zhang et al., 2009). (3) Combination of (1) and (2): each stem cell can divide either symmetrically or asymmetrically, but overall rates of self-renewal and differentiation are balanced as a population within a tissue (the idea of "population asymmetry") (Klein & Simons, 2011, 2015) (Fig. 5C). In recent studies, mammalian tissues such as the skin IFE (Clayton et al., 2007; Mascre et al., 2012), testes (Hara et al., 2014; Klein, Nakagawa, Ichikawa, Yoshida, & Simons, 2010) and the intestine (Lopez-Garcia, Klein, Simons, & Winton, 2010; Snippert, van der Flier, et al., 2010) favor the model of population asymmetry, in which stochastic loss and replacement of stem cells, so-called the "neutral drift" behavior, are observed (Krieger & Simons, 2015).

4.2 Slow-cycling nature and stemness

A hierarchical stem/TA cell model suggests that slow-cycling cells act as long-lived stem cells and place at the top of lineage hierarchy, generating short-lived progenitors, called TA cells (Fuchs, 2009; Sada & Tumbar, 2013) (Fig. 6). TA cells are characterized as follows: (1) the commitment to terminal-differentiation, (2) the lack of self-renewal capacity, (3) rapidly-dividing, and (4) progeny of stem cells. The slow-cycling nature is considered to be a mechanism to protect stem cells from replication-induced DNA damage, accumulation of mutations, telomere-shortening, all of which can lead to tumorigenesis and aging. In fact, hair follicle stem cells are known to be highly quiescent and they divide only ~50–100 times in a life (Waghmare et al., 2008). However, true biological functions of slow-cycling cells remain controversial and may be context-dependent (Sada & Tumbar, 2013). According to the stem/TA cell model, slow-cycling cells or LRCs are unique, long-lived stem cells (Blanpain et al., 2004; Cotsarelis et al., 1990; Foudi et al., 2009; Montgomery et al., 2011; Sangiorgi & Capecchi, 2008; Tumbar et al., 2004). Studies in the mouse intestine have challenged the generality of the stem/TA model, and found that the rapidly-dividing Lgr5+ cells act as stem cells

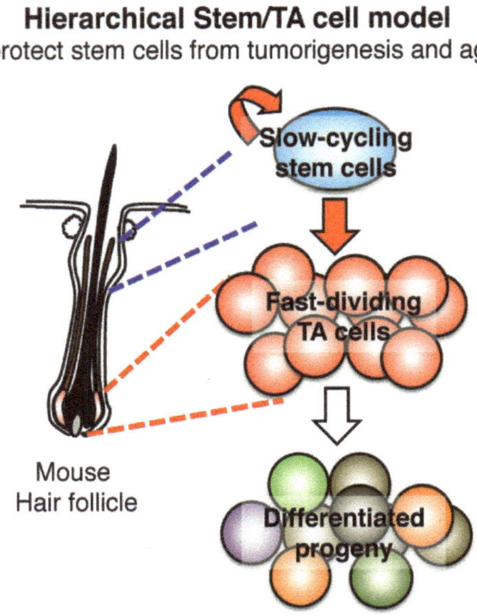

Fig. 6 Hierarchical stem/transit-amplifying (TA) cell model. Stem cells are thought to divide more slowly than their differentiating progeny, called TA cells, e.g., the mouse hair follicle.

(Barker et al., 2007). Other possibilities have been proposed. Slow-cycling cells could be a stem cell population co-existing with fast-dividing stem cells (Li & Clevers, 2010; Metcalfe, Kljavin, Ybarra, & de Sauvage, 2014; Takeda et al., 2011). Slow-cycling cells might be reserved or dormant stem cells that are required for injury-driven tissue regeneration (Li & Clevers, 2010; Tian et al., 2011; Wilson et al., 2008). LRCs are identified as differentiating stem cell progeny in the intestine (Buczacki et al., 2013). Slow-cycling cells within tumor can survive after drug treatments and contribute to tumor relapse (Sanchez-Danes et al., 2018; Schillert, Trumpp, & Sprick, 2013). On the other hand, the observed cell heterogeneity in cell division frequency, marker expressions or cellular behavior could also be explained by distinct cellular status of an equipotent cell population, not by the existence of functionally discrete cell populations (Krieger & Simons, 2015).

4.3 Hierarchical stem/transit-amplifying cell model

In the skin, earlier studies employing BrdU pulse-chase assay suggested a presence of LRCs and non-LRCs within the basal layer of the mouse IFE and that stem cells replicate slowly compared to their progeny, supporting the stem/TA model (Bickenbach, 1981) (Fig. 7A). Concurrently, an epidermal proliferative unit (EPU) model suggested that epidermal cells that reside at the center of EPU to be slow-cycling stem cells that divide and differentiate, and give rise to surrounding TA cells, which in turn differentiate into cells in the upper layer (Mackenzie, 1970; Potten, 1974). Later, when the fate of the labeled slow-cycling cells were analyzed, it provided direct evidence that the progeny of the labeled cells was not strictly contained within its own EPU, rather migrating to the adjacent EPUs (Doupe, Klein, Simons, & Jones, 2010; Kameda et al., 2003; Ro & Rannala, 2004, 2005). Thus, although the IFE contains slow-cycling LRCs, their clonal units do not always resemble classical EPUs, contradicting the idea of the EPU hypothesis.

More recently, the stem/TA cell model was re-visited and further tested by a quantitative lineage analysis by using K14- and Involucrin (Inv)-CreER (Mascre et al., 2012; Sanchez-Danes et al., 2016) (Fig. 7A). These studies proposed the existence of hierarchically related, two discrete basal populations. One is a slow-cycling and longer-lived stem cell population that is preferentially marked by K14-CreER. The second population is marked by an infrequent activation of Inv-CreER in the basal layer, normally active in differentiated suprabasal cells. Based on the lower rates of clone survival, cellular dynamics and more differentiated molecular phenotype of the Inv-CreER+ population, a hierarchical relationship between the K14-CreER+ stem cells and Inv-CreER+ committed progenitors (or TA cells)

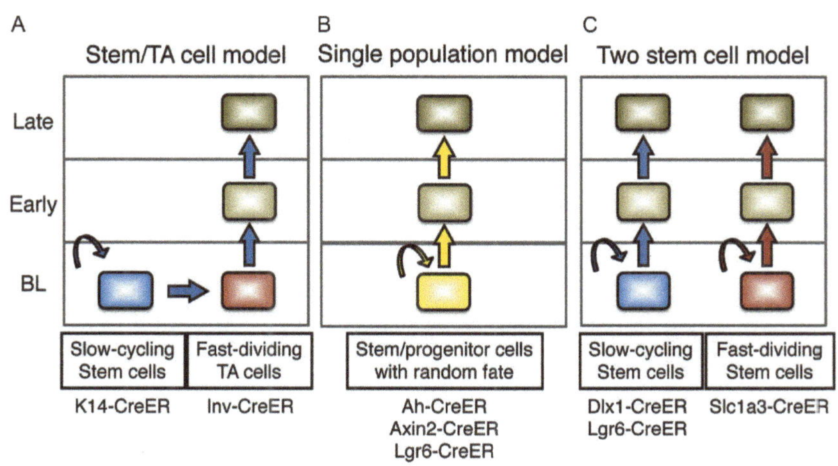

Fig. 7 Three stem cell models proposed in the interfollicular epidermis. (A) Hierarchical stem/transit-amplifying (TA) cell model. (B) Single population model. (C) Two-stem cell population model. The color in arrows represents cellular lineage of stem cells. Lineage markers are shown. BL, basal layer.

was suggested (Mascre et al., 2012; Sanchez-Danes et al., 2016). Interestingly, when territorial differences of the tail IFE, i.e., interscale and scale, were taken into consideration, the clonal data well-fit to the stem/TA cell model in the interscale region, but not in the scale (Sanchez-Danes et al., 2016). In the interscale IFE, division rates of stem and TA cells were calculated as 1 div/~15 days and 1 div/~4 days, respectively. In contrast, in the scale IFE, K14- and Inv-CreER+ cells showed similar clonal dynamics, indicating that scale is maintained by a single, equipotent progenitor population with a division rate of 1 div/~4 days (Sanchez-Danes et al., 2016).

4.4 Single population model

Contradicting to the stem/TA cell model, a single population model is proposed that all proliferating basal cells are identical, and that the behavior of any individual basal cell is stochastic irrespective of their cell division history (Clayton et al., 2007; Doupe et al., 2010) (Fig. 7B). This model was proposed based on a mathematical analysis of the size distribution of clones observed in lineage tracing experiments. In this study, Ah (also known as Cyp1a1)-CreER mice were used to label basal cells of the tail and ear IFE, to study the fate of the labeled cells over the course of 1 year period. The study observed that all epidermal basal cells follow a stochastic fate, possess equal potential to undergo symmetric or asymmetric cell divisions and the chances of generating a basal cell or a differentiated cell is balanced (Clayton et al., 2007; Doupe et al., 2010). This model was further supported by using other Cre lines such as Axin2- or Lgr6-CreER, suggesting that these Cre+basal cells undergoes probabilistic stem cell fate and no behavioral differences between Cre-positive and -negative cells (Fullgrabe et al., 2015; Lim et al., 2013). More recently, researchers utilized live imaging for in vivo tracking of the individual basal cells and their short-term fates in the ear and paw IFE (Rompolas et al., 2016). The study suggested that individual basal cells undergo seemingly random fates with no sign of lineage hierarchy, division heterogeneity or asymmetry. Although basal cells undergo differentiation in a stochastic manner, the trajectory of the differentiated cells is aligned in such a way that it occupies the position of a previous differentiated cell. This leads to a model that rather than EPUs, the IFE is maintained by epidermal differentiation units (EDU), where they gather differentiated progeny from basal cells and function as a buffer zone to moderate the transition of the cells from basal layer to terminally-differentiated cornified layer (Rompolas et al., 2016). A crucial point of the hierarchical stem/TA model of epidermal homeostasis revolves around the existence of a long-lived slow-cycling stem cells giving rise to TA cells, but the single population model denies such existence of slow-cycling cells as a top of lineage hierarchy. If the stochastic model is true, how self-renewal and differentiation of basal cells are balanced? Recent report points out that differentiation itself is a cue to induce the self-renewal division of basal cells, but not vice versa (Mesa et al., 2018). Hence, the rate of proliferation and differentiation are coordinated locally and tissue generates new cells on demand.

4.5 Two distinct stem cell population model

A third model is proposed that distinct populations of infrequently- and frequently-dividing basal cells might independently produce two distinct differentiation lineages (Fig. 7C). Gomez et al. showed that basal cells of scale IFE divided more rapidly than ones in the interscale IFE (Gomez et al., 2013). A lineage tracing using K14-CreER showed that clones didn't expand beyond the scale and interscale boundaries, indicating that scale/interscale are maintained locally by unipotent cells within each region.

A defining answer regarding the proliferative heterogeneity and the stemness of epidermal basal cells were reported more recently (Sada et al., 2016). Sada et al. identified novel molecular markers that enable us to distinguish epidermal LRCs (slow-cycling cells) and non-LRCs (fast-dividing cells) in both tail and dorsal skin and, thereby, to study their localization, kinetics and long-term stem cell potential during homeostasis and regeneration. By analyzing the proliferative history of basal cells using H2B-GFP pulse chase system, the study indicated the presence of spatially segregated LRC and non-LRC clusters in both dorsal and tail IFE (Fig. 3B). In the case of tail, LRC and non-LRC territories matched with interscale and scale IFE, whereas in the dorsal skin the size of LRC/non-LRC clusters were much smaller. Notably, the H2B-GFP label retention was quite uniform within the LRC territory, contradicting the idea of presence of rare, quiescent cells in the center of clonal units. Another interesting observation was that LRC clusters consisted of both basal and suprabasal cells, suggesting a possibility that basal LRCs directly differentiate into suprabasal LRCs after no or only a few rounds of TA cell divisions (Sada et al., 2016).

In order to quantitatively analyze H2B-GFP dilution data, Sada et al. collected basal and suprabasal cells from the IFE by FACS and obtained data was deconvolved to count the number of cell divisions after different chase periods (Sada et al., 2016). This experiment showed that during 3 days of doxycycline (doxy) chase, basal cells had divided several times and been transported into early and late differentiation layers. The unexpected rapid cell division and cellular transport was also confirmed later by live imaging (Rompolas et al., 2016) and could be explained by the existence of the fast-dividing population. The H2B-GFP data also implied a second, slower-cycling population that have stayed in the basal and suprabasal layers after longer chases (Mascre et al., 2012; Sada et al., 2016). Mathematical modeling showed that previous models, such as single population (Clayton et al., 2007) or stem/TA (Mascre et al., 2012) models could not fully explain these data. The best-fit was obtained with two independent population model that: the slow-cycling population represents ~35% of basal cells, divides on every 5–7 days; and the fast-dividing population represents ~65% of basal cells, divides every ~2 days in dorsal skin. This ratio was conserved between dorsal and tail skin, but overall turnover rates was 20–25% slower in tail.

Next, Sada et al. performed FACS-isolation and transcriptome analysis of basal LRCs and non-LRCs by using quadruple transgenic system and revealed the distinct molecular property and putative markers for LRCs and non-LRCs within the basal layer of the IFE (Sada et al., 2016). Genes involved in apoptosis inhibition, Ras-GTPase and insulin signaling pathways were significantly enriched in LRCs, whereas chromatin regulation, blood vessel development and MAPK (mitogen-activated protein kinase) pathway genes were over-represented in non-LRCs. The signature genes were also involved in metabolism, development, mitosis and cell migration.

To directly address whether slow-cycling and fast-dividing cells act as long-lived stem cells, new genetic tools Dlx1- and Slc1a3-CreER were selected based on the microarray data to specifically label slow-cycling and fast-dividing cells, respectively. Lineage tracing up to 1 year showed that Dlx1-CreER clones were enriched in the interscale IFE (or LRC territory in dorsal skin), whereas Slc1a3-CreER clones were preferentially found in scale IFE (or non-LRC territory in dorsal skin) (Sada et al., 2016). There was no apparent migration from interscale IFE to scale IFE during homeostasis, excluding the classical model of slow-cycling cells as stem cells, giving rise to fast-dividing TA cells. Similarly, Lgr6-CreER+ cells were largely restricted to the interscale IFE and contributed to the interscale, but not to the scale (Kretzschmar, Weber, Driskell, Calonje, & Watt, 2016). Thus, these experiments demonstrated that epidermal basal LRCs and non-LRCs are two distinct, spatially separated stem cell populations and that they replenish distinct epidermal territories during normal homeostasis (Sada et al., 2016) (Fig. 8).

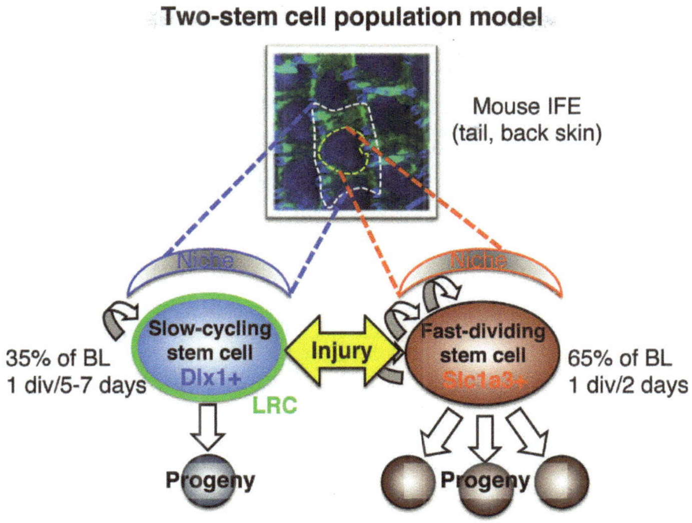

Fig. 8 Schematic representation of two-stem cell population model in the interfollicular epidermis. LRC, label-retaining cell; IFE, interfollicular epidermis; BL, basal layer.

These two stem cell populations can interconvert following injury, which demonstrates flexibility in their potential for robust skin regeneration, which we will discuss more details in injury repair session.

A remaining question is what is the relationship between new and previously-identified IFE populations marked by different CreER lines. The cells marked by K14- and Inv-CreER seems to be biased toward the interscale region (Mascre et al., 2012; Sanchez-Danes et al., 2016), implying that these two populations would be hierarchically-related stem/TA cells within the LRC territory (Sada et al., 2016). In fact, Sada et al. observed a short-lived population, aligned in the interscale "line" region, retained H2B-GFP (=slow-cycling), but marked by Slc1a3-CreER (Sada et al., 2016). This population might be the same population as those marked by Inv- (Mascre et al., 2012; Sanchez-Danes et al., 2016) and Ah- (Clayton et al., 2007) CreER. On the other hands, Slc1a3-CreER+ clones in the scale were long-lived, territory-specific stem cells, which matched with the idea that the cellular kinetics of scale IFE is balanced without having hierarchically-related subpopulations (Sanchez-Danes et al., 2016). A non-biased single cell RNA sequencing also supported molecular heterogeneity of both mouse and human epidermis (Joost et al., 2016; Tan et al., 2013), but information needs to be integrated and examined further to establish an unified model of stem cells in the IFE.

4.6 Context-dependent behavior of epidermal stem cells

The mouse skin, not only hair follicles but also other compartments in the epidermis and the dermis along with environmental signals, changes dramatically during the hair cycle. A question arose regarding whether such hair cycle-dependent changes can influence stem/progenitor cells in the IFE. A study by Roy et al. used a multicolor reporter with K14-CreER to track the clonal fate of individual basal cells in the IFE of dorsal skin (Roy et al., 2016). The study pointed out that during anagen, there was an increase in the size of clones as well as proliferation of IFE cells in the vicinity of hair follicles. In contrast, cells away from hair follicles were found to be slow-cycling. A recent study further indicated that the activation of IFE cells during anagen was controlled by transient up-regulation of Slc1a3, a glutamate transporter (Reichenbach et al., 2018; Roy & Khosrotehrani, 2018). Given that Slc1a3 was identified as a fast-dividing stem cell marker (Sada et al., 2016), it will be interesting to further address how growth activating signals and metabolic status of cells are dynamically changed during hair cycles within or across different epidermal compartments.

A recent study demonstrated that during pregnancy, rapidly-dividing Tbx3+ cells appeared in the basal IFE, divided parallely to the basement membrane and contributed to pregnancy-dependent expansion of skin (Ichijo et al., 2017). This process is controlled by the secretion of Wnt inhibitory factors, Sfrp1 and Igfbp2, from the dermis that would activate the Axin2+ basal cells to give rise to Tbx3+ population by suppressing autocrine, self-renewing signal of Wnt (Lim et al., 2013).

5. Epidermal stem cells in injury repair

As skin forms the outer most layer of the body, it is under a constant threat that it may break the barrier and affect the underlying cells and tissues. Given the reasons, it is crucial for skin to be equipped with complex mechanisms that would restore the integrity of the tissue quickly, decreasing the chances of infections.

5.1 Process of wound healing

Repair of skin tissue after wounding is heavily dependent on synchronized cross-talks between epidermal, dermal, immune and vascular cells which reside in close proximity to the wound. Full thickness wound models of skin suggest that wound healing generally follows three steps, (1) inflammation,

(2) re-epithelization, and (3) resolution phases (Gurtner, Werner, Barrandon, & Longaker, 2008; Sun, Siprashvili, & Khavari, 2014). In inflammatory phase, the immune cells are transported to the wound site and blood clots begin to form. A rapid re-epithelization involves covering the wound surface by migration, proliferation and stratification of epidermal cells. During resolution phase, the tissue remodeling of the epidermis, the dermis and ECM takes place.

5.2 Stem cell dynamics during injury repair

Advances in the technologies in labeling cells with high precision and various transgenic tools led us to the greater understanding of how different epidermal stem populations interact closely to re-instate epidermal homeostasis following wounding. In the epidermis, stem cell compartments identified to date includes the basal layer of the IFE, the bulge area of hair follicles and the junctional zone. Under homeostasis, stem cells in these niches repopulate the cells in their respective compartments. When skin is damaged, territorial barrier of these epidermal compartments seems to be canceled and stem cells display dramatic plasticity (also see reviews by Belokhvostova et al., 2018; Dekoninck & Blanpain, 2019).

Recent studies by intravital imaging (Park et al., 2017) and gene profiling (Aragona et al., 2017) have shown that epidermal cells are highly compartmentalized around the wound areas that are divided into (1) a proliferation zone, (2) a migration zone (the leading edge), and (3) a mixed zone. Cells near the wound rapidly respond to the signal and migrate toward the wound without proliferation, whereas cells located away from the wound divide and supply new cells. In the mixed zone, both proliferation and migration take place. The cells in the leading edge function as a scaffold and exhibit an upregulation of cell and ECM remodeling factors and pro-inflammatory molecules (Aragona et al., 2017). Cellular and molecular changes in the migration zone also resemble to the process of epithelial-to-mesenchymal transition (EMT), including upregulations of signature markers of EMT, which is recently reviewed in (Haensel & Dai, 2018).

Recent studies combined lineage tracing and live imaging of epidermal cells to understand individual cellular dynamics during injury repair. To investigate the contribution of cells within the IFE during wound healing, Mascre et al. employed a two Cre to mark the epidermal basal cells, K14- and Inv-CreER (Mascre et al., 2012). The mouse tail IFE was subjected to full thickness excisional wound by punch biopsy. They found that in the repaired tissue, clones were mostly derived from the K14-CreER labeled population, and few clones were obtained from Inv-CreER labeled population. This indicated that stem cells present in the basal layer marked by K14 are capable of tissue regeneration whereas the contribution of committed progenitors are negligible. Similarly, Axin2+ populations contributed robustly to the wound site (Lim et al., 2013). When Dlx1- and Slc1a3-CreER transgenic systems were used to study injury repair (in this case, epithelial surface wounding by scraping was used), the stem cells in the IFE migrated out of their specialized niche and once the injury was repaired, they were maintained only in their own environments (Sada et al., 2016). This suggests that these two stem cell populations have cell-intrinsic differences to better adapt to their typical territory. Sada et al. performed ablation of slow-cycling and fast-dividing cells using Dlx1- and Slc1a3-CreER crossing with transgenic mice carrying the Rosa-DTA and showed that these two stem cell populations were interchangeable. In contrast, Gata6+ cells of the sebaceous ducts were reported to migrate into the IFE in response to wounding, de-differentiate as stem cells and to contribute to the repair and long-term maintenance of the IFE (Donati et al., 2017).

6. Current knowledge on human epidermal stem cells

6.1 Epidermal cell clonogenicity

In primary human keratinocytes, three clonal subtypes are observed: stem cell-like holoclones; TA cell-like paraclones; and intermediate meroclones (Barrandon & Green, 1987). Holoclones contain a large population of small, regular and highly proliferative keratinocytes, which are thought to give rise to meroclones and paraclones. These colonies are enriched in fractions of cells that express β1-integrin, K14 and p63, indicating their self-renewal abilities (Jones, Harper, & Watt, 1995; Jones & Watt, 1993; Pellegrini et al., 2001). TA-cell like paraclones possess a limited proliferative ability but highly express differentiation markers such as Involucrin. Meroclones are mixture of cells with the different growth potential. The holoclone-forming population might correspond to α6 integrin+/CD71− infrequently-dividing cells in vivo, which have the greatest regenerative capacity after transplantation (Schluter, Paquet-Fifield, Gangatirkar, Li, & Kaur, 2011). In vivo clonal analysis in human further supports the idea that the human IFE contains heterogeneous cell populations with the different stem cell potential (Hirsch et al., 2017). Hence, these studies support a hierarchical model that the slow-cycling cells might represent a longer-lived stem cell population of the IFE.

In contrast, the clonal culture of human keratinocyte revealed that instead of exhibiting a hierarchy of stem and progenitor cells, human keratinocytes show two modes of proliferation (Roshan et al., 2016). Once keratinocytes reach to confluency, cells switch from expanding mode to balanced mode. In injury condition, cells in balanced mode are able to reversibly change to expanding mode to promote wound healing process. This in vitro cellular behavior supports the idea of dynamic stem cell heterogeneity or single population model, in which observed heterogeneity may reflect distinct cellular status of single population (Krieger & Simons, 2015).

6.2 Human epidermal architecture and stem cell location

At the tissue level, the human skin has characteristic tissue structures distinct from the mouse skin, named rete ridges and dermal papillae, which are the projection of epithelial layer into underlaying connective tissues and an upward projection of mesenchyme into the epithelium, respectively (Lawlor & Kaur, 2015). The localization of epidermal stem cells in human skin has been shown to correlate with the undulating pattern of the tissue: slow-cycling cells are preferentially located at the tip of dermal papillae, which are marked by β1-integrin (Jones et al., 1995) and MCSP (melanoma chondroitin sulfate proteoglycan) (Legg, Jensen, Broad, Leigh, & Watt, 2003), whereas fast-dividing populations labeled by K15 are enriched in the tip of rete ridges (Webb, Li, & Kaur, 2004). Interestingly, a recent study has shown that the topographical architecture of tissue itself could partially induce the localization pattern of epidermal stem cells and their progeny (Viswanathan et al., 2016). With the recent discovery of two populations of stem cells in the mouse IFE (Sada et al., 2016), further studies are needed to address the link between stem cell compartments and tissue architecture and to elucidate universal stem cell models between human and mice.

6.3 Clinical view

With abundant self-renewal potential, cultured human keratinocytes have been employed for reconstitution of epidermal grafts for decades. The first cultured epidermal autografts was successfully engrafted in severe burn victims, followed by several clinical cases in which the epidermal sheets were engrafted on patients with massive burns (Compton et al., 1989; Gallico, O'Connor, Compton, Kehinde, & Green, 1984; O'Connor, Mulliken,

Banks-Schlegel, Kehinde, & Green, 1981). The recent success in gene therapy has opened an avenue for the treatment of junctional epidermolysis bullosa by using genetically-corrected epidermal stem cells (Hirsch et al., 2017). However, this technique still holds some challenges related to stability of engraftments, unpredictable integration of transgenes, age of patients and the dermal conditions of patient's skin.

7. Epidermal stem cells in cancer and aging

Adult tissue confronts a variety of temporary or permanent physiological changes and stresses that may modify the overall structure and function of the tissue as well as character/behavior of stem cells. Here we discuss about contributions of heterogeneous stem cell populations to the cancer and aging.

7.1 Cancer

Basal cell carcinoma (BCC) and squamous cell carcinoma (SCC) are two major types of skin keratinocyte carcinomas in humans (Gordon, 2013; Nagarajan et al., 2019). In this review, we primary focus on current knowledge of stem cell dynamics in the context of the BCC formation and progression. Activation of the hedgehog signaling pathway, either by gain of function mutations of Smoothened or loss of function mutations of an inhibitor Patched, is often observed in these patients (Epstein, 2008; Kasper, Jaks, Hohl, & Toftgard, 2012; Yang, Xie, Fan, & Xie, 2010). p53 is also mutated in some patients with sporadic BCC. Recent studies suggest that the combinatorial treatments with vismodegib, a Smoothened inhibitor, and a Wnt pathway inhibitor would be a promising strategy to target slow-cycling vismodegib-resistant populations and suppress tumor relapse in BCC (Biehs et al., 2018; Sanchez-Danes et al., 2018). Which cells are the cells of (Yang et al., 2017) origin of BCC? How different epidermal compartments contribute to the formation of BCC? In mouse model, BCC can be induced by the activation of hedgehog pathway by overexpressing the mutant form of Smoothened or conditional knockout of Patched1 (Youssef et al., 2010). Lineage specific activation of hedgehog signaling showed that BCC can arise from the infundibulum and the IFE, but not from hair follicle stem or transit-amplifying matrix cells (Sanchez-Danes et al., 2016; Youssef et al., 2010). Within the IFE, BCC can arise from K14-CreER+ stem/progenitor cells, especially in the interscale region of the tail, but not from Inv-CreER+ committed progenitor cells (Sanchez-Danes et al., 2016).

Given that interscale and scale accommodate distinct stem and progenitor cell populations (Gomez et al., 2013; Sada et al., 2016; Sanchez-Danes et al., 2016), it is fascinating to address whether regional specificity of stem cells and their molecular differences contribute to the initiation and/or progression of BCC.

Skin is highly proliferative tissue, which may increase the risk for cancer-causing mutations, hijacking the physiological control and leading to uncontrolled growth. However, the healthy human skin seems to be able to retain harmful oncogenic mutations, indicating the innate ability of epidermal cells to withstand oncogenic loads without progressing to cancer (Martincorena et al., 2015). Studies suggested that non-mutated healthy epidermal cells have a mechanism to recognize and out-compete the mutant clones, preventing them to spread throughout the epidermis; therefore, the epidermal homeostasis can be restored (Brown et al., 2017; Murai et al., 2018). The oncogenic activation of PI3K (Phosphoinositide 3-kinase) caused the loss of mutant clones by promoting differentiation of epidermal basal cells (Ying, Sandoval, & Beronja, 2018). Such natural endurance for harmful mutations and the ability of the non-mutated cells to "tame" the mutated clones back to homeostasis are advantageous to the tissues that possess high cell-turnover rate.

7.2 Aging

Stem cell aging theory suggests that age-dependent functional alterations of stem cells play a critical role in driving tissue aging (Goodell & Rando, 2015; Lopez-Otin, Blasco, Partridge, Serrano, & Kroemer, 2013; Oh, Lee, & Wagers, 2014). It has been shown that hair follicle stem cells in the bulge area undergo transepidermal elimination with aging, resulting in the step-wise miniaturization of hair follicles and eventually hair loss (Matsumura et al., 2016). When studying the fate of hair follicle stem cells during the process of aging, hair follicle stem cells are shown to switch their fates and terminally differentiate as IFE lineages. Moreover, DNA damage and genomic instability increases in hair follicle stem cells with aging. This leads to proteolysis and depletion of type XVII collagen, which causes the hair follicle stem cells to be aged (Matsumura et al., 2016).

In human skin, the rete ridges, the epidermal invaginations to dermis, become flattened with age, along with decreased expression of epidermal stem cell markers (Giangreco, Goldie, Failla, Saintigny, & Watt, 2010). Studies in the human epidermis also point out that even though epidermal stem cells are conserved with aging, the number of TA cells increases after in vivo repopulation and colony forming assays, despite the decreased proliferative capacity and cell cycle duration (Charruyer et al., 2009). In mice, whole-mount immunostaining suggests that even though there are age-associated changes in the epidermal proliferation, the stem cells in the epidermis are overall retained through aging (Giangreco, Qin, Pintar, & Watt, 2008), and epidermal stem cells from young and old mice exhibit similar characteristics in vitro (Stern & Bickenbach, 2007). In the paw skin, aging leads to the hyperproliferation of IFE, along with the loss of non-hemidesmosomal type XVII collagen (Watanabe et al., 2017). The age-related phenotypes become more apparent after injury and aged epidermal stem cells show a delayed wound healing due to an impaired cell-cell communication between epidermal stem and immune cells (Keyes et al., 2016). Existing work does not delineate how the heterogeneous populations of epidermal stem cells react to aging and how their crosstalk with each other during aging, and further studies will shed light into the complexity of these relevant processes in aging.

8. Conclusions

Recent findings provide a perspective as to how epidermal stem cells maintain the tissue and how they respond to tissue damage and various changes that tissue encounters. Their regulatory mechanisms, including the cell-intrinsic or extrinsic factors, interaction of different cell types and components of the niche are currently being updated. It will be interesting to further unravel the biological meaning of highly-compartmentalized and distinct populations of stem cells. With emerging new experimental techniques, our understanding of the epidermal stem cells and its niches will be accelerated.

Acknowledgments

This work is supported in part by AMED-PRIME, AMED (JP18gm6110016), Grant-in-Aid for Scientific Research on Innovative Areas "Stem Cell Aging and Disease" (17H05631), Grant-in-Aid for Early-Career Scientists (18K14709), the Nanotech Research Professional (N.R.P.) course of the Nanotech Career-up Alliance in Nanotech (CUPAL) project to A.S., and research grants from The Hoyu Science Foundation, The Nakatomi Foundation, The Takeda Foundation, The Astellas Foundation for Research on Metabolic Disorders to A.S.

References

Adam, R. C., Yang, H., Ge, Y., Lien, W. H., Wang, P., Zhao, Y., et al. (2018). Temporal layering of signaling effectors drives chromatin remodeling during hair follicle stem cell lineage progression. *Cell Stem Cell, 22*(3), 398–413.e7. https://doi.org/10.1016/j.stem.2017.12.004.

Adam, R. C., Yang, H., Rockowitz, S., Larsen, S. B., Nikolova, M., Oristian, D. S., et al. (2015). Pioneer factors govern super-enhancer dynamics in stem cell plasticity and lineage choice. *Nature, 521*(7552), 366–370. https://doi.org/10.1038/nature14289.

Alcolea, M. P., & Jones, P. H. (2014). Lineage analysis of epidermal stem cells. *Cold Spring Harbor Perspectives in Medicine, 4*(1), a015206. https://doi.org/10.1101/cshperspect.a015206.

Ali, N., Zirak, B., Rodriguez, R. S., Pauli, M. L., Truong, H. A., Lai, K., et al. (2017). Regulatory T cells in skin facilitate epithelial stem cell differentiation. *Cell, 169*(6), 1119–1129.e11. https://doi.org/10.1016/j.cell.2017.05.002.

Alonso, L., & Fuchs, E. (2006). The hair cycle. *Journal of Cell Science, 119*(Pt. 3), 391–393. https://doi.org/10.1242/jcs02793.

Aragona, M., Dekoninck, S., Rulands, S., Lenglez, S., Mascre, G., Simons, B. D., et al. (2017). Defining stem cell dynamics and migration during wound healing in mouse skin epidermis. *Nature Communications, 8*, 14684. https://doi.org/10.1038/ncomms14684.

Barker, N., van Es, J. H., Kuipers, J., Kujala, P., van den Born, M., Cozijnsen, M., et al. (2007). Identification of stem cells in small intestine and colon by marker gene Lgr5. *Nature, 449*(7165), 1003–1007. https://doi.org/10.1038/nature06196.

Barrandon, Y., & Green, H. (1987). Three clonal types of keratinocyte with different capacities for multiplication. *Proceedings of the National Academy of Sciences of the United States of America, 84*(8), 2302–2306.

Belokhvostova, D., Berzanskyte, I., Cujba, A. M., Jowett, G., Marshall, L., Prueller, J., et al. (2018). Homeostasis, regeneration and tumour formation in the mammalian epidermis. *The International Journal of Developmental Biology, 62*(6-7-8), 571–582. https://doi.org/10.1387/ijdb.170341fw.

Beronja, S., Janki, P., Heller, E., Lien, W. H., Keyes, B. E., Oshimori, N., et al. (2013). RNAi screens in mice identify physiological regulators of oncogenic growth. *Nature, 501*(7466), 185–190. https://doi.org/10.1038/nature12464.

Beronja, S., Livshits, G., Williams, S., & Fuchs, E. (2010). Rapid functional dissection of genetic networks via tissue-specific transduction and RNAi in mouse embryos. *Nature Medicine, 16*(7), 821–827. https://doi.org/10.1038/nm.2167.

Betschinger, J., & Knoblich, J. A. (2004). Dare to be different: Asymmetric cell division in Drosophila, *C. elegans* and vertebrates. *Current Biology, 14*(16), R674–R685. https://doi.org/10.1016/j.cub.2004.08.017.

Bickenbach, J. R. (1981). Identification and behavior of label-retaining cells in oral-mucosa and skin. *Journal of Dental Research, 60*, 1611–1620. https://doi.org/10.1177/002203458106000311011.

Biehs, B., Dijkgraaf, G. J. P., Piskol, R., Alicke, B., Boumahdi, S., Peale, F., et al. (2018). A cell identity switch allows residual BCC to survive Hedgehog pathway inhibition. *Nature, 562*(7727), 429–433. https://doi.org/10.1038/s41586-018-0596-y.

Blanpain, C., Lowry, W. E., Geoghegan, A., Polak, L., & Fuchs, E. (2004). Self-renewal, multipotency, and the existence of two cell populations within an epithelial stem cell niche. *Cell, 118*(5), 635–648. https://doi.org/10.1016/j.cell.2004.08.012.

Blanpain, C., & Simons, B. D. (2013). Unravelling stem cell dynamics by lineage tracing. *Nature Reviews. Molecular Cell Biology, 14*(8), 489–502. https://doi.org/10.1038/nrm3625.

Braun, K. M., Niemann, C., Jensen, U. B., Sundberg, J. P., Silva-Vargas, V., & Watt, F. M. (2003). Manipulation of stem cell proliferation and lineage commitment: Visualisation of label-retaining cells in wholemounts of mouse epidermis. *Development*, *130*(21), 5241–5255. https://doi.org/10.1242/dev.00703.

Brown, S., Pineda, C. M., Xin, T., Boucher, J., Suozzi, K. C., Park, S., et al. (2017). Correction of aberrant growth preserves tissue homeostasis. *Nature*, *548*(7667), 334–337. https://doi.org/10.1038/nature23304.

Brownell, I., Guevara, E., Bai, C. B., Loomis, C. A., & Joyner, A. L. (2011). Nerve-derived sonic hedgehog defines a niche for hair follicle stem cells capable of becoming epidermal stem cells. *Cell Stem Cell*, *8*(5), 552–565. https://doi.org/10.1016/j.stem.2011.02.021.

Buch, T., Heppner, F. L., Tertilt, C., Heinen, T. J., Kremer, M., Wunderlich, F. T., et al. (2005). A Cre-inducible diphtheria toxin receptor mediates cell lineage ablation after toxin administration. *Nature Methods*, *2*(6), 419–426. https://doi.org/10.1038/nmeth762.

Buczacki, S. J., Zecchini, H. I., Nicholson, A. M., Russell, R., Vermeulen, L., Kemp, R., et al. (2013). Intestinal label-retaining cells are secretory precursors expressing Lgr5. *Nature*, *495*(7439), 65–69. https://doi.org/10.1038/nature11965.

Buenrostro, J. D., Giresi, P. G., Zaba, L. C., Chang, H. Y., & Greenleaf, W. J. (2013). Transposition of native chromatin for fast and sensitive epigenomic profiling of open chromatin, DNA-binding proteins and nucleosome position. *Nature Methods*, *10*(12), 1213–1218. https://doi.org/10.1038/nmeth.2688.

Candi, E., Schmidt, R., & Melino, G. (2005). The cornified envelope: A model of cell death in the skin. *Nature Reviews. Molecular Cell Biology*, *6*(4), 328–340. https://doi.org/10.1038/nrm1619.

Chang, H., & Nathans, J. (2013). Responses of hair follicle-associated structures to loss of planar cell polarity signaling. *Proceedings of the National Academy of Sciences of the United States of America*, *110*(10), E908–E917. https://doi.org/10.1073/pnas.1301430110.

Charruyer, A., Barland, C. O., Yue, L., Wessendorf, H. B., Lu, Y., Lawrence, H. J., et al. (2009). Transit-amplifying cell frequency and cell cycle kinetics are altered in aged epidermis. *The Journal of Investigative Dermatology*, *129*(11), 2574–2583. https://doi.org/10.1038/jid.2009.127.

Chen, T., Heller, E., Beronja, S., Oshimori, N., Stokes, N., & Fuchs, E. (2012). An RNA interference screen uncovers a new molecule in stem cell self-renewal and long-term regeneration. *Nature*, *485*(7396), 104–108. https://doi.org/10.1038/nature10940.

Cheng, C. C., Tsutsui, K., Taguchi, T., Sanzen, N., Nakagawa, A., Kakiguchi, K., et al. (2018). Hair follicle epidermal stem cells define a niche for tactile sensation. *eLife*, 7, e38883. https://doi.org/10.7554/eLife.38883.

Chermnykh, E., Kalabusheva, E., & Vorotelyak, E. (2018). Extracellular matrix as a regulator of epidermal stem cell fate. *International Journal of Molecular Sciences*, *19*(4), E1003. https://doi.org/10.3390/ijms19041003.

Clayton, E., Doupe, D. P., Klein, A. M., Winton, D. J., Simons, B. D., & Jones, P. H. (2007). A single type of progenitor cell maintains normal epidermis. *Nature*, *446*(7132), 185–189. https://doi.org/10.1038/nature05574.

Compton, C. C., Gill, J. M., Bradford, D. A., Regauer, S., Gallico, G. G., & O'Connor, N. E. (1989). Skin regenerated from cultured epithelial autografts on full-thickness burn wounds from 6 days to 5 years after grafting. A light, electron microscopic and immunohistochemical study. *Laboratory Investigation*, *60*(5), 600–612.

Cotsarelis, G., Sun, T. T., & Lavker, R. M. (1990). Label-retaining cells reside in the bulge area of pilosebaceous unit: Implications for follicular stem cells, hair cycle, and skin carcinogenesis. *Cell*, *61*(7), 1329–1337.

Dekoninck, S., & Blanpain, C. (2019). Stem cell dynamics, migration and plasticity during wound healing. *Nature Cell Biology*, *21*(1), 18–24. https://doi.org/10.1038/s41556-018-0237-6.

Diamond, I., Owolabi, T., Marco, M., Lam, C., & Glick, A. (2000). Conditional gene expression in the epidermis of transgenic mice using the tetracycline-regulated transactivators tTA and rTA linked to the keratin 5 promoter. *The Journal of Investigative Dermatology*, *115*(5), 788–794. https://doi.org/10.1046/j.1523-1747.2000.00144.x.
Donati, G., Rognoni, E., Hiratsuka, T., Liakath-Ali, K., Hoste, E., Kar, G., et al. (2017). Wounding induces dedifferentiation of epidermal Gata6(+) cells and acquisition of stem cell properties. *Nature Cell Biology*, *19*(6), 603–613. https://doi.org/10.1038/ncb3532.
Doucet, Y. S., Woo, S. H., Ruiz, M. E., & Owens, D. M. (2013). The touch dome defines an epidermal niche specialized for mechanosensory signaling. *Cell Reports*, *3*(6), 1759–1765. https://doi.org/10.1016/j.celrep.2013.04.026.
Doupe, D. P., Klein, A. M., Simons, B. D., & Jones, P. H. (2010). The ordered architecture of murine ear epidermis is maintained by progenitor cells with random fate. *Developmental Cell*, *18*(2), 317–323. https://doi.org/10.1016/j.devcel.2009.12.016.
Driskell, R. R., Clavel, C., Rendl, M., & Watt, F. M. (2011). Hair follicle dermal papilla cells at a glance. *Journal of Cell Science*, *124*(Pt. 8), 1179–1182. https://doi.org/10.1242/jcs.082446.
Driskell, R. R., Lichtenberger, B. M., Hoste, E., Kretzschmar, K., Simons, B. D., Charalambous, M., et al. (2013). Distinct fibroblast lineages determine dermal architecture in skin development and repair. *Nature*, *504*(7479), 277–281. https://doi.org/10.1038/nature12783.
Epstein, E. H. (2008). Basal cell carcinomas: Attack of the hedgehog. *Nature Reviews. Cancer*, *8*(10), 743–754. https://doi.org/10.1038/nrc2503.
Erturk, A., Becker, K., Jahrling, N., Mauch, C. P., Hojer, C. D., Egen, J. G., et al. (2012). Three-dimensional imaging of solvent-cleared organs using 3DISCO. *Nature Protocols*, 7(11), 1983–1995. https://doi.org/10.1038/nprot.2012.119.
Festa, E., Fretz, J., Berry, R., Schmidt, B., Rodeheffer, M., Horowitz, M., et al. (2011). Adipocyte lineage cells contribute to the skin stem cell niche to drive hair cycling. *Cell*, *146*(5), 761–771. https://doi.org/10.1016/j.cell.2011.07.019.
Foudi, A., Hochedlinger, K., Van Buren, D., Schindler, J. W., Jaenisch, R., Carey, V., et al. (2009). Analysis of histone 2B-GFP retention reveals slowly cycling hematopoietic stem cells. *Nature Biotechnology*, *27*(1), 84–90. https://doi.org/10.1038/nbt.1517.
Fuchs, E. (2009). The tortoise and the hair: Slow-cycling cells in the stem cell race. *Cell*, *137*(5), 811–819. https://doi.org/10.1016/j.cell.2009.05.002.
Fuchs, E. V., Coppock, S. M., Green, H., & Cleveland, D. W. (1981). Two distinct classes of keratin genes and their evolutionary significance. *Cell*, *27*(1 Pt. 2), 75–84.
Fuchs, E., & Green, H. (1980). Changes in keratin gene expression during terminal differentiation of the keratinocyte. *Cell*, *19*(4), 1033–1042.
Fujiwara, H., Ferreira, M., Donati, G., Marciano, D. K., Linton, J. M., Sato, Y., et al. (2011). The basement membrane of hair follicle stem cells is a muscle cell niche. *Cell*, *144*(4), 577–589. https://doi.org/10.1016/j.cell.2011.01.014.
Fujiwara, H., Tsutsui, K., & Morita, R. (2018). Multi-tasking epidermal stem cells: Beyond epidermal maintenance. *Development, Growth & Differentiation*, *60*, 531–541. https://doi.org/10.1111/dgd.12577.
Fullgrabe, A., Joost, S., Are, A., Jacob, T., Sivan, U., Haegebarth, A., et al. (2015). Dynamics of Lgr6(+) progenitor cells in the hair follicle, sebaceous gland, and interfollicular epidermis. *Stem Cell Reports*, *5*(5), 843–855. https://doi.org/10.1016/j.stemcr.2015.09.013.
Gallico, G. G., 3rd, O'Connor, N. E., Compton, C. C., Kehinde, O., & Green, H. (1984). Permanent coverage of large burn wounds with autologous cultured human epithelium. *The New England Journal of Medicine*, *311*(7), 448–451. https://doi.org/10.1056/NEJM198408163110706.
Ge, Y., Gomez, N. C., Adam, R. C., Nikolova, M., Yang, H., Verma, A., et al. (2017). Stem cell lineage infidelity drives wound repair and cancer. *Cell*, *169*(4), 636–650.e14. https://doi.org/10.1016/j.cell.2017.03.042.

Giangreco, A., Goldie, S. J., Failla, V., Saintigny, G., & Watt, F. M. (2010). Human skin aging is associated with reduced expression of the stem cell markers beta1 integrin and MCSP. *The Journal of Investigative Dermatology*, *130*(2), 604–608. https://doi.org/10.1038/jid.2009.297.

Giangreco, A., Qin, M., Pintar, J. E., & Watt, F. M. (2008). Epidermal stem cells are retained in vivo throughout skin aging. *Aging Cell*, 7(2), 250–259. https://doi.org/10.1111/j.1474-9726.2008.00372.x.

Goldstein, J., & Horsley, V. (2012). Home sweet home: Skin stem cell niches. *Cellular and Molecular Life Sciences*, *69*(15), 2573–2582. https://doi.org/10.1007/s00018-012-0943-3.

Gomez, C., Chua, W., Miremadi, A., Quist, S., Headon, D. J., & Watt, F. M. (2013). The interfollicular epidermis of adult mouse tail comprises two distinct cell lineages that are differentially regulated by Wnt, Edaradd, and Lrig 1. *Stem Cell Reports*, *1*(1), 19–27. https://doi.org/10.1016/j.stemcr.2013.04.001.

Goodell, M. A., & Rando, T. A. (2015). Stem cells and healthy aging. *Science*, *350*(6265), 1199–1204. https://doi.org/10.1126/science.aab3388.

Gordon, R. (2013). Skin cancer: An overview of epidemiology and risk factors. *Seminars in Oncology Nursing*, *29*(3), 160–169. https://doi.org/10.1016/j.soncn.2013.06.002.

Gurtner, G. C., Werner, S., Barrandon, Y., & Longaker, M. T. (2008). Wound repair and regeneration. *Nature*, *453*(7193), 314–321. https://doi.org/10.1038/nature07039.

Haensel, D., & Dai, X. (2018). Epithelial-to-mesenchymal transition in cutaneous wound healing: Where we are and where we are heading. *Developmental Dynamics*, *247*(3), 473–480. https://doi.org/10.1002/dvdy.24561.

Hara, K., Nakagawa, T., Enomoto, H., Suzuki, M., Yamamoto, M., Simons, B. D., et al. (2014). Mouse spermatogenic stem cells continually interconvert between equipotent singly isolated and syncytial states. *Cell Stem Cell*, *14*(5), 658–672. https://doi.org/10.1016/j.stem.2014.01.019.

Heath, W. R., & Carbone, F. R. (2013). The skin-resident and migratory immune system in steady state and memory: Innate lymphocytes, dendritic cells and T cells. *Nature Immunology*, *14*(10), 978–985. https://doi.org/10.1038/ni.2680.

Hennings, H., Michael, D., Cheng, C., Steinert, P., Holbrook, K., & Yuspa, S. H. (1980). Calcium regulation of growth and differentiation of mouse epidermal cells in culture. *Cell*, *19*(1), 245–254.

Hirsch, T., Rothoeft, T., Teig, N., Bauer, J. W., Pellegrini, G., De Rosa, L., et al. (2017). Regeneration of the entire human epidermis using transgenic stem cells. *Nature*, *551*(7680), 327–332. https://doi.org/10.1038/nature24487.

Hoeck, J. D., Biehs, B., Kurtova, A. V., Kljavin, N. M., de Sousa, E. M. F., Alicke, B., et al. (2017). Stem cell plasticity enables hair regeneration following Lgr5(+) cell loss. *Nature Cell Biology*, *19*(6), 666–676. https://doi.org/10.1038/ncb3535.

Horsley, V., Aliprantis, A. O., Polak, L., Glimcher, L. H., & Fuchs, E. (2008). NFATc1 balances quiescence and proliferation of skin stem cells. *Cell*, *132*(2), 299–310. https://doi.org/10.1016/j.cell.2007.11.047.

Horsley, V., O'Carroll, D., Tooze, R., Ohinata, Y., Saitou, M., Obukhanych, T., et al. (2006). Blimp1 defines a progenitor population that governs cellular input to the sebaceous gland. *Cell*, *126*(3), 597–609. https://doi.org/10.1016/j.cell.2006.06.048.

Hsu, Y. C., Li, L., & Fuchs, E. (2014). Transit-amplifying cells orchestrate stem cell activity and tissue regeneration. *Cell*, *157*(4), 935–949. https://doi.org/10.1016/j.cell.2014.02.057.

Hsu, Y. C., Pasolli, H. A., & Fuchs, E. (2011). Dynamics between stem cells, niche, and progeny in the hair follicle. *Cell*, *144*(1), 92–105. https://doi.org/10.1016/j.cell.2010.11.049.

Huang, S. X., & Rompolas, P. (2017). Two-photon microscopy for intracutaneous imaging of stem cell activity in mice. *Experimental Dermatology*, *26*(5), 379–383. https://doi.org/10.1111/exd.13221.

Ichijo, R., Kobayashi, H., Yoneda, S., Iizuka, Y., Kubo, H., Matsumura, S., et al. (2017). Tbx3-dependent amplifying stem cell progeny drives interfollicular epidermal expansion during pregnancy and regeneration. *Nature Communications*, *8*(1), 508. https://doi.org/10.1038/s41467-017-00433-7.

Ito, M., Liu, Y., Yang, Z., Nguyen, J., Liang, F., Morris, R. J., et al. (2005). Stem cells in the hair follicle bulge contribute to wound repair but not to homeostasis of the epidermis. *Nature Medicine*, *11*(12), 1351–1354. https://doi.org/10.1038/nm1328.

Ivanova, A., Signore, M., Caro, N., Greene, N. D., Copp, A. J., & Martinez-Barbera, J. P. (2005). In vivo genetic ablation by Cre-mediated expression of diphtheria toxin fragment A. *Genesis*, *43*(3), 129–135. https://doi.org/10.1002/gene.20162.

Jaks, V., Barker, N., Kasper, M., van Es, J. H., Snippert, H. J., Clevers, H., et al. (2008). Lgr5 marks cycling, yet long-lived, hair follicle stem cells. *Nature Genetics*, *40*(11), 1291–1299. https://doi.org/10.1038/ng.239.

Jaks, V., Kasper, M., & Toftgard, R. (2010). The hair follicle-a stem cell zoo. *Experimental Cell Research*, *316*(8), 1422–1428. https://doi.org/10.1016/j.yexcr.2010.03.014.

Jensen, K. B., Collins, C. A., Nascimento, E., Tan, D. W., Frye, M., Itami, S., et al. (2009). Lrig1 expression defines a distinct multipotent stem cell population in mammalian epidermis. *Cell Stem Cell*, *4*(5), 427–439. https://doi.org/10.1016/j.stem.2009.04.014.

Jensen, U. B., Yan, X., Triel, C., Woo, S. H., Christensen, R., & Owens, D. M. (2008). A distinct population of clonogenic and multipotent murine follicular keratinocytes residing in the upper isthmus. *Journal of Cell Science*, *121*(Pt. 5), 609–617. https://doi.org/10.1242/jcs.025502.

Jones, P. H., Harper, S., & Watt, F. M. (1995). Stem cell patterning and fate in human epidermis. *Cell*, *80*(1), 83–93.

Jones, P. H., & Watt, F. M. (1993). Separation of human epidermal stem cells from transit amplifying cells on the basis of differences in integrin function and expression. *Cell*, *73*(4), 713–724.

Joost, S., Jacob, T., Sun, X., Annusver, K., La Manno, G., Sur, I., et al. (2018). Single-cell transcriptomics of traced epidermal and hair follicle stem cells reveals rapid adaptations during wound healing. *Cell Reports*, *25*(3), 585–597.e7. https://doi.org/10.1016/j.celrep.2018.09.059.

Joost, S., Zeisel, A., Jacob, T., Sun, X., La Manno, G., Lonnerberg, P., et al. (2016). Single-cell transcriptomics reveals that differentiation and spatial signatures shape epidermal and hair follicle heterogeneity. *Cell Systems*, *3*(3), 221–237.e9. https://doi.org/10.1016/j.cels.2016.08.010.

Kameda, T., Nakata, A., Mizutani, T., Terada, K., Iba, H., & Sugiyama, T. (2003). Analysis of the cellular heterogeneity in the basal layer of mouse ear epidermis: An approach from partial decomposition in vitro and retroviral cell marking in vivo. *Experimental Cell Research*, *283*(2), 167–183. https://doi.org/10.1016/S0144-4827(02)00031-9.

Kasper, M., Jaks, V., Hohl, D., & Toftgard, R. (2012). Basal cell carcinoma—Molecular biology and potential new therapies. *The Journal of Clinical Investigation*, *122*(2), 455–463. https://doi.org/10.1172/JCI58779.

Keyes, B. E., Liu, S., Asare, A., Naik, S., Levorse, J., Polak, L., et al. (2016). Impaired epidermal to dendritic T cell signaling slows wound repair in aged skin. *Cell*, *167*(5), 1323–1338.e14. https://doi.org/10.1016/j.cell.2016.10.052.

Keyes, B. E., Segal, J. P., Heller, E., Lien, W. H., Chang, C. Y., Guo, X., et al. (2013). Nfatc1 orchestrates aging in hair follicle stem cells. *Proceedings of the National Academy of Sciences of the United States of America*, *110*(51), E4950–E4959. https://doi.org/10.1073/pnas.1320301110.

Klein, A. M., Nakagawa, T., Ichikawa, R., Yoshida, S., & Simons, B. D. (2010). Mouse germ line stem cells undergo rapid and stochastic turnover. *Cell Stem Cell*, *7*(2), 214–224. https://doi.org/10.1016/j.stem.2010.05.017.

Klein, A. M., & Simons, B. D. (2011). Universal patterns of stem cell fate in cycling adult tissues. *Development, 138*(15), 3103–3111. https://doi.org/10.1242/dev.060103.

Kretzschmar, K., Cottle, D. L., Donati, G., Chiang, M. F., Quist, S. R., Gollnick, H. P., et al. (2014). BLIMP1 is required for postnatal epidermal homeostasis but does not define a sebaceous gland progenitor under steady-state conditions. *Stem Cell Reports, 3*(4), 620–633. https://doi.org/10.1016/j.stemcr.2014.08.007.

Kretzschmar, K., Weber, C., Driskell, R. R., Calonje, E., & Watt, F. M. (2016). Compartmentalized epidermal activation of beta-catenin differentially affects lineage reprogramming and underlies tumor heterogeneity. *Cell Reports, 14*(2), 269–281. https://doi.org/10.1016/j.celrep.2015.12.041.

Krieger, T., & Simons, B. D. (2015). Dynamic stem cell heterogeneity. *Development, 142*(8), 1396–1406. https://doi.org/10.1242/dev.101063.

Lawlor, K. T., & Kaur, P. (2015). Dermal contributions to human interfollicular epidermal architecture and self-renewal. *International Journal of Molecular Sciences, 16*(12), 28098–28107. https://doi.org/10.3390/ijms161226078.

Lay, K., Kume, T., & Fuchs, E. (2016). FOXC1 maintains the hair follicle stem cell niche and governs stem cell quiescence to preserve long-term tissue-regenerating potential. *Proceedings of the National Academy of Sciences of the United States of America, 113*(11), E1506–E1515. https://doi.org/10.1073/pnas.1601569113.

Lechler, T., & Fuchs, E. (2005). Asymmetric cell divisions promote stratification and differentiation of mammalian skin. *Nature, 437*(7056), 275–280. https://doi.org/10.1038/nature03922.

Lee, J., Kang, S., Lilja, K. C., Colletier, K. J., Scheitz, C. J., Zhang, Y. V., et al. (2016). Signalling couples hair follicle stem cell quiescence with reduced histone H3 K4/K9/K27me3 for proper tissue homeostasis. *Nature Communications, 7*, 11278. https://doi.org/10.1038/ncomms11278.

Lee, S. E., Sada, A., Zhang, M., McDermitt, D. J., Lu, S. Y., Kemphues, K. J., et al. (2014). High Runx1 levels promote a reversible, more-differentiated cell state in hair-follicle stem cells during quiescence. *Cell Reports, 6*(3), 499–513. https://doi.org/10.1016/j.celrep.2013.12.039.

Lee, J., & Tumbar, T. (2012). Hairy tale of signaling in hair follicle development and cycling. *Seminars in Cell & Developmental Biology, 23*(8), 906–916. https://doi.org/10.1016/j.semcdb.2012.08.003.

Legg, J., Jensen, U. B., Broad, S., Leigh, I., & Watt, F. M. (2003). Role of melanoma chondroitin sulphate proteoglycan in patterning stem cells in human interfollicular epidermis. *Development, 130*(24), 6049–6063. https://doi.org/10.1242/dev.00837.

Levy, V., Lindon, C., Harfe, B. D., & Morgan, B. A. (2005). Distinct stem cell populations regenerate the follicle and interfollicular epidermis. *Developmental Cell, 9*(6), 855–861. https://doi.org/10.1016/j.devcel.2005.11.003.

Li, L., & Clevers, H. (2010). Coexistence of quiescent and active adult stem cells in mammals. *Science, 327*(5965), 542–545. https://doi.org/10.1126/science.1180794.

Liakath-Ali, K., Vancollie, V. E., Heath, E., Smedley, D. P., Estabel, J., Sunter, D., et al. (2014). Novel skin phenotypes revealed by a genome-wide mouse reverse genetic screen. *Nature Communications, 5*, 3540. https://doi.org/10.1038/ncomms4540.

Liang, H., Akladios, B., Canales, C. P., Francis, R., Hardeman, E. H., & Beverdam, A. (2016). CUBIC protocol visualizes protein expression at single cell resolution in whole mount skin preparations. *Journal of Visualized Experiments*, (114), e54401. https://doi.org/10.3791/54401.

Lichti, U., Anders, J., & Yuspa, S. H. (2008). Isolation and short-term culture of primary keratinocytes, hair follicle populations and dermal cells from newborn mice and keratinocytes from adult mice for in vitro analysis and for grafting to immunodeficient mice. *Nature Protocols, 3*(5), 799–810. https://doi.org/10.1038/nprot.2008.50.

Lien, W. H., Guo, X., Polak, L., Lawton, L. N., Young, R. A., Zheng, D., et al. (2011). Genome-wide maps of histone modifications unwind in vivo chromatin states of the hair follicle lineage. *Cell Stem Cell*, *9*(3), 219–232. https://doi.org/10.1016/j.stem.2011.07.015.

Lim, X. H., Tan, S. H., Koh, W. L. C., Chau, R. M. W., Yan, K. S., Kuo, C. J., et al. (2013). Interfollicular epidermal stem cells self-renew via autocrine Wnt signaling. *Science*, *342*(6163), 1226–1230. https://doi.org/10.1126/science.1239730.

Liu, Y., Lyle, S., Yang, Z., & Cotsarelis, G. (2003). Keratin 15 promoter targets putative epithelial stem cells in the hair follicle bulge. *The Journal of Investigative Dermatology*, *121*(5), 963–968. https://doi.org/10.1046/j.1523-1747.2003.12600.x.

Livet, J., Weissman, T. A., Kang, H., Draft, R. W., Lu, J., Bennis, R. A., et al. (2007). Transgenic strategies for combinatorial expression of fluorescent proteins in the nervous system. *Nature*, *450*(7166), 56–62. https://doi.org/10.1038/nature06293.

Lopez-Garcia, C., Klein, A. M., Simons, B. D., & Winton, D. J. (2010). Intestinal stem cell replacement follows a pattern of neutral drift. *Science*, *330*(6005), 822–825. https://doi.org/10.1126/science.1196236.

Lopez-Otin, C., Blasco, M. A., Partridge, L., Serrano, M., & Kroemer, G. (2013). The hallmarks of aging. *Cell*, *153*(6), 1194–1217. https://doi.org/10.1016/j.cell.2013.05.039.

Lu, C., & Fuchs, E. (2014). Sweat gland progenitors in development, homeostasis, and wound repair. *Cold Spring Harbor Perspectives in Medicine*, *4*(2), a015222. https://doi.org/10.1101/cshperspect.a015222.

Lu, C. P., Polak, L., Keyes, B. E., & Fuchs, E. (2016). Spatiotemporal antagonism in mesenchymal-epithelial signaling in sweat versus hair fate decision. *Science*, *354*(6319), aah6102. https://doi.org/10.1126/science.aah6102.

Lu, C. P., Polak, L., Rocha, A. S., Pasolli, H. A., Chen, S. C., Sharma, N., et al. (2012). Identification of stem cell populations in sweat glands and ducts reveals roles in homeostasis and wound repair. *Cell*, *150*(1), 136–150. https://doi.org/10.1016/j.cell.2012.04.045.

Mackenzie, I. C. (1970). Relationship between mitosis and the ordered structure of the stratum corneum in mouse epidermis. *Nature*, *226*(5246), 653–655.

Maksimovic, S., Nakatani, M., Baba, Y., Nelson, A. M., Marshall, K. L., Wellnitz, S. A., et al. (2014). Epidermal Merkel cells are mechanosensory cells that tune mammalian touch receptors. *Nature*, *509*(7502), 617–621. https://doi.org/10.1038/nature13250.

Malissen, B., Tamoutounour, S., & Henri, S. (2014). The origins and functions of dendritic cells and macrophages in the skin. *Nature Reviews. Immunology*, *14*(6), 417–428. https://doi.org/10.1038/nri3683.

Martincorena, I., Roshan, A., Gerstung, M., Ellis, P., Van Loo, P., McLaren, S., et al. (2015). Tumor evolution. High burden and pervasive positive selection of somatic mutations in normal human skin. *Science*, *348*(6237), 880–886. https://doi.org/10.1126/science.aaa6806.

Mascre, G., Dekoninck, S., Drogat, B., Youssef, K. K., Brohee, S., Sotiropoulou, P. A., et al. (2012). Distinct contribution of stem and progenitor cells to epidermal maintenance. *Nature*, *489*(7415), 257–262. https://doi.org/10.1038/nature11393.

Matsumura, H., Mohri, Y., Binh, N. T., Morinaga, H., Fukuda, M., Ito, M., et al. (2016). Hair follicle aging is driven by transepidermal elimination of stem cells via COL17A1 proteolysis. *Science*, *351*(6273), aad4395. https://doi.org/10.1126/science.aad4395.

Mesa, K. R., Kawaguchi, K., Cockburn, K., Gonzalez, D., Boucher, J., Xin, T., et al. (2018). Homeostatic epidermal stem cell self-renewal is driven by local differentiation. *Cell Stem Cell*, *23*(5), 677–686.e4. https://doi.org/10.1016/j.stem.2018.09.005.

Mesa, K. R., Rompolas, P., Zito, G., Myung, P., Sun, T. Y., Brown, S., et al. (2015). Niche-induced cell death and epithelial phagocytosis regulate hair follicle stem cell pool. *Nature*, *522*(7554), 94–97. https://doi.org/10.1038/nature14306.

Metcalfe, C., Kljavin, N. M., Ybarra, R., & de Sauvage, F. J. (2014). Lgr5+ stem cells are indispensable for radiation-induced intestinal regeneration. *Cell Stem Cell*, *14*(2), 149–159. https://doi.org/10.1016/j.stem.2013.11.008.

Mok, K. W., Saxena, N., & Rendl, M. (2018). More than the sum of its parts: Single-cell transcriptomics reveals epidermal cell states. *Cell Reports*, *25*(4), 823–824. https://doi.org/10.1016/j.celrep.2018.10.041.

Montgomery, R. K., Carlone, D. L., Richmond, C. A., Farilla, L., Kranendonk, M. E., Henderson, D. E., et al. (2011). Mouse telomerase reverse transcriptase (mTert) expression marks slowly cycling intestinal stem cells. *Proceedings of the National Academy of Sciences of the United States of America*, *108*(1), 179–184. https://doi.org/10.1073/pnas.1013004108.

Morris, R. J., Liu, Y., Marles, L., Yang, Z., Trempus, C., Li, S., et al. (2004). Capturing and profiling adult hair follicle stem cells. *Nature Biotechnology*, *22*(4), 411–417. https://doi.org/10.1038/nbt950.

Morrison, S. J., & Kimble, J. (2006). Asymmetric and symmetric stem-cell divisions in development and cancer. *Nature*, *441*(7097), 1068–1074. https://doi.org/10.1038/nature04956.

Muller-Rover, S., Handjiski, B., van der Veen, C., Eichmuller, S., Foitzik, K., McKay, I. A., et al. (2001). A comprehensive guide for the accurate classification of murine hair follicles in distinct hair cycle stages. *The Journal of Investigative Dermatology*, *117*(1), 3–15. https://doi.org/10.1046/j.0022-202x.2001.01377.x.

Murai, K., Skrupskelyte, G., Piedrafita, G., Hall, M., Kostiou, V., Ong, S. H., et al. (2018). Epidermal tissue adapts to restrain progenitors carrying clonal p53 mutations. *Cell Stem Cell*, *23*(5), 687–699.e8. https://doi.org/10.1016/j.stem.2018.08.017.

Nagarajan, P., Asgari, M. M., Green, A. C., Guhan, S. M., Arron, S. T., Proby, C. M., et al. (2019). Keratinocyte carcinomas: Current concepts and future research priorities. *Clinical Cancer Research*, *25*(8), 2379–2391. https://doi.org/10.1158/1078-0432.CCR-18-1122.

Naik, S., Larsen, S. B., Cowley, C. J., & Fuchs, E. (2018). Two to Tango: Dialog between immunity and stem cells in health and disease. *Cell*, *175*(4), 908–920. https://doi.org/10.1016/j.cell.2018.08.071.

Nakagawa, T., Nabeshima, Y., & Yoshida, S. (2007). Functional identification of the actual and potential stem cell compartments in mouse spermatogenesis. *Developmental Cell*, *12*(2), 195–206. https://doi.org/10.1016/j.devcel.2007.01.002.

Nijhof, J. G., Braun, K. M., Giangreco, A., van Pelt, C., Kawamoto, H., Boyd, R. L., et al. (2006). The cell-surface marker MTS24 identifies a novel population of follicular keratinocytes with characteristics of progenitor cells. *Development*, *133*(15), 3027–3037. https://doi.org/10.1242/dev.02443.

Nishimura, E. K. (2011). Melanocyte stem cells: A melanocyte reservoir in hair follicles for hair and skin pigmentation. *Pigment Cell & Melanoma Research*, *24*(3), 401–410. https://doi.org/10.1111/j.1755-148X.2011.00855.x.

O'Connor, N., Mulliken, J., Banks-Schlegel, S., Kehinde, O., & Green, H. (1981). Grafting of burns with cultured epithelium prepared from autologous epidermal cells. *Lancet*, *1*(8211), 75–78.

Oh, J., Lee, Y. D., & Wagers, A. J. (2014). Stem cell aging: Mechanisms, regulators and therapeutic opportunities. *Nature Medicine*, *20*(8), 870–880. https://doi.org/10.1038/nm.3651.

Osorio, K. M., Lee, S. E., McDermitt, D. J., Waghmare, S. K., Zhang, Y. V., Woo, H. N., et al. (2008). Runx1 modulates developmental, but not injury-driven, hair follicle stem cell activation. *Development*, *135*(6), 1059–1068. https://doi.org/10.1242/dev.012799.

Page, M. E., Lombard, P., Ng, F., Gottgens, B., & Jensen, K. B. (2013). The epidermis comprises autonomous compartments maintained by distinct stem cell populations. *Cell Stem Cell*, *13*(4), 471–482. https://doi.org/10.1016/j.stem.2013.07.010.

Park, S., Gonzalez, D. G., Guirao, B., Boucher, J. D., Cockburn, K., Marsh, E. D., et al. (2017). Tissue-scale coordination of cellular behaviour promotes epidermal wound repair in live mice. *Nature Cell Biology*, *19*(2), 155–163. https://doi.org/10.1038/ncb3472.

Pellegrini, G., Dellambra, E., Golisano, O., Martinelli, E., Fantozzi, I., Bondanza, S., et al. (2001). p63 identifies keratinocyte stem cells. *Proceedings of the National Academy of Sciences of the United States of America*, *98*(6), 3156–3161. https://doi.org/10.1073/pnas.061032098.

Philippeos, C., Telerman, S. B., Oules, B., Pisco, A. O., Shaw, T. J., Elgueta, R., et al. (2018). Spatial and single-cell transcriptional profiling identifies functionally distinct human dermal fibroblast subpopulations. *The Journal of Investigative Dermatology*, *138*(4), 811–825. https://doi.org/10.1016/j.jid.2018.01.016.

Pineda, C. M., Park, S., Mesa, K. R., Wolfel, M., Gonzalez, D. G., Haberman, A. M., et al. (2015). Intravital imaging of hair follicle regeneration in the mouse. *Nature Protocols*, *10*(7), 1116–1130. https://doi.org/10.1038/nprot.2015.070.

Potten, C. S. (1974). The epidermal proliferative unit: The possible role of the central basal cell. *Cell and Tissue Kinetics*, 7(1), 77–88.

Potten, C. S., Saffhill, R., & Maibach, H. I. (1987). Measurement of the transit time for cells through the epidermis and stratum corneum of the mouse and guinea-pig. *Cell and Tissue Kinetics*, *20*(5), 461–472.

Reichenbach, B., Classon, J., Aida, T., Tanaka, K., Genander, M., & Goritz, C. (2018). Glutamate transporter Slc1a3 mediates inter-niche stem cell activation during skin growth. *The EMBO Journal*, *37*(9), e98280. https://doi.org/10.15252/embj.201798280.

Rheinwald, J. G., & Green, H. (1975a). Formation of a keratinizing epithelium in culture by a cloned cell line derived from a teratoma. *Cell*, *6*(3), 317–330.

Rheinwald, J. G., & Green, H. (1975b). Serial cultivation of strains of human epidermal keratinocytes: The formation of keratinizing colonies from single cells. *Cell*, *6*(3), 331–343.

Rinkevich, Y., Walmsley, G. G., Hu, M. S., Maan, Z. N., Newman, A. M., Drukker, M., et al. (2015). Skin fibrosis. Identification and isolation of a dermal lineage with intrinsic fibrogenic potential. *Science*, *348*(6232), aaa2151. https://doi.org/10.1126/science.aaa2151.

Ro, S., & Rannala, B. (2004). A stop-EGFP transgenic mouse to detect clonal cell lineages generated by mutation. *EMBO Reports*, *5*(9), 914–920. https://doi.org/10.1038/sj.embor.7400218.

Ro, S., & Rannala, B. (2005). Evidence from the stop-EGFP mouse supports a niche-sharing model of epidermal proliferative units. *Experimental Dermatology*, *14*(11), 838–843. https://doi.org/10.1111/j.1600-0625.2005.00366.x.

Rognoni, E., & Watt, F. M. (2018). Skin cell heterogeneity in development, wound healing, and cancer. *Trends in Cell Biology*, *28*(9), 709–722. https://doi.org/10.1016/j.tcb.2018.05.002.

Rompolas, P., Deschene, E. R., Zito, G., Gonzalez, D. G., Saotome, I., Haberman, A. M., et al. (2012). Live imaging of stem cell and progeny behaviour in physiological hair-follicle regeneration. *Nature*, *487*(7408), 496–499. https://doi.org/10.1038/nature11218.

Rompolas, P., & Greco, V. (2014). Stem cell dynamics in the hair follicle niche. *Seminars in Cell & Developmental Biology*, *25–26*, 34–42. https://doi.org/10.1016/j.semcdb.2013.12.005.

Rompolas, P., Mesa, K. R., Kawaguchi, K., Park, S., Gonzalez, D., Brown, S., et al. (2016). Spatiotemporal coordination of stem cell commitment during epidermal homeostasis. *Science*, *352*(6292), 1471–1474. https://doi.org/10.1126/science.aaf7012.

Roshan, A., Murai, K., Fowler, J., Simons, B. D., Nikolaidou-Neokosmidou, V., & Jones, P. H. (2016). Human keratinocytes have two interconvertible modes of proliferation. *Nature Cell Biology*, *18*(2), 145–156. https://doi.org/10.1038/ncb3282.

Roy, E., & Khosrotehrani, K. (2018). Past stem cells and finally in transit: SLC1A3 instructs skin niche coupling. *The EMBO Journal*, *37*(9), e99393. https://doi.org/10.15252/embj.201899393.

Roy, E., Neufeld, Z., Cerone, L., Wong, H. Y., Hodgson, S., Livet, J., et al. (2016). Bimodal behaviour of interfollicular epidermal progenitors regulated by hair follicle position and cycling. *The EMBO Journal*, *35*(24), 2658–2670. https://doi.org/10.15252/embj.201693806.

Roy, E., Neufeld, Z., Livet, J., & Khosrotehrani, K. (2014). Concise review: Understanding clonal dynamics in homeostasis and injury through multicolor lineage tracing. *Stem Cells*, *32*(12), 3046–3054. https://doi.org/10.1002/stem.1804.

Rulands, S., & Simons, B. D. (2016). Tracing cellular dynamics in tissue development, maintenance and disease. *Current Opinion in Cell Biology*, *43*, 38–45. https://doi.org/10.1016/j.ceb.2016.07.001.

Sada, A., Jacob, F., Leung, E., Wang, S., White, B. S., Shalloway, D., et al. (2016). Defining the cellular lineage hierarchy in the interfollicular epidermis of adult skin. *Nature Cell Biology*, *18*(6), 619–631. https://doi.org/10.1038/ncb3359.

Sada, A., Jain, P., Wang, S., Leung, E., & Tumbar, T. (2017). Slc1a3-CreER as a targeting tool for the K6+ epithelial stem cell niche and its precursors during mouse hair follicle cycle. *The Journal of Investigative Dermatology*, *137*(7), 1569–1571. https://doi.org/10.1016/j.jid.2017.02.974.

Sada, A., & Tumbar, T. (2013). New insights into mechanisms of stem cell daughter fate determination in regenerative tissues. *International Review of Cell and Molecular Biology*, *300*, 1–50. https://doi.org/10.1016/B978-0-12-405210-9.00001-1.

Sanchez-Danes, A., Hannezo, E., Larsimont, J. C., Liagre, M., Youssef, K. K., Simons, B. D., et al. (2016). Defining the clonal dynamics leading to mouse skin tumour initiation. *Nature*, *536*(7616), 298–303. https://doi.org/10.1038/nature19069.

Sanchez-Danes, A., Larsimont, J. C., Liagre, M., Munoz-Couselo, E., Lapouge, G., Brisebarre, A., et al. (2018). A slow-cycling LGR5 tumour population mediates basal cell carcinoma relapse after therapy. *Nature*, *562*(7727), 434–438. https://doi.org/10.1038/s41586-018-0603-3.

Sangiorgi, E., & Capecchi, M. R. (2008). Bmi1 is expressed in vivo in intestinal stem cells. *Nature Genetics*, *40*(7), 915–920. https://doi.org/10.1038/ng.165.

Sanjana, N. E., Shalem, O., & Zhang, F. (2014). Improved vectors and genome-wide libraries for CRISPR screening. *Nature Methods*, *11*(8), 783–784. https://doi.org/10.1038/nmeth.3047.

Schillert, A., Trumpp, A., & Sprick, M. R. (2013). Label retaining cells in cancer—The dormant root of evil? *Cancer Letters*, *341*(1), 73–79. https://doi.org/10.1016/j.canlet.2013.04.019.

Schluter, H., Paquet-Fifield, S., Gangatirkar, P., Li, J., & Kaur, P. (2011). Functional characterization of quiescent keratinocyte stem cells and their progeny reveals a hierarchical organization in human skin epidermis. *Stem Cells*, *29*(8), 1256–1268. https://doi.org/10.1002/stem.675.

Schweizer, J., & Marks, F. (1977). A developmental study of the distribution and frequency of langerhans cells in relation to formation of patterning in mouse tail epidermis. *The Journal of Investigative Dermatology*, *69*(2), 198–204.

Shinin, V. V., & Panteleyev, A. A. (2017). Targeting of stem cell niches using unconventional genetic tools. *The Journal of Investigative Dermatology*, *137*(7), 1413–1414. https://doi.org/10.1016/j.jid.2017.03.022.

Simpson, C. L., Patel, D. M., & Green, K. J. (2011). Deconstructing the skin: Cytoarchitectural determinants of epidermal morphogenesis. *Nature Reviews. Molecular Cell Biology*, *12*(9), 565–580. https://doi.org/10.1038/nrm3175.

Snippert, H. J., Haegebarth, A., Kasper, M., Jaks, V., van Es, J. H., Barker, N., et al. (2010). Lgr6 marks stem cells in the hair follicle that generate all cell lineages of the skin. *Science*, *327*(5971), 1385–1389. https://doi.org/10.1126/science.1184733.

Snippert, H. J., van der Flier, L. G., Sato, T., van Es, J. H., van den Born, M., Kroon-Veenboer, C., et al. (2010). Intestinal crypt homeostasis results from neutral competition between symmetrically dividing Lgr5 stem cells. *Cell*, *143*(1), 134–144. https://doi.org/10.1016/j.cell.2010.09.016.

Sorrell, J. M., & Caplan, A. I. (2004). Fibroblast heterogeneity: More than skin deep. *Journal of Cell Science*, *117*(Pt. 5), 667–675. https://doi.org/10.1242/jcs.01005.

Soteriou, D., Kostic, L., Sedov, E., Yosefzon, Y., Steller, H., & Fuchs, Y. (2016). Isolating hair follicle stem cells and epidermal keratinocytes from dorsal mouse skin. *Journal of Visualized Experiments*, (110), e53931. https://doi.org/10.3791/53931.

Stern, M. M., & Bickenbach, J. R. (2007). Epidermal stem cells are resistant to cellular aging. *Aging Cell*, *6*(4), 439–452. https://doi.org/10.1111/j.1474-9726.2007.00318.x.

Sun, B. K., Siprashvili, Z., & Khavari, P. A. (2014). Advances in skin grafting and treatment of cutaneous wounds. *Science*, *346*(6212), 941–945. https://doi.org/10.1126/science.1253836.

Susaki, E. A., Tainaka, K., Perrin, D., Kishino, F., Tawara, T., Watanabe, T. M., et al. (2014). Whole-brain imaging with single-cell resolution using chemical cocktails and computational analysis. *Cell*, *157*(3), 726–739. https://doi.org/10.1016/j.cell.2014.03.042.

Takeda, N., Jain, R., Leboeuf, M. R., Padmanabhan, A., Wang, Q., Li, L., et al. (2013). Hopx expression defines a subset of multipotent hair follicle stem cells and a progenitor population primed to give rise to K6+ niche cells. *Development*, *140*(8), 1655–1664. https://doi.org/10.1242/dev.093005.

Takeda, N., Jain, R., LeBoeuf, M. R., Wang, Q., Lu, M. M., & Epstein, J. A. (2011). Interconversion between intestinal stem cell populations in distinct niches. *Science*, *334*(6061), 1420–1424. https://doi.org/10.1126/science.1213214.

Tan, D. W., Jensen, K. B., Trotter, M. W., Connelly, J. T., Broad, S., & Watt, F. M. (2013). Single-cell gene expression profiling reveals functional heterogeneity of undifferentiated human epidermal cells. *Development*, *140*(7), 1433–1444. https://doi.org/10.1242/dev.087551.

Tian, H., Biehs, B., Warming, S., Leong, K. G., Rangell, L., Klein, O. D., et al. (2011). A reserve stem cell population in small intestine renders Lgr5-positive cells dispensable. *Nature*, *478*(7368), 255–259. https://doi.org/10.1038/nature10408.

Trempus, C. S., Morris, R. J., Bortner, C. D., Cotsarelis, G., Faircloth, R. S., Reece, J. M., et al. (2003). Enrichment for living murine keratinocytes from the hair follicle bulge with the cell surface marker CD34. *The Journal of Investigative Dermatology*, *120*(4), 501–511. https://doi.org/10.1046/j.1523-1747.2003.12088.x.

Tumbar, T., Guasch, G., Greco, V., Blanpain, C., Lowry, W. E., Rendl, M., et al. (2004). Defining the epithelial stem cell niche in skin. *Science*, *303*(5656), 359–363. https://doi.org/10.1126/science.1092436.

Veltri, A., Lang, C., & Lien, W. H. (2018). Concise review: Wnt signaling pathways in skin development and epidermal stem cells. *Stem Cells*, *36*(1), 22–35. https://doi.org/10.1002/stem.2723.

Viswanathan, P., Guvendiren, M., Chua, W., Telerman, S. B., Liakath-Ali, K., Burdick, J. A., et al. (2016). Mimicking the topography of the epidermal-dermal interface with elastomer substrates. *Integrative Biology*, *8*(1), 21–29. https://doi.org/10.1039/c5ib00238a.

Waghmare, S. K., Bansal, R., Lee, J., Zhang, Y. V., McDermitt, D. J., & Tumbar, T. (2008). Quantitative proliferation dynamics and random chromosome segregation of hair follicle stem cells. *The EMBO Journal*, *27*(9), 1309–1320. https://doi.org/10.1038/emboj.2008.72.

Watanabe, M., Natsuga, K., Nishie, W., Kobayashi, Y., Donati, G., Suzuki, S., et al. (2017). Type XVII collagen coordinates proliferation in the interfollicular epidermis. *eLife*, *6*, e26635. https://doi.org/10.7554/eLife.26635.

Watt, F. M., & Fujiwara, H. (2011). Cell-extracellular matrix interactions in normal and diseased skin. *Cold Spring Harbor Perspectives in Biology*, *3*(4), a005124. https://doi.org/10.1101/cshperspect.a005124.

Watt, F. M., & Jones, P. H. (1993). Expression and function of the keratinocyte integrins. *Development Supplement*, *119*, 185–192.

Webb, A., Li, A., & Kaur, P. (2004). Location and phenotype of human adult keratinocyte stem cells of the skin. *Differentiation*, *72*(8), 387–395. https://doi.org/10.1111/j.1432-0436.2004.07208005.x.

Wilson, A., Laurenti, E., Oser, G., van der Wath, R. C., Blanco-Bose, W., Jaworski, M., et al. (2008). Hematopoietic stem cells reversibly switch from dormancy to self-renewal during homeostasis and repair. *Cell*, *135*(6), 1118–1129. https://doi.org/10.1016/j.cell.2008.10.048.

Xin, T., Gonzalez, D., Rompolas, P., & Greco, V. (2018). Flexible fate determination ensures robust differentiation in the hair follicle. *Nature Cell Biology*, *20*(12), 1361–1369. https://doi.org/10.1038/s41556-018-0232-y.

Yang, H., Adam, R. C., Ge, Y., Hua, Z. L., & Fuchs, E. (2017). Epithelial-mesenchymal micro-niches govern stem cell lineage choices. *Cell*, *169*(3), 483–496.e3. https://doi.org/10.1016/j.cell.2017.03.038.

Yang, L., Xie, G., Fan, Q., & Xie, J. (2010). Activation of the hedgehog-signaling pathway in human cancer and the clinical implications. *Oncogene*, *29*(4), 469–481. https://doi.org/10.1038/onc.2009.392.

Yi, R. (2017). Concise review: Mechanisms of quiescent hair follicle stem cell regulation. *Stem Cells*, *35*(12), 2323–2330. https://doi.org/10.1002/stem.2696.

Ying, Z., Sandoval, M., & Beronja, S. (2018). Oncogenic activation of PI3K induces progenitor cell differentiation to suppress epidermal growth. *Nature Cell Biology*, *20*(11), 1256–1266. https://doi.org/10.1038/s41556-018-0218-9.

Youssef, K. K., Van Keymeulen, A., Lapouge, G., Beck, B., Michaux, C., Achouri, Y., et al. (2010). Identification of the cell lineage at the origin of basal cell carcinoma. *Nature Cell Biology*, *12*(3), 299–305. https://doi.org/10.1038/ncb2031.

Yu, Z., Jiang, K., Xu, Z., Huang, H., Qian, N., Lu, Z., et al. (2018). Hoxc-dependent mesenchymal niche heterogeneity drives regional hair follicle regeneration. *Cell Stem Cell*, *23*(4), 487–500.e6. https://doi.org/10.1016/j.stem.2018.07.016.

Zhang, Y. V., Cheong, J., Ciapurin, N., McDermitt, D. J., & Tumbar, T. (2009). Distinct self-renewal and differentiation phases in the niche of infrequently dividing hair follicle stem cells. *Cell Stem Cell*, *5*(3), 267–278. https://doi.org/10.1016/j.stem.2009.06.004.

Zhang, B., & Hsu, Y. C. (2017). Emerging roles of transit-amplifying cells in tissue regeneration and cancer. *Wiley Interdisciplinary Reviews: Developmental Biology*, *6*(5), e282. https://doi.org/10.1002/wdev.282.

CHAPTER THREE

Molecular aspects governing epidermal stem cell niches

Paola Kuri, Gabriella Rice, Panteleimon Rompolas*
Department of Dermatology, Institute for Regenerative Medicine, University of Pennsylvania Perelman School of Medicine, Philadelphia, PA, United States
*Corresponding author: e-mail address: rompolas@pennmedicine.upenn.edu

Contents

1. Introduction

The skin is a large external organ that serves multiple functions, including thermoregulation and sensation. Additionally, the skin represents an ultimate barrier that shields the body from the environmental insults and defends it from pathogens. To perform these vital functions, the skin consists two layers, the outermost of which is the epidermis. This layer is organized as stratified sheets of cells that cover the surface skin and form a watertight

Advances in Stem Cells and their Niches, Volume 3
ISSN 2468-5097
https://doi.org/10.1016/bs.asn.2019.05.001

barrier. Interspersed within the epidermis are structures called appendages. These include hair follicles and eccrine sweat glands, which are specified during development and vary in numbers and function depending on the body site. There are specialized structures that are associated with the hair follicle including the isthmus, infundibulum and sebaceous gland, which together form the pilosebaceous unit (reviewed in Gonzales & Fuchs, 2017).

All epithelial compartments are formed during embryonic development from cells of ectodermal origin. In the fully developed adult skin, epithelial structures exist in a state of dynamic equilibrium and display some of the highest cellular turnovers in the mammalian body (for specific numbers, see Hoath & Leahy, 2003). Cells are shed on a daily basis from the surface of the epidermis due to normal wear from the external environment. The hair follicle generates an extremely large number of cells, which undergo a concerted process of terminal differentiation with the sole purpose of building a fiber that is ultimately dispensable and only transiently retains attachment to the skin (for a detailed description see Müller-Röver et al., 2001; Saxena, Mok, & Rendl, 2019). In the sebaceous glands, epithelial cells undergo a maturation process that involves the production of large deposits of triglycerides and wax esters. These deposits help lubricate and protect the skin, and are released through a secretory mechanism, in which the enlarged sebocytes rupture and secrete their contents into the hair canal (reviewed in Schneider & Paus, 2010). A common denominator of all these processes is the continuous loss of cells, which is inherent to the physiological function of the skin. Throughout life, the skin relies on stem cells that have the capacity to self-renew and differentiate to all the specialized cell types of the epithelium (Gonzales & Fuchs, 2017). These processes support the demand for new cells and in order to maintain homeostatic balance.

Stem cells in the skin have been at the center of research for many decades, due to their importance in understanding tissue physiology and for their tremendous potential for use in new therapies. However, many questions remain unanswered, including (1) how are stem cells specified during development, (2) which genes determine their identity, (3) what are the intrinsic and extrinsic factors that regulate their activity, (4) what is their contribution to normal tissue physiology and regeneration and (5) how does their function change in disease states such as cancer. Genetic mouse models have been tremendously valuable to understand stem cell regulation and function. These models have enabled the identification and characterization of key genes and signaling pathways that specify and control the activity of stem cells in adult skin. Despite this progress, it is becoming apparent that stem cell biology is far more complex than what can be resolved with traditional, static, experimental approaches. This is primarily due to the fact that stem cell identity and function are highly heterogeneous and dynamic and that stem cells are subject to multiple levels of regulation (reviewed in Rognoni & Watt, 2018). A prime example is the contrasting behavior that stem cells exhibit between tissue homeostasis and wound healing. It is generally accepted that each skin compartment is exclusively maintained by its own resident stem cells. However, experimental evidence suggests that in the event of injury to the skin, stem cells are capable of changing their identity and fate to contribute to the repair and long-term maintenance of a different skin structure. Therefore, the mechanisms that control cellular identity and plasticity *in vivo* are still not fully understood.

To address such challenges, new experimental approaches have been developed and widely adopted to study how stem cells function in health and disease. Due to the lack of a definitive "stemness" gene, researchers can only identify stem cells based on their capacity to satisfy two basic requirements. Namely (1) their ability to self-renew and (2) their ability differentiate into all the specialized lineages in the skin. Lineage tracing, in which individual stem cells are irreversibly marked by a fluorescent reporter that is inherited by all of the stem cell's progeny, has been the main assay to test these properties *in vivo* (described in Blanpain & Simons, 2013). However, the spatiotemporal resolution of the assay is limited by static histological analysis and the specificity of the genetic driver used. Intravital imaging effectively solved these shortcomings, since the same cells can be imaged and their activity tracked over time within their intact native tissue environment. With this approach, the fate decisions of individual cells can be directly captured and the environmental factors and associations that regulate these decisions can be better understood. Furthermore, live imaging has revealed the immense heterogeneity of stem and progenitor cell populations in the skin and highlighted the dynamic nature of their interactions with the respective tissue microenvironment, commonly known as the niche (Mesa et al., 2018; Rompolas et al., 2012, 2016; Rompolas, Mesa, & Greco, 2013). Concurrent to the recent advances in imaging technologies, an unprecedented growth in genomic approaches has illuminated the complexity of the regulatory gene networks and epigenetic mechanisms that control stem cell identity and fate. In this chapter we discuss stem cell activity and regulation in the skin and how our knowledge of the latter has reached greater depths with the experimental evidence that these powerful approaches have produced over the past decade.

2. Function and architecture of the epidermis

The epidermis is the most external layer of the mammalian skin and constitutes the main barrier between the environment and the internal organs. It is a stratified epithelium, composed of several layers of morphologically distinct cells (Fig. 1A). At the bottom, classical extracellular matrix

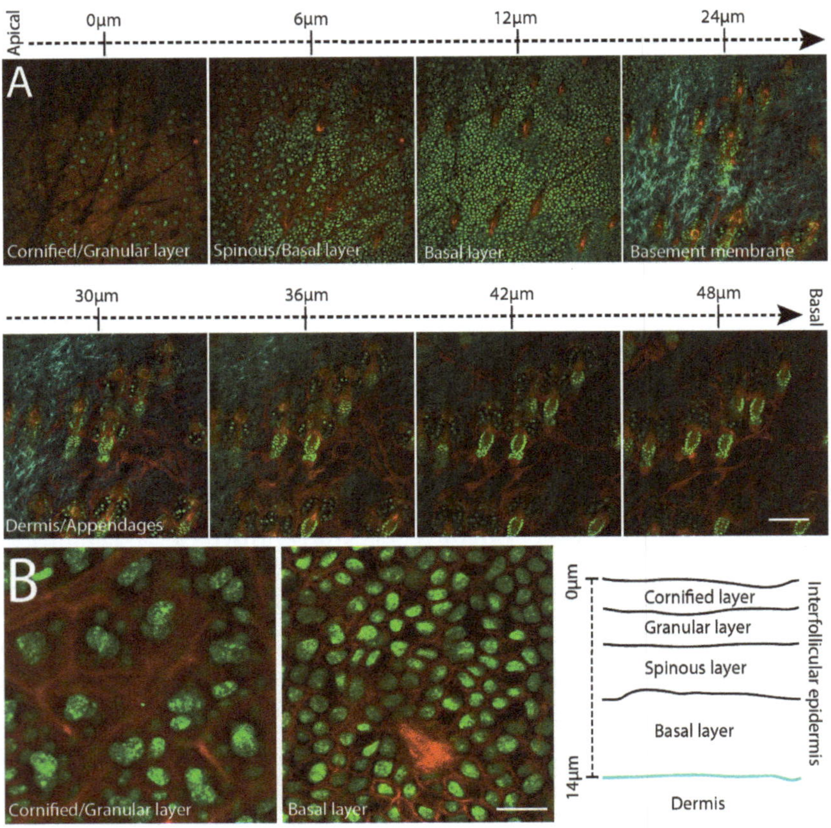

Fig. 1 Live imaging enables noninvasive visualization of the entire epidermis *in vivo* at the single cell level. Selected single planes from skin in telogen stage of K14H2BGFP; R26::mTom-LSL-mGFP mouse imaged using two-photon microscopy. Outer to innermost layers of the interfollicular epidermis as well as the entire hair follicle and sebaceous gland are shown (A). Higher magnification of interfollicular epidermis to highlight differences in cell morphology in each layer (B). Nuclear expression of GFP is driven by the K14 promoter, membrane Tomato is ubiquitously expressed. Collagen (cyan) was acquired using second harmonic generation. Total acquired stack was 60–70 μm thick, with single planes acquired every 2 μm. Depth of each layer is shown along dashed line. Scale bar, 100 μm (A) or 20 μm (B).

(ECM) components as well as specialized proteins make up the underlying layer of basement membrane (reviewed in Has & Nyström, 2015). This constitutes the boundary between the epidermis and the dermis. The structure of the epidermis is maintained by the activity of resident stem cell populations. The distribution of appendages, as well as the thickness and organization of the epidermal regions that lack appendages (called interfollicular epidermis) varies between body sites (Porter, Reichelt, Lunny, Magin, & Lane, 1998). In the interfollicular epidermis, cells that occupy the basal layer are the only ones capable of dividing. This layer is composed of polarized basal keratinocytes that are anchored to the basement membrane. Upon cell division, keratinocytes either remain part of the basal layer, expanding their numbers delaminate and moving toward the superficial layers. During this upward-directed transit to more superficial layers, keratinocytes undergo significant morphological and gene expression changes to differentiate into the spinous, granular, and cornified layers (Fig. 1B). Ultimately, this terminal differentiation constitutes a type of programmed cell death known as cornification (reviewed in Eckhart, Lippens, Tschachler, & Declercq, 2013). During cornification, the intracellular environment and intercellular junctions undergo extensive remodeling to yield a water-impermeable flat sheath of biologically inert, tightly linked cells. Therefore, basal cell division must be regulated to offset the loss of cells in the cornified layer through shedding.

The epidermis provides the body with crucial protection from potentially harmful environmental insults, including radiation, dehydration and infection. Epidermal appendages carry out specific functions that contribute to this protection, including thermoregulation and lubrication. Additionally, the epidermis must be able to withstand mechanical stress. Remarkably, the organ's constant turnover does not disrupt any of these functions. The loss of differentiated cells is also a feature of epidermal appendages, and, as in the interfollicular epidermis, stem cells with high proliferative potential is responsible for their replenishment. In the interfollicular epidermis, stem cells reside exclusively in the basal layer. In contrast, the hair follicle contains heterogeneous groups of stem cells residing in particular niches that have diverse contributions within the pilosebaceous unit. A niche is the specialized local microenvironment in which a stem cell resides that is crucial in regulating its activity. Because of these properties and the accessibility for direct observation of cell activity, the epidermis and its appendages have been used as paradigms to understand the relationship between stem cells and their niches.

3. Models and experimental approaches to study skin stem cell physiology

The mouse is the most common and best-characterized model to understand the role of stem cells in skin physiology and regeneration. This is especially true given the number of available molecular and genetic tools that have been generated over many decades of research. However, important differences between mouse and human skin should be noted (reviewed in Wong, Sorkin, Glotzbach, Longaker, & Gurtner, 2011; Zomer & Trentin, 2018).

An evolutionary change that expanded the body coverage of eccrine sweat glands allowed for greater thermoregulation in a range of climates but compromised the number of hair follicles specified during development (Kamberov et al., 2015; Lu, Polak, Keyes, & Fuchs, 2016). For this reason, human epidermis has a lower hair follicle density and a higher proportion of interfollicular epidermis compared to mouse epidermis. This also means that human skin is significantly more exposed to the environment than the skin of a furry rodent, which may explain why human epidermis is on average four times thicker than mouse epidermis. Another morphological difference is that the basal layer of the interfollicular epidermis, which is flat in mice, is characterized by downward undulations in humans (commonly known as rete ridges). An aspect that is particularly relevant for studying wound healing is that wound closure in mice is strongly driven by tissue contraction, promoted by the panniculus carnosus muscle which mostly absent in humans (Zomer & Trentin, 2018). One last consideration when extrapolating results from mouse studies to make claims about overall skin physiology is that different body sites vary in hair density, thickness and structure of the epidermis (Porter et al., 1998), which could explain some discrepancies in the literature. Overall, however, mice remain the best animal models for the study of skin physiology.

Another model to study human epidermis is to use primary human keratinocytes. These cells are seeded onto substrates that serve as dermal equivalents and cultivated at a liquid-air interface to generate an organotypic culture (Gangatirkar, Paquet-Fifield, Li, Rossi, & Kaur, 2007; Oh, Hsi, Guerrero-Juarez, Ramos, & Plikus, 2013; Ridky, Chow, Wong, & Khavari, 2010; Simpson, Kojima, & Getsios, 2010). Organotypic culture systems lack appendages, but fully differentiate to generate stratified skin and thus represent an experimental platform that allows for genetic manipulations in a tissue that closely recapitulates human skin. Engineered human skin can be studied long term if grafted onto immunocompromised recipient mice (reviewed in Salgado, Ng, Koh, Goh, & Common, 2017). Furthermore, skin organoids derived from human induced pluripotent stem cells (iPSCs) have also been established (Itoh et al., 2013). Recent exciting developments have led to the successful recapitulation of appendage morphogenesis, including hair follicle and sebaceous glands, in skin spheroid organoids generated from mouse pluripotent stem cells (Lee et al., 2018). These reductive approaches test the capability of epidermal stem cells to self-organize, by distilling the requirements for tissue morphogenesis to the most essential components. This is a tremendously powerful tool to interrogate the cellular and molecular changes that occur during morphogenesis. It has also been used to explore how aging affects the regenerative capacity of adult skin stem cells (Lei et al., 2017).

3.1 Lineage tracing: Seeing is believing

Lineage tracing is one of the most powerful approaches to both identify stem cell populations and understand how they behave under homeostatic conditions. Distinctly labeling single clones with the use of markers, genetic barcoding or deep sequencing can define the extent to which a population's progeny contributes to the maintenance of specific differentiated cell types (reviewed in Wells & Watt, 2018). In the skin, analyzing the progeny of labeled clones *in situ* is facilitated by the fact that newly generated keratinocytes stay within close proximity of each other, making the accurate quantification of clonal expansion feasible. Additionally, the accessibility of the skin makes it an advantageous model to study stem cells and their interactions with the niche in a noninvasive manner. This aspect has become particularly relevant with the advent of live imaging and the use of two-photon microscopy. This technique using infrared light to increase the imaging depth of a tissue with diminished scattering, enabling the real-time observation of stem cells with in their native environment (Huang & Rompolas, 2017; Pineda et al., 2015) (Fig. 2). Compared to classical methods to trace stem cells in the skin, lineage tracing can now be carried out much more efficiently by re-imaging the tissue in the same animal. The same cells can be visualized and tracked over long term and their fate precisely captured and quantified by live imaging (Mesa et al., 2018; Rompolas et al., 2012, 2013, 2016). This approach has also been used to determine the activity of epidermal stem cells in wound healing (Park et al., 2017).

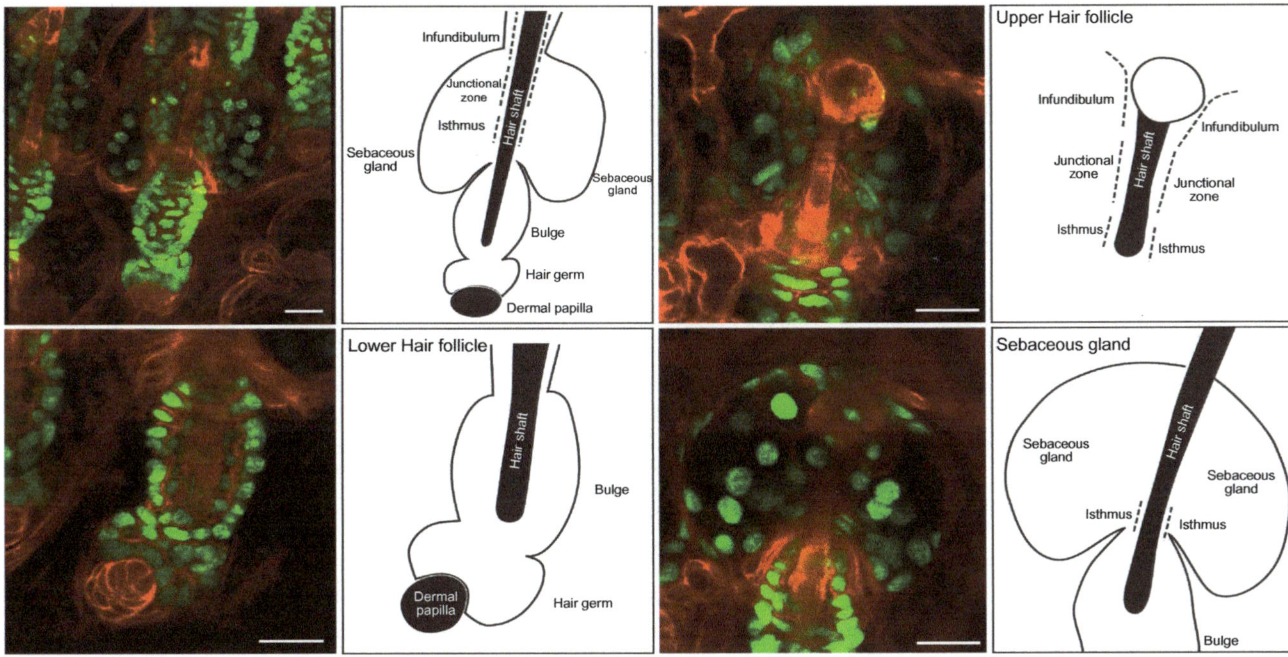

Fig. 2 Single stem cell populations in appendages can be visualized and monitored over time using live imaging. Telogen skin of K14H2BGFP; R26::mTom-LSL-mGFP mouse was imaged using two-photon microscopy. Three-dimensional rendering and diagram of hair follicle with sebaceous glands highlighting structural features of these appendages during telogen. Spatial localization of populations is shown by dashed lines (top left panels). Selected magnified regions of single planes allows monitoring of single cells belonging to specific progenitor populations at higher resolution: in the lower hair follicle (lower left panels), hair germ and bulge; in the upper hair follicle (top right panels), hair infundibulum, junctional zone and isthmus; and, close to the sebaceous glands (lower right panels), isthmus and bulge. Total acquired stack was 60–70 μm thick, with single planes acquired every 2 μm. Scale bar, 20 μm.

4. Skin morphogenesis and the emergence of stem cell niches

All epidermal stem cell populations arise from a single ectodermal layer of unspecified progenitors (described in Blanpain & Fuchs, 2006). Wnt signaling pushes these cells toward an epidermal fate though bone morphogenetic protein (BMP) signaling (Wilson & Hemmati-Brivanlou, 1995; Wilson et al., 2001). These progenitors will then switch expression from Keratin 8 and 18 to Keratin 5 and 14, two major markers that are specifically expressed in undifferentiated stem cell and progenitor populations in the skin (Blanpain, Lowry, Geoghegan, Polak, & Fuchs, 2004; Byrne, Tainsky, & Fuchs, 1994; Fuchs, 1993). This transition is driven by isoforms of the transcription factor p63, which drives expression of Keratin 5 and 14 and other genes required for epidermal fate commitment and stratification (Medawar et al., 2008; Romano, Ortt, Birkaya, Smalley, & Sinha, 2009; Shalom-Feuerstein et al., 2011). Recent work has begun to shed some more light on the epigenetic transitions that underlie this switch (Bao et al., 2015; Fan et al., 2018; Li et al., 2019). By using genetic profiling of *in vitro*-differentiated human pluripotent stem cells, it was shown that TFAP2C drives the chromatin rearrangements behind epidermal lineage commitment. TFAP2C increases chromatin of p63 binding sites and activates expression of p63. Activity from p63 can itself increase accessibility to keratinocyte maturation genes (Li et al., 2019). Chromatin analysis of Keratin 8-expressing epidermal progenitors during early mouse development have also shown how p63 activation to promote epidermal fate specification through extensive chromatin rearrangements (Fan et al., 2018).

The specification of appendages, such as hair follicles, in embryonic skin involves a complex cascade of interactions and signaling between mesenchymal and epithelial cells. To disentangle the very early stages of this process, researchers have turned to both live imaging and single-cell transcriptional analysis (reviewed in Saxena et al., 2019). Signaling from the underlying dermis causes pockets of epidermal cells at specific intervals to express higher Wnt levels and assemble into evenly spaced structures called hair placodes (Ahtiainen et al., 2014; Cetera, Leybova, Joyce, & Devenport, 2018; Gupta et al., 2019; Mok et al., 2019). Owing to the differential Wnt signaling received, these cells develop into hair follicles following divisions perpendicular to the basement membrane that drive downward growth into the dermis (Ouspenskaia, Matos, Mertz, Fiore, & Fuchs, 2016). Furthermore, variable combinations of Wnt, BMP and sonic Hedgehog (SHH) signals specify the cell populations in each of the epithelial compartments that form the hair follicle. At this stage, stem cell markers such as Sox9, Lgr6 and Lrig1 segregate into distinct domains (Jensen et al., 2009; Nowak, Polak, Pasolli, & Fuchs, 2008; Snippert, Haegebarth, et al., 2010). Within these domains are cells that will become hair follicle stem cells and Sox9 appears to be at the top of a hierarchy of regulatory factors that determine this quiescent population's identity (Adam et al., 2015). Because Wnt signaling represses Sox9 expression, a reduction in Wnt is needed for the specification of the hair follicle stem cell population. The attenuation of Wnt occurs only in a specific section of the embryonic niche, highlighting how spatial localization is crucial for the establishment of a stem cell identity (Xu et al., 2015).

Around the time when the hair placodes are established, other ectodermal progenitor cells have begun to stratify upwards and form the basis of what will become the adult interfollicular epidermis. In addition to Wnt signaling, p63 is involved in epidermal stratification (Koster, Kim, Mills, DeMayo, & Roop, 2004; Koster et al., 2007; Mills et al., 1999). During stratification, keratin expression switches from Keratin 14 and 5, to Keratin 1 and 10. These, together with integrin α6 are lifelong basal cell markers. Additional differentiation markers that are expressed in the most superficial layers during the process of stratification include components of the cornified envelope, including involucrin and loricrin (reviewed in Simpson, Patel, & Green, 2011).

5. Adult stem cell niches

As described above, the interfollicular epidermis and the skin's appendages contain populations of stem cells that are responsible for their maintenance. We will now discuss how stem cell activity is regulated to maintain homeostasis in each compartment.

5.1 Interfollicular epidermis: Models of homeostatic maintenance

When referring to the fate of the daughter cells, stem cells can be said to divide symmetrically, and generate two cells that acquire the same fate, or divide asymmetrically, resulting in one stem cell and one terminally differentiated cell (discussed in Morrison & Kimble, 2006). These decisions are crucial for maintaining homeostatic balance and enabling regeneration of the tissue. Therefore, characterizing basal cell behavior in the interfollicular epidermis to determine their regenerative potential and fate choices has been a subject of significant interest. Traditionally, this type of analysis is performed by lineage tracing and clonal analysis. These studies aim to answer how symmetric and asymmetric fates are balanced in the tissue and whether all cells in the basal layer have the same regenerative capacity. Furthermore, data from such experiments can be used to apply these principles to construct models that explain the clonal diversity seen in tissues *in vivo*.

One model suggests that homeostatic maintenance of the interfollicular epidermis is hierarchical and relies on the concerted action of a largely quiescent population that divides very rarely only to give rise to a second population, composed of short-lived committed progenitors. According to this model, transit amplifying cells divide more frequently to replenish the tissue but have limited long-term potential and eventually terminally differentiate (Mascré et al., 2012). Other studies (Gomez et al., 2013; Sada et al., 2016) have proposed that independent stem cell populations occupy discrete sites in the tail skin, named scale and interscale regions. The basal populations maintaining each of these regions express unique sets of genes and display distinct homeostatic maintenance dynamics. Two single-cell RNAseq studies using either human epidermal cells in culture (Tan et al., 2013) or mouse epidermis (Joost et al., 2016) have shown there is a high level of transcriptional heterogeneity within the interfollicular epidermis. Nonetheless, specific transcriptional signatures that allow the clustering of interfollicular stem cells were found in both cases. Whether and how these molecular signatures define stem cell identity and translate into a functional difference is unclear.

An alternative model proposes that all cells in the basal layer are uniform in their potency and their fate is stochastically determined (Clayton et al., 2007; Doupé, Klein, Simons, & Jones, 2010; Lim et al., 2013). Based on this model, each of the daughter cells that are the product of a cell division in the basal layer adopts a random fate, with the likelihood of each outcome varying between different areas in the skin. Two recent studies have provided new insights into the mechanism of epidermal homeostasis by turning to intravital imaging. In one, the movements of individual labeled nuclei of basal cells were followed over time to track their fate, as they either divided or moved upward in the interfollicular epidermis. This study showed that asymmetric and symmetric fates were randomly distributed and that all daughter cells had equal probability to differentiate or undergo another cell division (Rompolas et al., 2016). Furthermore, unbiased marking and tracing of single basal cells in the interfollicular epidermis using a genetically encoded photoactivatable reporter, confirmed these results. A subsequent study looked at the mechanisms regulating homeostatic self-renewal by analyzing division and differentiation events in adjacent individual cells. During homeostasis and under perturbation, it was determined that basal cells are driven to self-renew by local increases in cell area. This transient increase in size occurs when a cell expands to cover the recently vacated space of a differentiating neighboring cell and drives the cell cycle to enter into S/G2 (Mesa et al., 2018). In this model, proliferation is not required for differentiation, but rather, happens as a consequence of the latter. Together, these data support the notion that clonal diversity in a tissue may be the result of neutral competition and are consistent with an equipotency model of stem cell fate determination.

There are several factors that could explain the discrepancies in the interpretation of the data from studies that use mouse genetic models to understand mechanisms of epidermal homeostatic maintenance. Most of these experiments rely on the use of an inducible Cre-recombinase (CreER) that is under the control of a specific promoter and drives expression of a reporter gene. An important factor to consider is the inherent biases introduced by differences in the activity of a given genetic driver as well as the sensitivity of the Cre-reporter alleles. Additionally, other morphological parameters may influence cell behavior. For example, the position of basal cells relative to cycling hair follicles can alter their proliferative capacities, resulting in populations of cells that may appear as fast- or slow-cycling without there being a hierarchy in their origins (Roy et al., 2016). Therefore, because different body sites in the mouse have variable hair densities, proximity to appendages could be another factor introducing bias. One last consideration is that not all studies determine stemness using the same parameters. Genetic markers, tissue location or cycling rates (i.e., quiescence) may not be definitive features of stem cells and can be tissue-specific (discussed in Clevers & Watt, 2018). Unbiased approaches, such as those that rely on live imaging of ubiquitously labeled populations, can circumvent the use of markers and are therefore better suited to resolve principles behind the coordination of stem cell fate choices in time and space (Mesa et al., 2018; Rompolas et al., 2016).

A groundbreaking study in which skin grafts were used to reconstruct almost the entire skin of a patient with a severe skin disorder represents a one-of-a-kind example to evaluate clonal expansion in human epidermis (Hirsch et al., 2017). The authors assayed clonal diversity of the cells in culture at 4 and 8 months following engraftment. Based on the significant loss of diversity observed over time, the authors concluded that a small number of long-lived stem cells are responsible for the long-term maintenance of the entire human epidermis. However, given that neutral competition also predicts the loss of diversity of time, further studies are necessary to corroborate either hypothesis.

5.2 Hair follicle

5.2.1 Anatomy and markers

A mature hair follicle is composed of a central keratinized fiber, called the hair shaft, that is directly surrounded by layers of terminally differentiated keratinocytes called the inner root sheath. The entire structure is lined externally by a single layer of undifferentiated basal cells that comprise the outer root sheath. The outer root sheath is continuous to the basal layer of the interfollicular epidermis and the cells within it retain many stem cell characteristics. The anatomical structure called the bulge, whose morphology varies during the hair cycle, is home to quiescent stem cells throughout the adult lifetime and was the first identified hair follicle stem cell niche (Cotsarelis, Sun, & Lavker, 1990). Hair follicle stem cells in the bulge can give rise to the entire hair follicle in transplantation assays (Jaks et al., 2008; Morris et al., 2004; Tumbar et al., 2004) and are distinguished by the expression of bulge-exclusive markers: Keratin 15 (Liu, Lyle, Yang, & Cotsarelis, 2003; Lyle et al., 1998; Morris et al., 2004), Keratin 19 (Means, Xu, Zhao, Ray, & Gu, 2008; Michel et al., 1996), Lgr5 (Jaks et al., 2008), CD34 (Blanpain et al., 2004; Trempus et al., 2003) and Sox9 (Nowak et al., 2008; Vidal et al., 2005). A second niche is located in the hair germ, a cluster of cells below the bulge that contain stem cells more prone to proliferation than those in the bulge (Greco et al., 2009; Ito, Kizawa, Hamada, & Cotsarelis, 2004). Located at the top of the bulge is a structure called the isthmus, an epithelial compartment occupied by a heterogeneous population of stem cells that, when transplanted, can give rise to all epidermal lineages (Jensen et al., 2008; Snippert, van der Flier, et al., 2010). Cells in the isthmus lack CD34 expression and instead express Lgr6 (Snippert, van der Flier, et al., 2010), MTS24 (also known as Plet1) (Nijhof et al., 2006) and Lrig1 (Jensen et al., 2009). The junctional zone, located right above the isthmus, also contains cells that express Lrig1 (Page, Lombard, Ng, Göttgens, & Jensen, 2013), Gli1 (Brownell, Guevara, Bai, Loomis, & Joyner, 2011) and Gata6 (Donati et al., 2017). The Lrig1-expressing population is responsible for the maintenance of the upper part of the hair follicle, called the infundibulum, which connects the pilosebaceous unit to the interfollicular epidermis (Page et al., 2013; Veniaminova et al., 2013).

5.2.2 Stem cell niche-specific activity in homeostasis

Unlike the interfollicular epidermis, which is continuously and constantly replenished, hair follicles in both humans and mice have stereotypical regenerative cycles during which they experience phases of growth (anagen), regression (catagen) and relative stability or quiescence (telogen). The hair follicle undergoes significant morphological changes during each phase, which also impacts the general physiology and appearance of the rest of the skin (described in Müller-Röver et al., 2001). The start of the hair cycle is in great part determined by signaling cross-talk with a group of compact mesenchymal cells at the base of the hair follicle, known as the dermal papilla (Chase, Montagna, & Malone, 1953). The dermal papilla constitutes a critical niche whose signaling is necessary for the regulation of hair follicle stem cells (Jahoda, Horne, & Oliver, 1984; Rendl, Lewis, & Fuchs, 2005).

The requirement of the dermal papilla for the initiation of hair growth was directly tested by intravital imaging, combined with laser-induced cell ablation (Chi, Wu, & Morgan, 2013; Rompolas et al., 2012). After complete elimination of the dermal papilla during the resting phase of the hair cycle (telogen) hair follicle stem cells fail to activate in order to start a new hair cycle. In the absence of a functional niche, stem cells still retain expression of all the major stem cell marker genes, indicating that they maintain their potency, but lack the instructions to implement a differentiation program. Furthermore, modulation of the number of dermal papilla cells by partial genetic ablation was shown to alter the type of hair fiber that the follicles generates, indicating that the size of the mesenchymal niche specifies the hair size and type, as well as its regenerative properties (Chi et al., 2013). The dermal papilla is populated every hair cycle by the progeny of a self-renewing population of dermal sheath cells. These cells are also involved in the regulation of the hair follicle as it cycles, since the ablation of these cells at the start of anagen disrupts hair growth, indicating they are also involved in the regulation of the hair follicle cycling (Rahmani et al., 2014). In addition to hair follicle growth, the dermal papilla triggers the regression of the hair follicle at the onset of catagen, at least partly through TGF-β signaling (Mesa et al., 2015). During this stage, the fate of cells is also spatially determined, since epithelial located closest to the DP are more likely to undergo cell death than those located further away. Overall, it is clear that the epithelial-mesenchymal crosstalk is crucial in regulating cell fate, ultimately enabling proper hair growth.

In the mouse, hair specification and initial growth begins during embryonic development. During this first cycle, the growth of all hair follicles is synchronized and comes to an end at about 3 weeks after birth, covering the young animal with a dense fur. The second hair cycle begins after a brief period of quiescence where no apparent cell activity is observed and the organ exists in its most anatomically reduced state. During the telogen phase the regenerating part of the hair follicle is divided into two anatomically and molecularly distinct compartments, the bulge and the hair germ (Greco et al., 2009; Ito et al., 2004). The growth phase of the second hair cycle begins with the mobilization of stem cells in the hair germ; a niche located at the most distal part of the hair follicle where the epithelial cells come into direct contact with the dermal papilla.

Activation of hair germ cells occurs due to dampening of repressive BMP signaling and concurrent activation of Wnt signaling (Hsu, Li, & Fuchs, 2014). Time-lapse intravital imaging revealed that cells in the hair germ are the first to divide (Rompolas et al., 2012). These cell divisions are oriented perpendicular to the plane of the dermal papilla and parallel to the long axis of follicle growth. This initial spur of cell proliferation results in the rapid enlargement of the hair germ, which quickly begins to envelop the dermal papilla (Legué & Nicolas, 2005; Rompolas et al., 2012). This process establishes the hair matrix, which represents a specific niche of epithelial progenitors that exclusively generate all the differentiated lineages that build up the new and growing hair fiber (Rompolas et al., 2013). A recent study showed that there are two distinct waves of cell differentiation that occurs in the matrix, with the first one occurring very early, even before the epithelial interface with the dermal papilla is fully formed (Mesler, Veniaminova, Lull, & Wong, 2017). The first established matrix progenitors give rise to descendants that form the hair follicle companion layer that is in contact with and expands along the outer root sheath. Early differentiation events are also important for establishing the hair follicle canal in preparation for the breakout of the newly formed hair shaft (Veniaminova et al., 2013).

Matrix progenitors undergo continuous asymmetric divisions that both replenish their own pool and produce cells that immediately initiate a hair differentiation program as they are displaced away from the dermal papilla, and progressively move along the length of the follicle, toward the surface of the skin (Legué, Sequeira, & Nicolas, 2010; Sequeira & Nicolas, 2012). Which distinct lineage these cells commit to is thought to be dictated by the interaction of the matrix progenitors with the dermal papilla and the relative positional information from their niche. Single-cell transcriptomic analysis revealed a high level of heterogeneity within the pool of epithelial progenitors in the matrix and supported the idea that unique genetic signatures emerge from spatially confined interactions within the niche (Yang, Adam, Ge, Hua, & Fuchs, 2017). Despite this highly stereotypical organization within the matrix and the spatial segregation of progenitors with unique gene expression profiles, their commitment to a specific hair lineage is not definitive but rather flexible. This was tested directly by live imaging combined with a lineage tracing strategy that enabled the visualization and direct tracking of the same marked progenitors in the matrix as hair growth commences (Xin, Gonzalez, Rompolas, & Greco, 2018). These experiments indicate that stem cells constantly tune their differentiation program based on environmental cues, which ensures robustness in the regeneration program.

Anagen is characterized by the rapid production of seven distinct types of differentiated cells that are organized in concentric layers at the core of the elongated hair follicle (Yang et al., 2017). These cells are generated in the hair matrix and expand longitudinally, toward the surface of the skin. The upward movement of the post-mitotic differentiated hair lineages causes the outward displacement of the still undifferentiated cells in the follicle. Lineage tracing combined with live imaging showed that stem cells in the bulge and upper hair also begin to divide with distinct temporal and spatial patterns (Rompolas et al., 2013). Specifically, during anagen, stem cells in the upper bulge exit their quiescence and briefly enter the cell cycle. A limited number of cell divisions in the bulge likely serves to replenish the long lived stem cell pool (Hsu et al., 2014). Conversely, cells in the lower bulge experience much more sustained proliferation, which establishes the outer root sheath that lines the external surface of the growing hair follicle. Taken together, the data indicate the spatial organization of stem cells within the telogen hair follicle niche determines their fate and contribution in the subsequent hair cycle (Rompolas et al., 2013). Thus, stem cells in the hair germ establish the matrix that builds the hair but are eventually lost through terminal differentiation at the end of anagen. Progenitors in the outer root sheath give rise to the new hair germ whereas the slower-cycling cells in the upper bulge repopulate this compartment by undergoing self-renewal.

Most epithelial cells in the hair follicle are eliminated during the regression stage of the hair cycle either by terminal differentiation or by apoptosis. Catagen begins by apoptotic cell death in the epithelial cells closest to the dermal papilla. TGF-β signaling is required for this process, highlighting this mesenchymal population's role in regulating hair cycle progression (Mesa et al., 2015). Surprisingly, live imaging also revealed that cellular debris is ultimately cleared by surviving epithelial cells through apoptosis

(Mesa et al., 2015). Single cell transcriptomic analysis was recently used to examine the heterogeneity of the stem cell population within the hair follicle niche. Joost et al. (2016) utilized single-cell RNAseq to examine telogen epidermis. Unbiased clustering yielded populations from which the authors reconstructed the interfollicular epidermis and hair follicle differentiation program. Additionally, the sources of stem cell heterogeneity were examined independently to the differentiation signature to yield a spatial ordering of the cells matching the proximal-distal axis of the hair follicle spanning form the inner bulge to the interfollicular epidermis. The fact that these lineage-identifying spatial gene signatures were expressed independent to the differentiation program underscores the influence that the local niche microenvironment has in shaping a cell's identity (Joost et al., 2016).

5.3 Sebaceous gland

Sebaceous glands are an epidermal appendage associated with the upper portion of the hair follicle in adult skin. Sebocytes are the major cell type in the sebaceous gland. The exact identity and point of origin of the epithelial progenitor that gives rise to these cells has yet to be defined. Sebocytes differentiate through the progressive accumulation of cytoplasmic lipids, leading to disruption and discharge of sebum, a waxy oily material, into the hair follicle shaft by an exocrine secretory mechanism termed holocrine secretion (Schneider & Paus, 2010; Thody & Shuster, 1989).

Compared to the interfollicular epidermis and hair follicle, less is known about the stem cell niche that maintains the sebaceous gland. One of the difficulties in studying stem cells of the upper pilosebaceous unit is the difficulty to separate overlapping stem cell populations (Horsley et al., 2006; Snippert, Haegebarth, et al., 2010). Early studies suggested Blimp1 to be a putative sebaceous gland stem cell marker (Horsley et al., 2006). However, Blimp1 was shown to also be expressed in the epidermis and differentiated sebaceous gland cells (Magnúsdóttir et al., 2007; Page et al., 2013). In steady-state epidermis, these cells seem to exhibit bipotency, replenishing cells of the interfollicular epidermis and sebaceous gland lineages (Jensen et al., 2009). However, lineage tracing experiments also showed that Lrig1$^+$ stem cells are capable of contributing to interfollicular epidermis, hair follicle, and sebaceous gland after injury or transplantation (Jensen et al., 2009; Page et al., 2013), suggesting that they are not lineage-restricted, but rather that their fate may dictated by the niche microenvironment.

Lgr6 is another gene that marks stem cells in the isthmus area, above the hair follicle bulge. Under homeostasis, Lgr6$^+$ cells contribute to the sebaceous glands, isthmus, and interfollicular epidermis. In transplantation assays into dorsal skin of nude mice, Lgr6 + are able to fully reconstitute the interfollicular epidermis and hair follicles (Füllgrabe et al., 2015; Snippert, Haegebarth, et al., 2010). A major challenge in using Lgr6 as a marker for presumptive sebaceous gland stem cells is its expression in multiple skin niches, including the interfollicular epidermis and hair follicle. Therefore, it is plausible that the labeled progeny in lineage-tracing studies originates from $Lgr6^+$ cells of different compartments rather than a single source in the isthmus. To address this issue, Füllgrabe et al. (2015) utilized multicolor lineage tracing to track $Lgr6^+$ epidermal cells in the interfollicular epidermis, isthmus, and sebaceous gland. Their results revealed that the progeny of $Lgr6^+$ remained confined in their resident niche, providing evidence in support of a compartmentalized stem cell organization. Lgr6 expression depends on β-catenin and is upregulated upon entry into anagen (Füllgrabe et al., 2015; Lien et al., 2014). However, β-catenin signaling suppresses sebaceous gland development, so it is highly unlikely that Lgr6 expression is being regulated by canonical Wnt signaling (Petersson et al., 2011; Silva-Vargas et al., 2005). The signaling and regulatory networks that control the expression of Lgr6 and how this relates to the function of Lgr6-expressing cells as putative sebaceous gland stem cells is complex and remains poorly understood and thus, further work is needed in this area.

Recently, the transcription factor Gata6 was shown to play a role in sebaceous lineage determination and identified as a marker of the previously uncharacterized sebaceous duct, which occupies the upper portion of the sebaceous gland connecting to the junctional zone (Donati et al., 2017). In the transgenic mouse line K14ΔNLef1, epidermal Wnt signaling is inhibited due to keratin 14-driven expression of a mutant Lef1 transgene which lacks the β-catenin binding site. Wnt inhibition results in cyst formation due to the conversion of hair follicle stem cells into ectopic sebocytes (Donati et al., 2014; Niemann, Owens, Hülsken, Birchmeier, & Watt, 2002). Additionally, K14ΔNLef1 transgenic mice have an expanded junctional zone with upregulated GATA6 expression compared to wild type mice (Donati et al., 2017). As previously mentioned, $Lgr6^+$ and $Lrig1^+$ junctional zone stem cells differentiate to produce progenitors of the sebaceous gland. GATA6 is expressed in a subpopulation of these stem cells and was also found to be expressed in the sebaceous gland duct (Donati et al., 2017). Furthermore, GATA6 loss-of-function results in the reduction of cells in the sebaceous duct, while GATA6 gain-of-function leads to the upregulation of sebaceous duct markers. More recent work by Oulès et al. (2019) has provided an interesting connection between the ΔNLef1 mutant and Gata6 (Oulès et al., 2019). Gata6, as well as the positive sebaceous gland regulators Pparγ and Edar (Chang, Jobling, Brennan, & Headon, 2009; Rosenfield, Deplewski, Kentsis, & Ciletti, 1998; Trivedi et al., 2006), were all found to be direct transcriptional targets of ΔNLef1 and exhibited elevated expression in K14ΔNLef1 epidermis (Oulès et al., 2019). However, while Gata6 is a target gene of mutant Lef1, Gata6 expression does not seem to be dependent on Lef1 nor β-catenin (Oulès et al., 2019). Oulès et al. (2019) also identified a potential role of GATA6 as a tumor suppressor in the context of sebaceous tumors, through regulation of DNA mismatch repair genes.

The sebaceous gland and duct niches can be defined by distinct transcription factor sub-networks, including the androgen receptor (Ar) in the sebaceous gland and Blimp1 in the duct. While Ar and Blimp1 are both Gata6 targets, Gata6 overexpression induces Blimp1 expression while suppressing Ar, suggesting that Gata6 defines the sebaceous duct. Under homeostatic conditions, Gata6^{+} cells are restricted to the junctional zone and sebaceous duct, but after full thickness wounding, Gata6^{+} cells of the sebaceous duct and junctional zone extended to the base of the sebaceous gland in hair follicles proximal to the wound site (Donati et al., 2017). Differentiated Gata6^{+} cells of the sebaceous duct can de-differentiate and mobilize to the interfollicular epidermis in response to wounding (Donati et al., 2017). The ability of differentiated cells to contribute to a wider range of epidermal lineages was previously unrecognized.

It is important to note that other factors outside the hair niche also play a role in regulating sebaceous gland physiology. Recent work described the cross-talk between the skin epithelium and the immune system in maintaining the microbial equilibrium of the skin barrier. The recent identification of skin-resident innate lymphoid cells along with the subsequent identification of a RORϒt^{+} innate lymphoid cells subset around the hair follicle near sebaceous glands, led Kobayashi et al. (2019) to investigate the role of innate lymphoid cells in regulating sebaceous gland function. Rag2$^{-/-}$ Il2rg$^{-/-}$ mice lack epidermal innate lymphoid cells, resulting in multi-lobulated hair follicles and sebaceous hyperplasia accounted for by increased Lrig1^{+} compartments and sebocyte progenitor proliferation (Kobayashi et al., 2019). Additionally, Rag2$^{-/-}$ Il2rg$^{-/-}$ mice produced an altered sebum-rich environment with significant increase in the free fatty acid, palmitoleic acid, which has been shown to possess antimicrobial activity (Zheng et al., 2005), and lead to restricted commensalism of Gram-positive bacterial species (Kobayashi et al., 2019). Notch signaling plays an essential role in sebaceous gland formation and maintenance. The expression of tumor necrosis factor (TNF) and lymphotoxin by epidermal innate lymphoid cells suppress sebocyte expression of JAG2, a Notch ligand. Thus, epidermal innate lymphoid cells regulate the skin microbiota under homeostatic conditions by controlling sebaceous gland function via TNF/TA-mediated negative regulation of Notch signaling, providing evidence for an immune-epithelia interaction facilitating host-microbial equilibrium.

5.4 Clonal dynamics in disease: The case for competition in the niche

The interest in understanding stem cell regulation in skin niches also stems from its relevance to disease, especially cancer (reviewed in Blanpain, 2013; Sánchez-Danés & Blanpain, 2018). There are many unknowns surrounding the activity of stem cells that acquire gene mutations and how they interact with their niche environment. Stem cells are subject to mutations caused by mistakes that occur during genome replication due to the effects of environmental agents. In the skin, cells are at a particular risk of the mutagenic effects of UV irradiation. At the tissue level, some of these mutations get eliminated as the cells that carry them undergo terminal differentiation. However, mutations that cause a stem cell's ability to expand in the basal layer can result in positive selection. Ultimately, such cells can generate progeny that outcompetes the wild type neighbors to effectively colonize the tissue.

Recently, the mutational profile of small biopsies from normal sun-exposed epidermis was examined by deep sequencing (Martincorena et al., 2015). The author's analysis showed that the density and average size of clones carrying mutations in Notch signaling genes and p53 were increased, suggesting that mutations targeting these pathways are under positive selection. Some have challenged this view (Simons, 2016), suggesting that the stochastic fate of the basal cells make the tissue robust to acquisition of cancer driver mutations. It is also possible that clones carrying cancer-driver mutations expand initially through positive selection, but then revert to neutral drift after a certain limit in growth advantage is reached (Martincorena, Jones, & Campbell, 2016).

More recent studies have used transgenic mouse models to shed light on the cell autonomous or non-cell autonomous tumor-suppressive mechanisms that may play a role in maintaining epidermal homeostasis. By studying the long-term behavior of basal clones carrying induced heterozygous TP53 mutations, Murai et al. (2018) observed that mutated cells do have an initial competitive advantage. However, this advantage is lost after

6 months and without impacting tissue integrity. Chronic exposure to UV led to the appearance of new mutations that limited the colonization of the p53 mutant clones, suggesting competitive equilibrium might contribute to homeostatic maintenance (Murai et al., 2018).

The notion that the epidermis is resilient to certain mutations is also supported by data from skin appendages. Live imaging of the hair follicle showed that hair follicle stem cell carrying oncogenic Wnt or Hras mutations would give rise to abnormal outgrowths. Strikingly, these regress thanks to the activity of neighboring wild type cells (Brown et al., 2017). Whether the outgrowths are completely eliminated or integrated into functional appendages depends on the compartment the outgrowths originate from. This suggests that the specific fate of these aberrant cells is niche-dependent. Based on results from computational modeling, Lynch et al. (2017) proposed a mechanism where stochastic secondary mutations that arise specifically at the boundary between expanding mutant clones are likelier to get fixated in the tissue, since they outcompete the adjacent wild type clones (Lynch et al., 2017).

Altogether, there's sufficient evidence to suggest that a combination of neutral drift, cell competition and niche/spatial interactions are involved in clonal expansion and cancer initiation. However, the specific molecular mechanisms that regulate the activity of stem cells that carry oncogenic mutations are still not fully understood. It is possible that cell-autonomous mechanisms are also at work. One recent example of a cell-autonomous mechanism is the oncogenic activation of PI3k-AKT. Although over-activation of PI3k-AKT confers enhanced proliferation, Ying, Sandoval, and Beronja (2018) found it cause a growth disadvantage *in vivo* because it triggered enforced differentiation and thus, restricted the ability of these mutant cells to colonize the epidermis (Ying et al., 2018). Because uncontrolled tissue expansion is one of the hallmarks of cancer, dissecting the mechanisms that restrict the expansion of mutated clones *in vivo* is of paramount importance. This knowledge will not only help us understand how cancer arises, but also how it evolves during tumorigenesis (Reeves, Kandyba, Harris, Del Rosario, & Balmain, 2018).

6. Regulation of the niche

6.1 The role of signaling pathways

Several key signaling pathways are involved in regulating skin stem cell fate. Notch signaling becomes activated in cells that are committed to epidermal differentiation (reviewed in Watt, Estrach, & Ambler, 2008) and is involved in hair follicle homeostasis, likely through regulation of bulge cells (Demehri & Kopan, 2009; Vauclair, Nicolas, Barrandon, & Radtke, 2005). Furthermore, Notch acts as a tumor suppressor in the skin (Nicolas et al., 2003; Proweller et al., 2006). The role of Wnt/β-catenin signaling in the skin is complex (reviewed in Veltri, Lang, & Lien, 2018). As mentioned earlier, Wnt signaling is involved in the commitment to epidermal identity during early development and also contributes to the specification of hair follicle placodes. During adulthood, specific Wnt ligands can regulate quiescence in hair follicle stem cells in an autocrine manner (Lim, Tan, Yu, Lim, & Nusse, 2016), whereas paracrine signaling by other factors activate the cells in the hair germ and initiate the hair cycle by stimulating cell proliferation (Choi et al., 2013; Greco et al., 2009; Lien et al., 2014). BMPs also contribute to maintaining bulge stem cell quiescence, and aberrant BMP signaling may be responsible for diminished hair growth with age (Keyes et al., 2013). Repression by BMP inhibitors is required for the activation of a new hair cycle (Oshimori & Fuchs, 2012; Plikus et al., 2008). FGF and TGF-β signaling factors generated from the dermal papilla also contribute to this transition (Hsu et al., 2014; Telerman et al., 2017). It was recently shown that Wnt and BMP signaling create signaling the microniches along the interface with the dermal papilla (Yang et al., 2017), so it is likely that ligands and inhibitors for other signaling pathways may create additional signaling microniches that generates the stereotypical hair follicle cell fate spatial organization (Rompolas et al., 2013).

6.2 Mechanical properties affecting the niche microenvironment

The stem cell niche has specific physical attributes that provide environmental cues, which can change over time and play an important role in influencing stem cell activity (reviewed in Kumar, Placone, & Engler, 2017). However, past research has mainly focused on classic signaling pathways to understand how epithelial cell fate is regulated and comparatively little is known about the mechanosensory systems. Detecting biophysical properties of the niche would allow a stem cell in the basal layer of the skin to detect population density changes in its local environment or alterations in the stiffness of the basement membrane and respond accordingly. Such cues from the immediate cellular environment may instruct a stem cell to initiate a cell cycle or to stop it. Changes in the physical and mechanical properties of the niche have special relevance in the context of wound healing.

There is significant evidence to support the notion that cytoskeletal and adhesive features of the epidermis have an impact on cellular fate. One particularly important connection for homeostasis and disease is the interaction with the basement membrane extracellular matrix (reviewed in Watt & Fujiwara, 2011). Integrins, whose expression is largely limited to the basal layer, are the main mediators of basement membrane anchoring. This attachment keeps cells undifferentiated (Levy, Broad, Diekmann, Evans, & Watt, 2000; Watt, Kubler, Hotchin, Nicholson, & Adams, 1993). Integrins bind to specific ECM components and link them to the actin cytoskeleton or via hemidesmosomes (in the case of α6β4 integrins), to intermediate filaments (reviewed in Watt, 2002). The link between integrins and cell fate has been known for decades; high integrin expression in basal cells correlates with higher proliferative potential (Jones & Watt, 1993) and proper anchorage through α6β4 integrin is crucial for cell survival (Dowling, Yu, & Fuchs, 1996). Integrins recruit regulatory proteins that mediate their binding to the actin cytoskeleton and can regulate the fate of basal cells in the interfollicular epidermis, but also of hair follicle stem cells in the bulge, as reported for integrin-linked kinase (ILK) (Morgner et al., 2015).

To understand how biophysical parameters can affect cell fate, several studies have used human epidermal stem cells cultured on artificial substrates. By using micropatterned surface to culture individual cells, Connelly et al. (2010) found that restricting cell spreading by culturing cells in small areas led to shape-induced differentiation. This happened as a result of serum response factor (SRF) signaling triggered by actin cytoskeleton depolymerization (Connelly et al., 2010). A subsequent study used substrates of different stiffness and found cells cultured in softer ECM failed to form stable adhesions, which decreased extracellular-signal-related kinase (ERK)/mitogen-activated protein kinase (MAPK) signaling and led to differentiation (Trappmann et al., 2012). Culturing human epidermal cells on stiffer substrates enables the formation of stable adhesions and higher actin cytoskeleton tension. This leads to activating epidermal growth factor (EGF) signaling, upstream of ERK, which drives cell proliferation without affecting cell fate (Kenny et al., 2018).

Epidermal cells are mechanically anchored through one another by epidermal adherens junctions, which physically link the cytoskeleton of neighboring cells (Simpson et al., 2011). Cell-cell contacts are also indicators of density and can determine proliferation rates of human epidermal cells in culture (Roshan et al., 2016). A recent study combined live imaging of mouse embryogenesis, epidermal progenitor cell monolayers and computational simulations to show how basal cell delamination and relieves compression generated from proliferation to reinstate homeostatic density (Miroshnikova et al., 2018). Differentiation was triggered by an increase in cell-cell adhesion forces through changing levels of E- and P-cadherin that led to diminished cortical tension. Mechanical signaling through E-cadherin is also crucial in the spatiotemporal coordination of epidermal barrier function establishment (Rübsam et al., 2017). These studies highlight how understanding how adhesion, membrane and cytoskeletal components generate specific mechanical properties help elucidate how stem cell decisions are made *in vivo*.

Other regulatory proteins associated to adherens junctions proteins have been shown to have specific roles in influencing cellular fate. Such is the case of αE-catenin, whose absence leads to altered cell polarity, a loss of contact-mediated inhibition of cell proliferation (Livshits, Kobielak, & Fuchs, 2012; Silvis et al., 2011; Vasioukhin, Bauer, Degenstein, Wise, & Fuchs, 2001). Both αE-catenin and integrins have been shown to be upstream regulators of YAP1 (Elbediwy et al., 2016; Schlegelmilch et al., 2011; Silvis et al., 2011; Zhang, Pasolli, & Fuchs, 2011). Yap/Taz is a mechanosensory signaling pathway that controls cell survival and self-renewal in many tissues (reviewed in Low et al., 2014), including the epidermis (Zhang et al., 2011). Because Yap/Taz is upstream of Notch signaling, however, it establishes a direct link between mechanosensing with cell fate (Totaro et al., 2017). Additionally, Yap/Taz signaling has been linked to wound healing (Lee, Byun, Furutani-Seiki, Hong, & Jung, 2014) and to cancer (reviewed in Zanconato, Cordenonsi, & Piccolo, 2016). Mechanoregulation of stem cell niches can also lead to structural chromatin rearrangements that affect gene expression and lineage commitment, highlighting the alternative ways in which the niche's physical properties can be coupled to cell behavior (Le et al., 2016).

Overall, it is clear that the mechanical properties of a stem cell's microenvironment can strongly affect its fate. To further characterize these niche biophysical parameters, approaches that are compatible with live imaging, such as optogenetics, laser ablation (to measure recoil) and FRET (Förster resonance energy transfer) could prove particularly useful.

6.3 Regulation of the niche by circadian rhythms

The circadian clock, which regulates a myriad of physiological and behavioral processes over 24-h cycles, also has an influence in homeostasis and function of many adult stem cells (reviewed in Weger, Diotel, Dorsemans, Dickmeis, & Weger, 2017). Circadian clock genes are expressed in the skin and their expression oscillates in distinctive phases over 24-h periods. In the interfollicular epidermis, this has been reported to bias a basal cell's likelihood to divide or differentiate (Janich et al., 2013). Opposite phase regulation of clock genes within bulge hair follicle stem cells results in a heterogeneous population with varying predisposition to activation stimuli (Janich et al., 2011). Circadian rhythms of gene expression are also involved in regulating hair follicle cycling, by controlling cell divisions in the matrix and dermal papilla during early anagen (Lin et al., 2009; Plikus et al., 2013). It is plausible that stem cells in the epidermal and appendage niches are scheduled to divide at times that minimizes exposure to UV irradiation and thus prevent damage to the genome when DNA is at its most susceptible (Geyfman et al., 2012; Janich et al., 2013; Plikus et al., 2013). The control of cell proliferation by circadian rhythms could also be influenced by daily changes in the cell's metabolic state (Stringari et al., 2015) to avoid exposure of DNA to higher ROS levels produced during oxidative phosphorylation. A recent study showed that circadian rhythms also regulate human skin gene expression, even at the population level (Wu et al., 2018), which underscores how the study of temporal variation of biological processes has significant medical implications.

6.4 Regulation of the stem cell niche from other cell types

The well-established role of the dermis in specifying appendage fate during development was briefly mentioned. Together with epithelial cells, fibroblasts are the most abundant cell type in the skin. They constitute most of the dermis and constantly provide growth factors and signaling cues to epidermal stem cells. However, many other cell types are found in the skin and associate with stem cell niches either transiently or permanently. These include innate and adaptive immune cells as well as endothelial cells, adipocytes, melanocytes and nerves (Ali et al., 2017; Brownell et al., 2011; Festa et al., 2011; Rabbani et al., 2011; Tanimura et al., 2011; Xiao et al., 2013; Yang et al., 2017). Interactions between these cell types and epithelial cells can have profound effects in stem cell identity and fate. A closer look at the role that non-epithelial cells play in maintenance of epidermal stem cell niches will help disentangle how diseases that impair their function affect epidermal homeostasis and wound healing (reviewed in Naik, Larsen, Cowley, & Fuchs, 2018).

7. Cell plasticity: Wound healing and beyond

De-differentiation refers to a committed cell's ability to revert into a cell of greater developmental potential (reviewed in Merrell & Stanger, 2016). Although under homeostatic conditions resident stem cell populations are responsible for the regeneration of specific compartments in the epidermis, changes in cellular identity or fate can occur under circumstances of physiological stress or injury. In the skin, cells from the hair follicle niche migrate to the interfollicular epidermis upon wounding and strongly contribute to re-epithelialization. This observation was consistent for K15 (Ito et al., 2005), Lgr5 (Jaks et al., 2008; Joost et al., 2018), Lrig1 (Page et al., 2013) and Gata6 populations (Donati et al., 2017). By combining single cell transcriptomics with lineage tracing, Joost et al. (2018) have provided a more detailed map of how wound adaptation occurs genetically in the Lgr5 bulge lineage. To contribute to reepithelialization, these cells switch to an interfollicular epidermis gene signature even before leaving the bulge by downregulating bulge signature genes and increasing the number of receptors to respond to ligand signaling from the wound environment at levels present in Lgr6+ interfollicular epidermis basal cells during homeostasis. This rapid transcriptional adaptation to wound signals highlights how readily plastic cells are when it comes to responding to an injury in order to promote repair.

The converse transition, cells from interfollicular epidermis contributing to appendage regeneration, appears to be possible. *De novo* hair formation was shown to occur in the center of a large re-epithelialized full-thickness wound in a manner that largely recapitulated embryonic hair morphogenesis (Ito et al., 2007). Lineage tracing showed that cells giving rise to these hair follicles did not come from hair follicle niches, suggesting epidermal cells outside the hair follicle stem cell niche acquired the ability to dedifferentiate to give rise to the appendages. This indicates that a reversion of commitment is possible and that cellular state is dictated at least partly by the microenvironment, instead of being solely determined by ancestry. An important aspect related to this is that hair neogenesis is not observed with smaller wounds, probably due to differences in dermal SHH signaling (Lim et al., 2018). Dedifferentiation has been visualized in real time by live imaging of the hair follicle (Rompolas et al., 2013). During homeostasis, the localization of a stem cell within the hair follicle niche was found to determine its fate in a highly stereotypical fashion. However, when bulge stem cells were targeted by laser ablation, surrounding cells would undergo de-differentiation to replace the lost cells, ultimately acquiring hair follicle stem cell identity. Only the disruption of signaling from the dermal papilla impaired the functional reconstitution of the hair stem cell niche, which indicates that while cells can be flexible, ultimately certain essential factors that define the niche need to remain intact for cells to initiate reprogramming. Overall, the nature and extent of the insult determines the changes in identity and fate that occur when a cell goes out of its way to fill a vacated niche in order to restore tissue function.

A lingering question is what epigenetic and gene expression changes control the ability of stem cells in the skin to display increased plasticity when niche conditions change. To address this, a recent study (Ge et al., 2017) compared chromatin accessibility profiles from resting and injured skin. Wounding was shown to initiate a transient genetic program that "hovers" between distinct cell lineages. This transient state is important for proper contribution to wound healing. Interestingly, skin cancer cells were found to be in a permanent "lineage infidelity" state, highlighting how plasticity at the genomic level may come at the high cost of increasing cancer susceptibility (Ge et al., 2017; Latil et al., 2017; Sánchez-Danés et al., 2016; Youssef et al., 2012). This work highlights how the chromatin landscape is crucial in determining cellular state, a concept that has gained traction in recent years.

A master epigenetic regulator that has been shown to have a role in skin homeostasis is the Polycomb-repressive complex (both PRC1 and 2). Previous work characterizing the role of PRC components, had shown Ezh1 and Ezh2, functionally redundant methyltransferases that catalyze H3K27 trimethylation, regulate the proliferative potential of basal cells by repressing genes involved in epidermal differentiation (Ezhkova et al., 2011, 2009). Further work using *in vivo* chromatin and transcriptional profiling has uncovered catalytic-dependent and independent roles for PRC1 in both activating and repressing genes required for epidermal stem cell specification and maintenance of tissue integrity (Cohen et al., 2018, 2019). Histone deacetylases (HDACs), a family of proteins which removes acetyl groups from lysins in histone tails, also have several roles in epidermal morphogenesis, homeostasis and disease (reviewed in Brennan-Crispi & Millar, 2018). These include HDAC1/2 regulation epidermal morphogenesis through control of p63 (LeBoeuf et al., 2010) and HDAC1 regulation of proliferation and as a tumor suppressor (Winter et al., 2013). PRC components and HDACs have been seen to change expression level or their activity in response to injury (Shaw & Martin, 2009; Spallotta et al., 2013), which is consistent with the idea that a change in chromatin landscape underlies the plasticity seen during wound healing.

What are the key environmental factors that lead to the genetic changes that alter the activity of stem cells? Acute inflammation plays an important role in shaping the chromatin landscape of epidermal stem cells by making key stress response genes more accessible. Such genes are then primed for robust expression that ultimately results in a faster response to subsequent insults (Naik et al., 2017). Despite recent progress, the precise signaling, genetic and epigenetic factors that regulate stem cell identity and plasticity in the regenerating skin remain unclear. Integrative experimental approaches that combine genetic mouse models, imaging and genomics will undoubtedly shine light to these lingering questions.

8. Conclusion

There are many aspects surrounding the activity of stem cells in the skin that have gained clarity with the advent of new technologies. *In vivo* imaging of the skin has allowed the study of tissue physiology in unprecedented detail. Together with genomic and transcriptional profiling approaches we have come to understand many of the extrinsic and intrinsic factors that determine cellular identity in the skim stem cell niches. However, many questions remain unanswered. We now know that the niche is influenced not only by classical signaling, but also by factors including biophysical properties, circadian rhythms and epigenetic changes. Altogether, the more we broaden our understanding of the niche as an extremely rich environment (or a mixture of microenvironments), the more it becomes clear that the complexity of signal integration that underlies decision making along a stem cell's fate is still far from being fully understood.

Cell plasticity, a property at the heart of regeneration but also of disease, is a central issue. How can we distinguish between intermittent cellular states and true cellular identity? What defines either? A combination of unbiased and single-cell profiling approaches will likely prove useful in answering these questions, and how cell intrinsic and extrinsic events are coordinated at the tissue level. Ultimately, a global understanding of what constitutes homeostasis and the heterogeneity that underlies it, from cell to tissue scale, is necessary to tease apart the consequences brought about by deviations from the norm. A major goal that stems from this is to shed light on what constitutes a pro-tumorigenic cellular state and environment and the conditions that bring about such a state.

References

Adam, R. C., Yang, H., Rockowitz, S., Larsen, S. B., Nikolova, M., Oristian, D. S., et al. (2015). Pioneer factors govern super-enhancer dynamics in stem cell plasticity and lineage choice. *Nature*, *521*(7552), 366–370. https://doi.org/10.1038/nature14289.

Ahtiainen, L., Lefebvre, S., Lindfors, P. H., Renvoisé, E., Shirokova, V., Vartiainen, M. K., et al. (2014). Directional cell migration, but not proliferation, drives hair placode morphogenesis. *Developmental Cell*, *28*(5), 588–602. https://doi.org/10.1016/j.devcel.2014.02.003.

Ali, N., Zirak, B., Rodriguez, R. S., Pauli, M. L., Truong, H.-A., Lai, K., et al. (2017). Regulatory T cells in skin facilitate epithelial stem cell differentiation. *Cell*, *169*(6), 1119–1129.e11. https://doi.org/10.1016/j.cell.2017.05.002.

Bao, X., Rubin, A. J., Qu, K., Zhang, J., Giresi, P. G., Chang, H. Y., et al. (2015). A novel ATAC-seq approach reveals lineage-specific reinforcement of the open chromatin landscape via cooperation between BAF and p63. *Genome Biology*, *16*(1), 284. https://doi.org/10.1186/s13059-015-0840-9.

Blanpain, C. (2013). Tracing the cellular origin of cancer. *Nature Cell Biology*, *15*(2), 126–134. https://doi.org/10.1038/ncb2657.

Blanpain, C., & Fuchs, E. (2006). Epidermal stem cells of the skin. *Annual Review of Cell and Developmental Biology*, *22*(1), 339–373. https://doi.org/10.1146/annurev.cellbio.22.010305.104357.

Blanpain, C., Lowry, W. E., Geoghegan, A., Polak, L., & Fuchs, E. (2004). Self-renewal, multipotency, and the existence of two cell populations within an epithelial stem cell niche. *Cell*, *118*(5), 635–648. https://doi.org/10.1016/j.cell.2004.08.012.

Blanpain, C., & Simons, B. D. (2013). Unravelling stem cell dynamics by lineage tracing. *Nature Reviews Molecular Cell Biology*, *14*(8), 489–502. https://doi.org/10.1038/nrm3625.

Brennan-Crispi, D. M., & Millar, S. E. (2018). Histone deacetylase functions in epidermal development, homeostasis and cancer. In *Vol. 61. Epigenetic regulation of skin development and regeneration* (pp. 121–157): https://doi.org/10.1007/978-3-319-16769-5_5.

Brown, S., Pineda, C. M., Xin, T., Boucher, J., Suozzi, K. C., Park, S., et al. (2017). Correction of aberrant growth preserves tissue homeostasis. *Nature*, *548*(7667), 334–337. https://doi.org/10.1038/nature23304.

Brownell, I., Guevara, E., Bai, C. B., Loomis, C. A., & Joyner, A. L. (2011). Nerve-derived sonic hedgehog defines a niche for hair follicle stem cells capable of becoming epidermal stem cells. *Cell Stem Cell*, *8*(5), 552–565. https://doi.org/10.1016/j.stem.2011.02.021.

Byrne, C., Tainsky, M., & Fuchs, E. (1994). Programming gene expression in developing epidermis. *Development*, *120*(9), 2369–2383.

Cetera, M., Leybova, L., Joyce, B., & Devenport, D. (2018). Counter-rotational cell flows drive morphological and cell fate asymmetries in mammalian hair follicles. *Nature Cell Biology*, *20*(5), 541–552. https://doi.org/10.1038/s41556-018-0082-7.

Chang, S. H., Jobling, S., Brennan, K., & Headon, D. J. (2009). Enhanced Edar signalling has pleiotropic effects on craniofacial and cutaneous glands. *PLoS One*, *4*(10). e7591.https://doi.org/10.1371/journal.pone.0007591.

Chase, H. B., Montagna, W., & Malone, J. D. (1953). Changes in the skin in relation to the hair growth cycle. *The Anatomical Record*, *116*(1), 75–81.

Chi, W., Wu, E., & Morgan, B. A. (2013). Dermal papilla cell number specifies hair size, shape and cycling and its reduction causes follicular decline. *Development*, *140*(8), 1676–1683. https://doi.org/10.1242/dev.090662.

Choi, Y. S., Zhang, Y., Xu, M., Yang, Y., Ito, M., Peng, T., et al. (2013). Distinct functions for Wnt/β-catenin in hair follicle stem cell proliferation and survival and interfollicular epidermal homeostasis. *Cell Stem Cell*, *13*(6), 720–733. https://doi.org/10.1016/j.stem.2013.10.003.

Clayton, E., Doupé, D. P., Klein, A. M., Winton, D. J., Simons, B. D., & Jones, P. H. (2007). A single type of progenitor cell maintains normal epidermis. *Nature, 446*(7132), 185–189. https://doi.org/10.1038/nature05574.

Clevers, H., & Watt, F. M. (2018). Defining adult stem cells by function, not by phenotype. *Annual Review of Biochemistry, 87*, 1015–1027. https://doi.org/10.1146/annurev-biochem-062917-012341.

Cohen, I., Zhao, D., Bar, C., Valdes, V. J., Dauber-Decker, K. L., Nguyen, M. B., et al. (2018). PRC1 fine-tunes gene repression and activation to safeguard skin development and stem cell specification. *Cell Stem Cell, 22*(5), 726–739.e7. https://doi.org/10.1016/j.stem.2018.04.005.

Cohen, I., Zhao, D., Menon, G., Nakayama, M., Koseki, H., Zheng, D., et al. (2019). PRC1 preserves epidermal tissue integrity independently of PRC2. *Genes & Development, 33*(1–2), 55–60. https://doi.org/10.1101/gad.319939.118.

Connelly, J. T., Gautrot, J. E., Trappmann, B., Tan, D. W.-M., Donati, G., Huck, W. T. S., et al. (2010). Actin and serum response factor transduce physical cues from the microenvironment to regulate epidermal stem cell fate decisions. *Nature Cell Biology, 12*(7), 711–718. https://doi.org/10.1038/ncb2074.

Cotsarelis, G., Sun, T. T., & Lavker, R. M. (1990). Label-retaining cells reside in the bulge area of pilosebaceous unit: Implications for follicular stem cells, hair cycle, and skin carcinogenesis. *Cell, 61*(7), 1329–1337.

Demehri, S., & Kopan, R. (2009). Notch signaling in bulge stem cells is not required for selection of hair follicle fate. *Development, 136*(6), 891–896. https://doi.org/10.1242/dev.030700.

Donati, G., Proserpio, V., Lichtenberger, B. M., Natsuga, K., Sinclair, R., Fujiwara, H., et al. (2014). Epidermal Wnt/β-catenin signaling regulates adipocyte differentiation via secretion of adipogenic factors. *Proceedings of the National Academy of Sciences of the United States of America, 111*(15), E1501–E1509. https://doi.org/10.1073/pnas.1312880111.

Donati, G., Rognoni, E., Hiratsuka, T., Liakath-Ali, K., Hoste, E., Kar, G., et al. (2017). Wounding induces dedifferentiation of epidermal Gata6+ cells and acquisition of stem cell properties. *Nature Cell Biology, 19*(6), 603–613. https://doi.org/10.1038/ncb3532.

Doupé, D. P., Klein, A. M., Simons, B. D., & Jones, P. H. (2010). The ordered architecture of murine ear epidermis is maintained by progenitor cells with random fate. *Developmental Cell, 18*(2), 317–323. https://doi.org/10.1016/j.devcel.2009.12.016.

Dowling, J., Yu, Q. C., & Fuchs, E. (1996). Beta4 integrin is required for hemidesmosome formation, cell adhesion and cell survival. *The Journal of Cell Biology, 134*(2), 559–572.

Eckhart, L., Lippens, S., Tschachler, E., & Declercq, W. (2013). Cell death by cornification. *Biochimica et Biophysica Acta, 1833*(12), 3471–3480. https://doi.org/10.1016/j.bbamcr.2013.06.010.

Elbediwy, A., Vincent-Mistiaen, Z. I., Spencer-Dene, B., Stone, R. K., Boeing, S., Wculek, S. K., et al. (2016). Integrin signalling regulates YAP and TAZ to control skin homeostasis. *Development, 143*(10), 1674–1687. https://doi.org/10.1242/dev.133728.

Ezhkova, E., Lien, W.-H., Stokes, N., Pasolli, H. A., Silva, J. M., & Fuchs, E. (2011). EZH1 and EZH2 cogovern histone H3K27 trimethylation and are essential for hair follicle homeostasis and wound repair. *Genes & Development, 25*(5), 485–498. https://doi.org/10.1101/gad.2019811.

Ezhkova, E., Pasolli, H. A., Parker, J. S., Stokes, N., Su, I.-H., Hannon, G., et al. (2009). Ezh2 orchestrates gene expression for the stepwise differentiation of tissue-specific stem cells. *Cell, 136*(6), 1122–1135. https://doi.org/10.1016/j.cell.2008.12.043.

Fan, X., Wang, D., Burgmaier, J. E., Teng, Y., Romano, R.-A., Sinha, S., et al. (2018). Single cell and open chromatin analysis reveals molecular origin of epidermal cells of the skin. *Developmental Cell, 47*(1), 21–37.e5. https://doi.org/10.1016/j.devcel.2018.08.010.

Festa, E., Fretz, J., Berry, R., Schmidt, B., Rodeheffer, M., Horowitz, M., et al. (2011). Adipocyte lineage cells contribute to the skin stem cell niche to drive hair cycling. *Cell, 146*(5), 761–771. https://doi.org/10.1016/j.cell.2011.07.019.

Fuchs, E. (1993). Epidermal differentiation and keratin gene expression. *Journal of Cell Science. Supplement, 17*, 197–208.

Füllgrabe, A., Joost, S., Are, A., Jacob, T., Sivan, U., Haegebarth, A., et al. (2015). Dynamics of Lgr6+ progenitor cells in the hair follicle, sebaceous gland, and interfollicular epidermis. *Stem Cell Reports, 5*(5), 843–855. https://doi.org/10.1016/j.stemcr.2015.09.013.

Gangatirkar, P., Paquet-Fifield, S., Li, A., Rossi, R., & Kaur, P. (2007). Establishment of 3D organotypic cultures using human neonatal epidermal cells. *Nature Protocols, 2*(1), 178–186. https://doi.org/10.1038/nprot.2006.448.

Ge, Y., Gomez, N. C., Adam, R. C., Nikolova, M., Yang, H., Verma, A., et al. (2017). Stem cell lineage infidelity drives wound repair and cancer. *Cell, 169*(4), 636–642.e14. https://doi.org/10.1016/j.cell.2017.03.042.

Geyfman, M., Kumar, V., Liu, Q., Ruiz, R., Gordon, W., Espitia, F., et al. (2012). Brain and muscle Arnt-like protein-1 (BMAL1) controls circadian cell proliferation and susceptibility to UVB-induced DNA damage in the epidermis. *Proceedings of the National Academy of Sciences of the United States of America, 109*(29), 11758–11763. https://doi.org/10.1073/pnas.1209592109.

Gomez, C., Chua, W., Miremadi, A., Quist, S., Headon, D. J., & Watt, F. M. (2013). The interfollicular epidermis of adult mouse tail comprises two distinct cell lineages that are differentially regulated by Wnt, Edaradd, and Lrig1. *Stem Cell Reports, 1*(1), 19–27. https://doi.org/10.1016/j.stemcr.2013.04.001.

Gonzales, K. A. U., & Fuchs, E. (2017). Skin and its regenerative powers: An alliance between stem cells and their niche. *Developmental Cell, 43*(4), 387–401. https://doi.org/10.1016/j.devcel.2017.10.001.

Greco, V., Chen, T., Rendl, M., Schober, M., Pasolli, H. A., Stokes, N., et al. (2009). A two-step mechanism for stem cell activation during hair regeneration. *Cell Stem Cell, 4*(2), 155–169. https://doi.org/10.1016/j.stem.2008.12.009.

Gupta, K., Levinsohn, J., Linderman, G., Chen, D., Sun, T. Y., Dong, D., et al. (2019). Single-cell analysis reveals a hair follicle dermal niche molecular differentiation trajectory that begins prior to morphogenesis. *Developmental Cell, 48*(1), 17–31.e6. https://doi.org/10.1016/j.devcel.2018.11.032.

Has, C., & Nyström, A. (2015). Epidermal basement membrane in health and disease. *Current Topics in Membranes, 76*, 117–170. https://doi.org/10.1016/bs.ctm.2015.05.003.

Hirsch, T., Rothoeft, T., Teig, N., Bauer, J. W., Pellegrini, G., De Rosa, L., et al. (2017). Regeneration of the entire human epidermis using transgenic stem cells. *Nature, 551*(7680), 327–332. https://doi.org/10.1038/nature24487.

Hoath, S. B., & Leahy, D. G. (2003). The organization of human epidermis: Functional epidermal units and phi proportionality. *The Journal of Investigative Dermatology, 121*(6), 1440–1446. https://doi.org/10.1046/j.1523-1747.2003.12606.x.

Horsley, V., O'Carroll, D., Tooze, R., Ohinata, Y., Saitou, M., Obukhanych, T., et al. (2006). Blimp1 defines a progenitor population that governs cellular input to the sebaceous gland. *Cell, 126*(3), 597–609. https://doi.org/10.1016/j.cell.2006.06.048.

Hsu, Y.-C., Li, L., & Fuchs, E. (2014). Transit-amplifying cells orchestrate stem cell activity and tissue regeneration. *Cell, 157*(4), 935–949. https://doi.org/10.1016/j.cell.2014.02.057.

Huang, S., & Rompolas, P. (2017). Two-photon microscopy for intracutaneous imaging of stem cell activity in mice. *Experimental Dermatology, 26*(5), 379–383. https://doi.org/10.1111/exd.13221.

Ito, M., Kizawa, K., Hamada, K., & Cotsarelis, G. (2004). Hair follicle stem cells in the lower bulge form the secondary germ, a biochemically distinct but functionally equivalent progenitor cell population, at the termination of catagen. *Differentiation; Research in Biological Diversity*, *72*(9–10), 548–557. https://doi.org/10.1111/j.1432-0436.2004.07209008.x.

Ito, M., Liu, Y., Yang, Z., Nguyen, J., Liang, F., Morris, R. J., et al. (2005). Stem cells in the hair follicle bulge contribute to wound repair but not to homeostasis of the epidermis. *Nature Medicine*, *11*(12), 1351–1354. https://doi.org/10.1038/nm1328.

Ito, M., Yang, Z., Andl, T., Cui, C., Kim, N., Millar, S. E., et al. (2007). Wnt-dependent de novo hair follicle regeneration in adult mouse skin after wounding. *Nature*, *447*(7142), 316–320. https://doi.org/10.1038/nature05766.

Itoh, M., Umegaki-Arao, N., Guo, Z., Liu, L., Higgins, C. A., & Christiano, A. M. (2013). Generation of 3D skin equivalents fully reconstituted from human induced pluripotent stem cells (iPSCs). *PLoS One*, *8*(10). e77673https://doi.org/10.1371/journal.pone.0077673.

Jahoda, C. A., Horne, K. A., & Oliver, R. F. (1984). Induction of hair growth by implantation of cultured dermal papilla cells. *Nature*, *311*(5986), 560–562.

Jaks, V., Barker, N., Kasper, M., van Es, J. H., Snippert, H. J., Clevers, H., et al. (2008). Lgr5 marks cycling, yet long-lived, hair follicle stem cells. *Nature Genetics*, *40*(11), 1291–1299. https://doi.org/10.1038/ng.239.

Janich, P., Pascual, G., Merlos-Suárez, A., Batlle, E., Ripperger, J., Albrecht, U., et al. (2011). The circadian molecular clock creates epidermal stem cell heterogeneity. *Nature*, *480*(7376), 209–214. https://doi.org/10.1038/nature10649.

Janich, P., Toufighi, K., Solanas, G., Luis, N. M., Minkwitz, S., Serrano, L., et al. (2013). Human epidermal stem cell function is regulated by circadian oscillations. *Cell Stem Cell*, *13*(6), 745–753. https://doi.org/10.1016/j.stem.2013.09.004.

Jensen, K. B., Collins, C. A., Nascimento, E., Tan, D. W., Frye, M., Itami, S., et al. (2009). Lrig1 expression defines a distinct multipotent stem cell population in mammalian epidermis. *Cell Stem Cell*, *4*(5), 427–439. https://doi.org/10.1016/j.stem.2009.04.014.

Jensen, U. B., Yan, X., Triel, C., Woo, S.-H., Christensen, R., & Owens, D. M. (2008). A distinct population of clonogenic and multipotent murine follicular keratinocytes residing in the upper isthmus. *Journal of Cell Science*, *121*(Pt. 5), 609–617. https://doi.org/10.1242/jcs.025502.

Jones, P. H., & Watt, F. M. (1993). Separation of human epidermal stem cells from transit amplifying cells on the basis of differences in integrin function and expression. *Cell*, *73*(4), 713–724. https://doi.org/10.1016/0092-8674(93)90251-K.

Joost, S., Jacob, T., Sun, X., Annusver, K., La Manno, G., Sur, I., et al. (2018). Single-cell transcriptomics of traced epidermal and hair follicle stem cells reveals rapid adaptations during wound healing. *Cell Reports*, *25*(3), 585–597.e7. https://doi.org/10.1016/j.celrep.2018.09.059.

Joost, S., Zeisel, A., Jacob, T., Sun, X., La Manno, G., Lönnerberg, P., et al. (2016). Single-cell transcriptomics reveals that differentiation and spatial signatures shape epidermal and hair follicle heterogeneity. *Cell Systems*, *3*(3), 221–237.e9. https://doi.org/10.1016/j.cels.2016.08.010.

Kamberov, Y. G., Karlsson, E. K., Kamberova, G. L., Lieberman, D. E., Sabeti, P. C., Morgan, B. A., et al. (2015). A genetic basis of variation in eccrine sweat gland and hair follicle density. *Proceedings of the National Academy of Sciences of the United States of America*, *112*(32), 9932–9937. https://doi.org/10.1073/pnas.1511680112.

Kenny, F. N., Drymoussi, Z., Delaine-Smith, R., Kao, A. P., Laly, A. C., Knight, M. M., et al. (2018). Tissue stiffening promotes keratinocyte proliferation through activation of epidermal growth factor signaling. *Journal of Cell Science*, *131*(10). jcs215780. https://doi.org/10.1242/jcs.215780.

Keyes, B. E., Segal, J. P., Heller, E., Lien, W.-H., Chang, C.-Y., Guo, X., et al. (2013). Nfatc1 orchestrates aging in hair follicle stem cells. *Proceedings of the National Academy of Sciences of the United States of America, 110*(51), E4950–E4959. https://doi.org/10.1073/pnas.1320301110.

Kobayashi, T., Voisin, B., Kim, D.-Y., Kennedy, E. A., Jo, J.-H., Shih, H.-Y., et al. (2019). Homeostatic control of sebaceous glands by innate lymphoid cells regulates commensal bacteria equilibrium. *Cell, 176*(5), 982–997.e16. https://doi.org/10.1016/j.cell.2018.12.031.

Koster, M. I., Dai, D., Marinari, B., Sano, Y., Costanzo, A., Karin, M., et al. (2007). p63 induces key target genes required for epidermal morphogenesis. *Proceedings of the National Academy of Sciences of the United States of America, 104*(9), 3255–3260. https://doi.org/10.1073/pnas.0611376104.

Koster, M. I., Kim, S., Mills, A. A., DeMayo, F. J., & Roop, D. R. (2004). p63 is the molecular switch for initiation of an epithelial stratification program. *Genes & Development, 18*(2), 126–131. https://doi.org/10.1101/gad.1165104.

Kumar, A., Placone, J. K., & Engler, A. J. (2017). Understanding the extracellular forces that determine cell fate and maintenance. *Development, 144*(23), 4261–4270. https://doi.org/10.1242/dev.158469.

Latil, M., Nassar, D., Beck, B., Boumahdi, S., Wang, L., Brisebarre, A., et al. (2017). Cell-type-specific chromatin states differentially prime squamous cell carcinoma tumor-initiating cells for epithelial to mesenchymal transition. *Cell Stem Cell, 20*(2), 191–204.e5. https://doi.org/10.1016/j.stem.2016.10.018.

Le, H. Q., Ghatak, S., Yeung, C.-Y. C., Tellkamp, F., Günschmann, C., Dieterich, C., et al. (2016). Mechanical regulation of transcription controls polycomb-mediated gene silencing during lineage commitment. *Nature Cell Biology, 18*(8), 864–875. https://doi.org/10.1038/ncb3387.

LeBoeuf, M., Terrell, A., Trivedi, S., Sinha, S., Epstein, J. A., Olson, E. N., et al. (2010). Hdac1 and Hdac2 act redundantly to control p63 and p53 functions in epidermal progenitor cells. *Developmental Cell, 19*(6), 807–818. https://doi.org/10.1016/j.devcel.2010.10.015.

Lee, J., Böscke, R., Tang, P.-C., Hartman, B. H., Heller, S., & Koehler, K. R. (2018). Hair follicle development in mouse pluripotent stem cell-derived skin organoids. *Cell Reports, 22*(1), 242–254. https://doi.org/10.1016/j.celrep.2017.12.007.

Lee, M.-J., Byun, M. R., Furutani-Seiki, M., Hong, J.-H., & Jung, H.-S. (2014). YAP and TAZ regulate skin wound healing. *Journal of Investigative Dermatology, 134*(2), 518–525. https://doi.org/10.1038/jid.2013.339.

Legué, E., & Nicolas, J.-F. (2005). Hair follicle renewal: Organization of stem cells in the matrix and the role of stereotyped lineages and behaviors. *Development, 132*(18), 4143–4154. https://doi.org/10.1242/dev.01975.

Legué, E., Sequeira, I., & Nicolas, J.-F. (2010). Hair follicle renewal: Authentic morphogenesis that depends on a complex progression of stem cell lineages. *Development, 137*(4), 569–577. https://doi.org/10.1242/dev.044123.

Lei, M., Schumacher, L. J., Lai, Y.-C., Juan, W.-T., Yeh, C.-Y., Wu, P., et al. (2017). Self-organization process in newborn skin organoid formation inspires strategy to restore hair regeneration of adult cells. *Proceedings of the National Academy of Sciences of the United States of America, 114*(34), E7101–E7110. https://doi.org/10.1073/pnas.1700475114.

Levy, L., Broad, S., Diekmann, D., Evans, R. D., & Watt, F. M. (2000). beta1 integrins regulate keratinocyte adhesion and differentiation by distinct mechanisms. *Molecular Biology of the Cell, 11*(2), 453–466. https://doi.org/10.1091/mbc.11.2.453.

Li, L., Wang, Y., Torkelson, J. L., Shankar, G., Pattison, J. M., Zhen, H. H., et al. (2019). TFAP2C- and p63-dependent networks sequentially rearrange chromatin landscapes to drive human epidermal lineage commitment. *Cell Stem Cell, 24*(2), 271–284.e8. https://doi.org/10.1016/j.stem.2018.12.012.

Lien, W.-H., Polak, L., Lin, M., Lay, K., Zheng, D., & Fuchs, E. (2014). In vivo transcriptional governance of hair follicle stem cells by canonical Wnt regulators. *Nature Cell Biology*, *16*(2), 179–190. https://doi.org/10.1038/ncb2903.

Lim, C. H., Sun, Q., Ratti, K., Lee, S.-H., Zheng, Y., Takeo, M., et al. (2018). Hedgehog stimulates hair follicle neogenesis by creating inductive dermis during murine skin wound healing. *Nature Communications*, *9*(1), 4903. https://doi.org/10.1038/s41467-018-07142-9.

Lim, X., Tan, S. H., Koh, W. L. C., Chau, R. M. W., Yan, K. S., Kuo, C. J., et al. (2013). Interfollicular epidermal stem cells self-renew via autocrine Wnt signaling. *Science (New York, N.Y.)*, *342*(6163), 1226–1230. https://doi.org/10.1126/science.1239730.

Lim, X., Tan, S. H., Yu, K. L., Lim, S. B. H., & Nusse, R. (2016). Axin 2 marks quiescent hair follicle bulge stem cells that are maintained by autocrine Wnt/β-catenin signaling. *Proceedings of the National Academy of Sciences of the United States of America*, *113*(11), E1498–E1505. https://doi.org/10.1073/pnas.1601599113.

Lin, K. K., Kumar, V., Geyfman, M., Chudova, D., Ihler, A. T., Smyth, P., et al. (2009). Circadian clock genes contribute to the regulation of hair follicle cycling. *PLoS Genetics*, *5*(7). e1000573https://doi.org/10.1371/journal.pgen.1000573.

Liu, Y., Lyle, S., Yang, Z., & Cotsarelis, G. (2003). Keratin 15 promoter targets putative epithelial stem cells in the hair follicle bulge. *The Journal of Investigative Dermatology*, *121*(5), 963–968. https://doi.org/10.1046/j.1523-1747.2003.12600.x.

Livshits, G., Kobielak, A., & Fuchs, E. (2012). Governing epidermal homeostasis by coupling cell-cell adhesion to integrin and growth factor signaling, proliferation, and apoptosis. *Proceedings of the National Academy of Sciences of the United States of America*, *109*(13), 4886–4891. https://doi.org/10.1073/pnas.1202120109.

Low, B. C., Pan, C. Q., Shivashankar, G. V., Bershadsky, A., Sudol, M., & Sheetz, M. (2014). YAP/TAZ as mechanosensors and mechanotransducers in regulating organ size and tumor growth. *FEBS Letters*, *588*(16), 2663–2670. https://doi.org/10.1016/j.febslet.2014.04.012.

Lu, C. P., Polak, L., Keyes, B. E., & Fuchs, E. (2016). Spatiotemporal antagonism in mesenchymal-epithelial signaling in sweat versus hair fate decision. *Science (New York, N.Y.)*, *354*(6319). aah6102. https://doi.org/10.1126/science.aah6102.

Lyle, S., Christofidou-Solomidou, M., Liu, Y., Elder, D. E., Albelda, S., & Cotsarelis, G. (1998). The C8/144B monoclonal antibody recognizes cytokeratin 15 and defines the location of human hair follicle stem cells. *Journal of Cell Science*, *111*(Pt. 21), 3179–3188.

Lynch, M. D., Lynch, C. N. S., Craythorne, E., Liakath-Ali, K., Mallipeddi, R., Barker, J. N., et al. (2017). Spatial constraints govern competition of mutant clones in human epidermis. *Nature Communications*, *8*(1), 1119. https://doi.org/10.1038/s41467-017-00993-8.

Magnúsdóttir, E., Kalachikov, S., Mizukoshi, K., Savitsky, D., Ishida-Yamamoto, A., Panteleyev, A. A., et al. (2007). Epidermal terminal differentiation depends on B lymphocyte-induced maturation protein-1. *Proceedings of the National Academy of Sciences of the United States of America*, *104*(38), 14988–14993. https://doi.org/10.1073/pnas.0707323104.

Martincorena, I., Jones, P. H., & Campbell, P. J. (2016). Constrained positive selection on cancer mutations in normal skin. *Proceedings of the National Academy of Sciences of the United States of America*, *113*(9), E1128–E1129. https://doi.org/10.1073/pnas.1600910113.

Martincorena, I., Roshan, A., Gerstung, M., Ellis, P., Van Loo, P., McLaren, S., et al. (2015). Tumor evolution. High burden and pervasive positive selection of somatic mutations in normal human skin. *Science (New York, N.Y.)*, *348*(6237), 880–886. https://doi.org/10.1126/science.aaa6806.

Mascré, G., Dekoninck, S., Drogat, B., Youssef, K. K., Broheé, S., Sotiropoulou, P. A., et al. (2012). Distinct contribution of stem and progenitor cells to epidermal maintenance. *Nature*, *489*(7415), 257–262. https://doi.org/10.1038/nature11393.

Means, A. L., Xu, Y., Zhao, A., Ray, K. C., & Gu, G. (2008). A CK19(CreERT) knockin mouse line allows for conditional DNA recombination in epithelial cells in multiple endodermal organs. *Genesis*, *46*(6), 318–323. https://doi.org/10.1002/dvg.20397.

Medawar, A., Virolle, T., Rostagno, P., De La Forest-Divonne, S., Gambaro, K., Rouleau, M., et al. (2008). DeltaNp63 is essential for epidermal commitment of embryonic stem cells. *PLoS One*, *3*(10), e3441. https://doi.org/10.1371/journal.pone.0003441.

Merrell, A. J., & Stanger, B. Z. (2016). Adult cell plasticity in vivo: De-differentiation and transdifferentiation are back in style. *Nature Reviews Molecular Cell Biology*, *17*(7), 413–425. https://doi.org/10.1038/nrm.2016.24.

Mesa, K. R., Kawaguchi, K., Cockburn, K., Gonzalez, D., Boucher, J., Xin, T., et al. (2018). Homeostatic epidermal stem cell self-renewal is driven by local differentiation. *Cell Stem Cell*, *23*(5), 677–686.e4. https://doi.org/10.1016/j.stem.2018.09.005.

Mesa, K. R., Rompolas, P., Zito, G., Myung, P., Sun, T. Y., Brown, S., et al. (2015). Niche-induced cell death and epithelial phagocytosis regulate hair follicle stem cell pool. *Nature*, *522*(7554), 94–97. https://doi.org/10.1038/nature14306.

Mesler, A. L., Veniaminova, N. A., Lull, M. V., & Wong, S. Y. (2017). Hair follicle terminal differentiation is orchestrated by distinct early and late matrix progenitors. *Cell Reports*, *19*(4), 809–821. https://doi.org/10.1016/j.celrep.2017.03.077.

Michel, M., Török, N., Godbout, M. J., Lussier, M., Gaudreau, P., Royal, A., et al. (1996). Keratin 19 as a biochemical marker of skin stem cells in vivo and in vitro: Keratin 19 expressing cells are differentially localized in function of anatomic sites, and their number varies with donor age and culture stage. *Journal of Cell Science*, *109*(Pt 5), 1017–1028.

Mills, A. A., Zheng, B., Wang, X. J., Vogel, H., Roop, D. R., & Bradley, A. (1999). p63 is a p53 homologue required for limb and epidermal morphogenesis. *Nature*, *398*(6729), 708–713. https://doi.org/10.1038/19531.

Miroshnikova, Y. A., Le, H. Q., Schneider, D., Thalheim, T., Rübsam, M., Bremicker, N., et al. (2018). Adhesion forces and cortical tension couple cell proliferation and differentiation to drive epidermal stratification. *Nature Cell Biology*, *20*(1), 69–80. https://doi.org/10.1038/s41556-017-0005-z.

Mok, K.-W., Saxena, N., Heitman, N., Grisanti, L., Srivastava, D., Muraro, M. J., et al. (2019). Dermal condensate niche fate specification occurs prior to formation and is placode progenitor dependent. *Developmental Cell*, *48*(1), 32–48.e5. https://doi.org/10.1016/j.devcel.2018.11.034.

Morgner, J., Ghatak, S., Jakobi, T., Dieterich, C., Aumailley, M., & Wickström, S. A. (2015). Integrin-linked kinase regulates the niche of quiescent epidermal stem cells. *Nature Communications*, *6*, 8198. https://doi.org/10.1038/ncomms9198.

Morris, R. J., Liu, Y., Marles, L., Yang, Z., Trempus, C., Li, S., et al. (2004). Capturing and profiling adult hair follicle stem cells. *Nature Biotechnology*, *22*(4), 411–417. https://doi.org/10.1038/nbt950.

Morrison, S. J., & Kimble, J. (2006). Asymmetric and symmetric stem-cell divisions in development and cancer. *Nature*, *441*(7097), 1068–1074. https://doi.org/10.1038/nature04956.

Müller-Röver, S., Handjiski, B., Van Der Veen, C., Eichmüller, S., Foitzik, K., McKay, I. A., et al. (2001). A comprehensive guide for the accurate classification of murine hair follicles in distinct hair cycle stages. *The Journal of Investigative Dermatology*, *117*(1), 3–15. https://doi.org/10.1046/j.0022-202x.2001.01377.x.

Murai, K., Skrupskelyte, G., Piedrafita, G., Hall, M., Kostiou, V., Ong, S. H., et al. (2018). Epidermal tissue adapts to restrain progenitors carrying clonal p53 mutations. *Cell Stem Cell*, *23*, 687–699. https://doi.org/10.1016/j.stem.2018.08.017.

Naik, S., Larsen, S. B., Cowley, C. J., & Fuchs, E. (2018). Two to tango: Dialog between immunity and stem cells in health and disease. *Cell*, *175*(4), 908–920. https://doi.org/10.1016/j.cell.2018.08.071.

Naik, S., Larsen, S. B., Gomez, N. C., Alaverdyan, K., Sendoel, A., Yuan, S., et al. (2017). Inflammatory memory sensitizes skin epithelial stem cells to tissue damage. *Nature*, *550*(7677), 475–480. https://doi.org/10.1038/nature24271.

Nicolas, M., Wolfer, A., Raj, K., Kummer, J. A., Mill, P., van Noort, M., et al. (2003). Notch1 functions as a tumor suppressor in mouse skin. *Nature Genetics*, *33*(3), 416–421. https://doi.org/10.1038/ng1099.

Niemann, C., Owens, D. M., Hülsken, J., Birchmeier, W., & Watt, F. M. (2002). Expression of DeltaNLef1 in mouse epidermis results in differentiation of hair follicles into squamous epidermal cysts and formation of skin tumours. *Development*, *129*(1), 95–109.

Nijhof, J. G. W., Braun, K. M., Giangreco, A., van Pelt, C., Kawamoto, H., Boyd, R. L., et al. (2006). The cell-surface marker MTS24 identifies a novel population of follicular keratinocytes with characteristics of progenitor cells. *Development*, *133*(15), 3027–3037. https://doi.org/10.1242/dev.02443.

Nowak, J. A., Polak, L., Pasolli, H. A., & Fuchs, E. (2008). Hair follicle stem cells are specified and function in early skin morphogenesis. *Cell Stem Cell*, *3*(1), 33–43. https://doi.org/10.1016/j.stem.2008.05.009.

Oh, J. W., Hsi, T.-C., Guerrero-Juarez, C. F., Ramos, R., & Plikus, M. V. (2013). Organotypic skin culture. *Journal of Investigative Dermatology*, *133*(11), 1–4. https://doi.org/10.1038/jid.2013.387.

Oshimori, N., & Fuchs, E. (2012). Paracrine TGF-β signaling counterbalances BMP-mediated repression in hair follicle stem cell activation. *Cell Stem Cell*, *10*(1), 63–75. https://doi.org/10.1016/j.stem.2011.11.005.

Oulès, B., Rognoni, E., Hoste, E., Goss, G., Fiehler, R., Natsuga, K., et al. (2019). Mutant Lef1 controls Gata6 in sebaceous gland development and cancer. *The EMBO Journal38*. e100526. https://doi.org/10.15252/embj.2018100526.

Ouspenskaia, T., Matos, I., Mertz, A. F., Fiore, V. F., & Fuchs, E. (2016). WNT-SHH antagonism specifies and expands stem cells prior to niche formation. *Cell*, *164*(1–2), 156–169. https://doi.org/10.1016/j.cell.2015.11.058.

Page, M. E., Lombard, P., Ng, F., Göttgens, B., & Jensen, K. B. (2013). The epidermis comprises autonomous compartments maintained by distinct stem cell populations. *Cell Stem Cell*, *13*(4), 471–482. https://doi.org/10.1016/j.stem.2013.07.010.

Park, S., Gonzalez, D. G., Guirao, B., Boucher, J. D., Cockburn, K., Marsh, E. D., et al. (2017). Tissue-scale coordination of cellular behaviour promotes epidermal wound repair in live mice. *Nature Cell Biology*, *19*(2), 155–163. https://doi.org/10.1038/ncb3472.

Petersson, M., Brylka, H., Kraus, A., John, S., Rappl, G., Schettina, P., et al. (2011). TCF/Lef1 activity controls establishment of diverse stem and progenitor cell compartments in mouse epidermis. *The EMBO Journal*, *30*(15), 3004–3018. https://doi.org/10.1038/emboj.2011.199.

Pineda, C. M., Park, S., Mesa, K. R., Wolfel, M., Gonzalez, D. G., Haberman, A. M., et al. (2015). Intravital imaging of hair follicle regeneration in the mouse. *Nature Protocols*, *10*(7), 1116–1130. https://doi.org/10.1038/nprot.2015.070.

Plikus, M. V., Mayer, J. A., de la Cruz, D., Baker, R. E., Maini, P. K., Maxson, R., et al. (2008). Cyclic dermal BMP signalling regulates stem cell activation during hair regeneration. *Nature*, *451*(7176), 340–344. https://doi.org/10.1038/nature06457.

Plikus, M. V., Vollmers, C., de la Cruz, D., Chaix, A., Ramos, R., Panda, S., et al. (2013). Local circadian clock gates cell cycle progression of transient amplifying cells during regenerative hair cycling. *Proceedings of the National Academy of Sciences of the United States of America*, *110*(23), E2106–E2115. https://doi.org/10.1073/pnas.1215935110.

Porter, R. M., Reichelt, J., Lunny, D. P., Magin, T. M., & Lane, E. B. (1998). The relationship between hyperproliferation and epidermal thickening in a mouse model for BCIE. *The Journal of Investigative Dermatology*, *110*(6), 951–957. https://doi.org/10.1046/j.1523-1747.1998.00218.x.

Proweller, A., Tu, L., Lepore, J. J., Cheng, L., Lu, M. M., Seykora, J., et al. (2006). Impaired notch signaling promotes de novo squamous cell carcinoma formation. *Cancer Research*, *66*(15), 7438–7444. https://doi.org/10.1158/0008-5472.CAN-06-0793.

Rabbani, P., Takeo, M., Chou, W., Myung, P., Bosenberg, M., Chin, L., et al. (2011). Coordinated activation of Wnt in epithelial and melanocyte stem cells initiates pigmented hair regeneration. *Cell*, *145*(6), 941–955. https://doi.org/10.1016/j.cell.2011.05.004.

Rahmani, W., Abbasi, S., Hagner, A., Raharjo, E., Kumar, R., Hotta, A., et al. (2014). Hair follicle dermal stem cells regenerate the dermal sheath, repopulate the dermal papilla, and modulate hair type. *Developmental Cell*, *31*(5), 543–558. https://doi.org/10.1016/j.devcel.2014.10.022.

Reeves, M. Q., Kandyba, E., Harris, S., Del Rosario, R., & Balmain, A. (2018). Multicolour lineage tracing reveals clonal dynamics of squamous carcinoma evolution from initiation to metastasis. *Nature Cell Biology*, *20*(6), 699–709. https://doi.org/10.1038/s41556-018-0109-0.

Rendl, M., Lewis, L., & Fuchs, E. (2005). Molecular dissection of mesenchymal-epithelial interactions in the hair follicle. *PLoS Biology*, *3*(11), e331. https://doi.org/10.1371/journal.pbio.0030331.

Ridky, T. W., Chow, J. M., Wong, D. J., & Khavari, P. A. (2010). Invasive three-dimensional organotypic neoplasia from multiple normal human epithelia. *Nature Medicine*, *16*(12), 1450–1455. https://doi.org/10.1038/nm.2265.

Rognoni, E., & Watt, F. M. (2018). Skin cell heterogeneity in development, wound healing, and cancer. *Trends in Cell Biology*, *28*(9), 709–722. https://doi.org/10.1016/j.tcb.2018.05.002.

Romano, R.-A., Ortt, K., Birkaya, B., Smalley, K., & Sinha, S. (2009). An active role of the DeltaN isoform of p63 in regulating basal keratin genes K5 and K14 and directing epidermal cell fate. *PLoS One*, *4*(5) e5623.

Rompolas, P., Deschene, E. R., Zito, G., Gonzalez, D. G., Saotome, I., Haberman, A. M., et al. (2012). Live imaging of stem cell and progeny behaviour in physiological hair-follicle regeneration. *Nature*, *487*(7408), 496–499. https://doi.org/10.1038/nature11218.

Rompolas, P., Mesa, K. R., & Greco, V. (2013). Spatial organization within a niche as a determinant of stem-cell fate. *Nature*, *502*(7472), 513–518. https://doi.org/10.1038/nature12602.

Rompolas, P., Mesa, K. R., Kawaguchi, K., Park, S., Gonzalez, D., Brown, S., et al. (2016). Spatiotemporal coordination of stem cell commitment during epidermal homeostasis. *Science (New York, N.Y.)*, *352*(6292), 1471–1474. https://doi.org/10.1126/science.aaf7012.

Rosenfield, R. L., Deplewski, D., Kentsis, A., & Ciletti, N. (1998). Mechanisms of androgen induction of sebocyte differentiation. *Dermatology (Basel, Switzerland)*, *196*(1), 43–46. https://doi.org/10.1159/000017864.

Roshan, A., Murai, K., Fowler, J., Simons, B. D., Nikolaidou-Neokosmidou, V., & Jones, P. H. (2016). Human keratinocytes have two interconvertible modes of proliferation. *Nature Cell Biology*, *18*(2), 145–156. https://doi.org/10.1038/ncb3282.

Roy, E., Neufeld, Z., Cerone, L., Wong, H. Y., Hodgson, S., Livet, J., et al. (2016). Bimodal behaviour of interfollicular epidermal progenitors regulated by hair follicle position and cycling. *The EMBO Journal*, *35*(24), 2658–2670. https://doi.org/10.15252/embj.201693806.

Rübsam, M., Mertz, A. F., Kubo, A., Marg, S., Jüngst, C., Goranci-Buzhala, G., et al. (2017). E-cadherin integrates mechanotransduction and EGFR signaling to control junctional tissue polarization and tight junction positioning. *Nature Communications*, *8*(1), 1250. https://doi.org/10.1038/s41467-017-01170-7.

Sada, A., Jacob, F., Leung, E., Wang, S., White, B. S., Shalloway, D., et al. (2016). Defining the cellular lineage hierarchy in the interfollicular epidermis of adult skin. *Nature Cell Biology*, *18*(6), 619–631. https://doi.org/10.1038/ncb3359.

Salgado, G., Ng, Y. Z., Koh, L. F., Goh, C. S. M., & Common, J. E. (2017). Human reconstructed skin xenografts on mice to model skin physiology. *Differentiation*, *98*, 14–24. https://doi.org/10.1016/j.diff.2017.09.004.

Sánchez-Danés, A., & Blanpain, C. (2018). Deciphering the cells of origin of squamous cell carcinomas. *Nature Reviews Cancer*, *18*(9), 549–561. https://doi.org/10.1038/s41568-018-0024-5.

Sánchez-Danés, A., Hannezo, E., Larsimont, J.-C., Liagre, M., Youssef, K. K., Simons, B. D., et al. (2016). Defining the clonal dynamics leading to mouse skin tumour initiation. *Nature*, *536*(7616), 298–303. https://doi.org/10.1038/nature19069.

Saxena, N., Mok, K.-W., & Rendl, M. (2019). An updated classification of hair follicle morphogenesis. *Experimental Dermatology*, *28*, 332–344. https://doi.org/10.1111/exd.13913.

Schlegelmilch, K., Mohseni, M., Kirak, O., Pruszak, J., Rodriguez, J. R., Zhou, D., et al. (2011). Yap1 acts downstream of α-catenin to control epidermal proliferation. *Cell*, *144*(5), 782–795. https://doi.org/10.1016/j.cell.2011.02.031.

Schneider, M. R., & Paus, R. (2010). Sebocytes, multifaceted epithelial cells: Lipid production and holocrine secretion. *The International Journal of Biochemistry & Cell Biology*, *42*(2), 181–185. https://doi.org/10.1016/j.biocel.2009.11.017.

Sequeira, I., & Nicolas, J.-F. (2012). Redefining the structure of the hair follicle by 3D clonal analysis. *Development*, *139*(20), 3741–3751. https://doi.org/10.1242/dev.081091.

Shalom-Feuerstein, R., Lena, A. M., Zhou, H., De La Forest Divonne, S., Van Bokhoven, H., Candi, E., et al. (2011). ΔNp63 is an ectodermal gatekeeper of epidermal morphogenesis. *Cell Death and Differentiation*, *18*(5), 887–896. https://doi.org/10.1038/cdd.2010.159.

Shaw, T., & Martin, P. (2009). Epigenetic reprogramming during wound healing: Loss of polycomb-mediated silencing may enable upregulation of repair genes. *EMBO Reports*, *10*(8), 881–886. https://doi.org/10.1038/embor.2009.102.

Silva-Vargas, V., Lo Celso, C., Giangreco, A., Ofstad, T., Prowse, D. M., Braun, K. M., et al. (2005). Beta-catenin and hedgehog signal strength can specify number and location of hair follicles in adult epidermis without recruitment of bulge stem cells. *Developmental Cell*, *9*(1), 121–131. https://doi.org/10.1016/j.devcel.2005.04.013.

Silvis, M. R., Kreger, B. T., Lien, W.-H., Klezovitch, O., Rudakova, G. M., Camargo, F. D., et al. (2011). α-Catenin is a tumor suppressor that controls cell accumulation by regulating the localization and activity of the transcriptional coactivator Yap1. *Science Signaling*, *4*(174), ra33. https://doi.org/10.1126/scisignal.2001823.

Simons, B. D. (2016). Deep sequencing as a probe of normal stem cell fate and preneoplasia in human epidermis. *Proceedings of the National Academy of Sciences of the United States of America*, *113*(1), 128–133. https://doi.org/10.1073/pnas.1516123113.

Simpson, C. L., Kojima, S.-I., & Getsios, S. (2010). RNA interference in keratinocytes and an organotypic model of human epidermis. *Methods in Molecular Biology (Clifton, NJ)*, *585*(2), 127–146. https://doi.org/10.1007/978-1-60761-380-0_10.

Simpson, C. L., Patel, D. M., & Green, K. J. (2011). Deconstructing the skin: Cytoarchitectural determinants of epidermal morphogenesis. *Nature Reviews Molecular Cell Biology, 12*(9), 565–580. https://doi.org/10.1038/nrm3175.
Snippert, H. J., Haegebarth, A., Kasper, M., Jaks, V., van Es, J. H., Barker, N., et al. (2010). Lgr6 marks stem cells in the hair follicle that generate all cell lineages of the skin. *Science (New York, N.Y.), 327*(5971), 1385–1389. https://doi.org/10.1126/science.1184733.
Snippert, H. J., van der Flier, L. G., Sato, T., van Es, J. H., van den Born, M., Kroon-Veenboer, C., et al. (2010). Intestinal crypt homeostasis results from neutral competition between symmetrically dividing Lgr5 stem cells. *Cell, 143*(1), 134–144. https://doi.org/10.1016/j.cell.2010.09.016.
Spallotta, F., Cencioni, C., Straino, S., Nanni, S., Rosati, J., Artuso, S., et al. (2013). A nitric oxide-dependent cross-talk between class I and III histone deacetylases accelerates skin repair. *The Journal of Biological Chemistry, 288*(16), 11004–11012. https://doi.org/10.1074/jbc.M112.441816.
Stringari, C., Wang, H., Geyfman, M., Crosignani, V., Kumar, V., Takahashi, J. S., et al. (2015). In vivo single-cell detection of metabolic oscillations in stem cells. *Cell Reports, 10*(1), 1–7. https://doi.org/10.1016/j.celrep.2014.12.007.
Tan, D. W., Jensen, K. B., Trotter, M. W., Connelly, J. T., Broad, S., & Watt, F. M. (2013). Single-cell gene expression profiling reveals functional heterogeneity of undifferentiated human epidermal cells. *Development, 140*(7), 1433–1444. https://doi.org/10.1242/dev.087551.
Tanimura, S., Tadokoro, Y., Inomata, K., Binh, N. T., Nishie, W., Yamazaki, S., et al. (2011). Hair follicle stem cells provide a functional niche for melanocyte stem cells. *Cell Stem Cell, 8*(2), 177–187. https://doi.org/10.1016/j.stem.2010.11.029.
Telerman, S. B., Rognoni, E., Sequeira, I., Pisco, A. O., Lichtenberger, B. M., Culley, O. J., et al. (2017). Dermal Blimp1 acts downstream of epidermal TGFβ and Wnt/β-catenin to regulate hair follicle formation and growth. *Journal of Investigative Dermatology, 137*(11), 2270–2281. https://doi.org/10.1016/j.jid.2017.06.015.
Thody, A. J., & Shuster, S. (1989). Control and function of sebaceous glands. *Physiological Reviews, 69*(2), 383–416. https://doi.org/10.1152/physrev.1989.69.2.383.
Totaro, A., Castellan, M., Battilana, G., Zanconato, F., Azzolin, L., Giulitti, S., et al. (2017). YAP/TAZ link cell mechanics to Notch signalling to control epidermal stem cell fate. *Nature Communications, 8*. 15206. https://doi.org/10.1038/ncomms15206.
Trappmann, B., Gautrot, J. E., Connelly, J. T., Strange, D. G. T., Li, Y., Oyen, M. L., et al. (2012). Extracellular-matrix tethering regulates stem-cell fate. *Nature Materials, 11*(7), 642–649. https://doi.org/10.1038/nmat3339.
Trempus, C. S., Morris, R. J., Bortner, C. D., Cotsarelis, G., Faircloth, R. S., Reece, J. M., et al. (2003). Enrichment for living murine keratinocytes from the hair follicle bulge with the cell surface marker CD34. *The Journal of Investigative Dermatology, 120*(4), 501–511. https://doi.org/10.1046/j.1523-1747.2003.12088.x.
Trivedi, N. R., Cong, Z., Nelson, A. M., Albert, A. J., Rosamilia, L. L., Sivarajah, S., et al. (2006). Peroxisome proliferator-activated receptors increase human sebum production. *The Journal of Investigative Dermatology, 126*(9), 2002–2009. https://doi.org/10.1038/sj.jid.5700336.
Tumbar, T., Guasch, G., Greco, V., Blanpain, C., Lowry, W. E., Rendl, M., et al. (2004). Defining the epithelial stem cell niche in skin. *Science (New York, N.Y.), 303*(5656), 359–363. https://doi.org/10.1126/science.1092436.
Vasioukhin, V., Bauer, C., Degenstein, L., Wise, B., & Fuchs, E. (2001). Hyperproliferation and defects in epithelial polarity upon conditional ablation of alpha-catenin in skin. *Cell, 104*(4), 605–617.

Vauclair, S., Nicolas, M., Barrandon, Y., & Radtke, F. (2005). Notch1 is essential for post-natal hair follicle development and homeostasis. *Developmental Biology, 284*(1), 184–193. https://doi.org/10.1016/j.ydbio.2005.05.018.

Veltri, A., Lang, C., & Lien, W.-H. (2018). Concise review: Wnt signaling pathways in skin development and epidermal stem cells. *Stem Cells (Dayton, Ohio), 36*(1), 22–35. https://doi.org/10.1002/stem.2723.

Veniaminova, N. A., Vagnozzi, A. N., Kopinke, D., Do, T. T., Murtaugh, L. C., Maillard, I., et al. (2013). Keratin 79 identifies a novel population of migratory epithelial cells that initiates hair canal morphogenesis and regeneration. *Development, 140*(24), 4870–4880. https://doi.org/10.1242/dev.101725.

Vidal, V. P. I., Chaboissier, M.-C., Lützkendorf, S., Cotsarelis, G., Mill, P., Hui, C.-C., et al. (2005). Sox 9 is essential for outer root sheath differentiation and the formation of the hair stem cell compartment. *Current Biology, 15*(15), 1340–1351. https://doi.org/10.1016/j.cub.2005.06.064.

Watt, F. M. (2002). Role of integrins in regulating epidermal adhesion, growth and differentiation. *The EMBO Journal, 21*(15), 3919–3926. https://doi.org/10.1093/emboj/cdf399.

Watt, F. M., Estrach, S., & Ambler, C. A. (2008). Epidermal notch signalling: Differentiation, cancer and adhesion. *Current Opinion in Cell Biology, 20*(2), 171–179. https://doi.org/10.1016/j.ceb.2008.01.010.

Watt, F. M., & Fujiwara, H. (2011). Cell-extracellular matrix interactions in normal and diseased skin. *Cold Spring Harbor Perspectives in Biology, 3*(4), a005124. https://doi.org/10.1101/cshperspect.a005124.

Watt, F. M., Kubler, M. D., Hotchin, N. A., Nicholson, L. J., & Adams, J. C. (1993). Regulation of keratinocyte terminal differentiation by integrin-extracellular matrix interactions. *Journal of Cell Science, 106*(Pt. 1), 175–182.

Weger, M., Diotel, N., Dorsemans, A.-C., Dickmeis, T., & Weger, B. D. (2017). Stem cells and the circadian clock. *Developmental Biology, 431*(2), 111–123. https://doi.org/10.1016/j.ydbio.2017.09.012.

Wells, J. M., & Watt, F. M. (2018). Diverse mechanisms for endogenous regeneration and repair in mammalian organs. *Nature, 557*(7705), 322–328. https://doi.org/10.1038/s41586-018-0073-7.

Wilson, P. A., & Hemmati-Brivanlou, A. (1995). Induction of epidermis and inhibition of neural fate by Bmp-4. *Nature, 376*(6538), 331–333. https://doi.org/10.1038/376331a0.

Wilson, S. I., Rydström, A., Trimborn, T., Willert, K., Nusse, R., Jessell, T. M., et al. (2001). The status of Wnt signalling regulates neural and epidermal fates in the chick embryo. *Nature, 411*(6835), 325–330. https://doi.org/10.1038/35077115.

Winter, M., Moser, M. A., Meunier, D., Fischer, C., Machat, G., Mattes, K., et al. (2013). Divergent roles of HDAC1 and HDAC2 in the regulation of epidermal development and tumorigenesis. *The EMBO Journal, 32*(24), 3176–3191. https://doi.org/10.1038/emboj.2013.243.

Wong, V. W., Sorkin, M., Glotzbach, J. P., Longaker, M. T., & Gurtner, G. C. (2011). Surgical approaches to create murine models of human wound healing. *Journal of Biomedicine & Biotechnology, 2011*, 969618. https://doi.org/10.1155/2011/969618.

Wu, G., Ruben, M. D., Schmidt, R. E., Francey, L. J., Smith, D. F., Anafi, R. C., et al. (2018). Population-level rhythms in human skin with implications for circadian medicine. *Proceedings of the National Academy of Sciences of the United States of America, 115*(48), 12313–12318. https://doi.org/10.1073/pnas.1809442115.

Xiao, Y., Woo, W.-M., Nagao, K., Li, W., Terunuma, A., Mukouyama, Y.-S., et al. (2013). Perivascular hair follicle stem cells associate with a venule annulus. *Journal of Investigative Dermatology, 133*(10), 2324–2331. https://doi.org/10.1038/jid.2013.167.

Xin, T., Gonzalez, D., Rompolas, P., & Greco, V. (2018). Flexible fate determination ensures robust differentiation in the hair follicle. *Nature Cell Biology, 20*(12), 1361–1369. https://doi.org/10.1038/s41556-018-0232-y.

Xu, Z., Wang, W., Jiang, K., Yu, Z., Huang, H., Wang, F., et al. (2015). Embryonic attenuated Wnt/β-catenin signaling defines niche location and long-term stem cell fate in hair follicle. *eLife, 4*. e10567. https://doi.org/10.7554/eLife.10567.

Yang, H., Adam, R. C., Ge, Y., Hua, Z. L., & Fuchs, E. (2017). Epithelial-mesenchymal micro-niches govern stem cell lineage choices. *Cell, 169*(3), 483–496.e13. https://doi.org/10.1016/j.cell.2017.03.038.

Ying, Z., Sandoval, M., & Beronja, S. (2018). Oncogenic activation of PI3K induces progenitor cell differentiation to suppress epidermal growth. *Nature Cell Biology, 20*(11), 1256–1266. https://doi.org/10.1038/s41556-018-0218-9.

Youssef, K. K., Lapouge, G., Bouvrée, K., Rorive, S., Broheé, S., Appelstein, O., et al. (2012). Adult interfollicular tumour-initiating cells are reprogrammed into an embryonic hair follicle progenitor-like fate during basal cell carcinoma initiation. *Nature Cell Biology, 14*(12), 1282–1294. https://doi.org/10.1038/ncb2628.

Zanconato, F., Cordenonsi, M., & Piccolo, S. (2016). YAP/TAZ at the roots of cancer. *Cancer Cell, 29*(6), 783–803. https://doi.org/10.1016/j.ccell.2016.05.005.

Zhang, H., Pasolli, H. A., & Fuchs, E. (2011). Yes-associated protein (YAP) transcriptional coactivator functions in balancing growth and differentiation in skin. *Proceedings of the National Academy of Sciences of the United States of America, 108*(6), 2270–2275. https://doi.org/10.1073/pnas.1019603108.

Zheng, C. J., Yoo, J.-S., Lee, T.-G., Cho, H.-Y., Kim, Y.-H., & Kim, W.-G. (2005). Fatty acid synthesis is a target for antibacterial activity of unsaturated fatty acids. *FEBS Letters, 579*(23), 5157–5162. https://doi.org/10.1016/j.febslet.2005.08.028.

Zomer, H. D., & Trentin, A. G. (2018). Skin wound healing in humans and mice: Challenges in translational research. *Journal of Dermatological Science, 90*(1), 3–12. https://doi.org/10.1016/j.jdermsci.2017.12.009.

CHAPTER FOUR

Dermal papilla regulation of hair growth and pigmentation

Jamien Lim[a,b], Krystle Joy Ng[a,b], Carlos Clavel[a,b,]*
[a]Hair & Pigmentation Development, Skin Research Institute of Singapore (SRIS), Singapore, Singapore
[b]Lee Kong Chian School of Medicine, Nanyang Technological University, Singapore, Singapore
*Corresponding author: e-mail address: carlos.clavel@sris.a-star.edu.sg

Contents

1. Stem cell niches

Tissues and organs are tightly regulated by stem cell (SC) activation and self-renewal to maintain normal homeostasis in our bodies. In most cases, this occurs in specialized anatomical structures, or SC niches, that provide a microenvironment for precise spatiotemporal communication between SCs and niche cells (Rezza, Sennett, & Rendl, 2014; Walker, Patel, & Stappenbeck, 2009). A thorough understanding of how SC niche cells become programmed and function during development to regulate SCs is a crucial requirement for future tissue-specific regenerative therapies. While much knowledge has been gained from studies in invertebrates, a molecular understanding of how niche cells in mammalian SC niches, such as the bone marrow, brain, intestine and hair follicles, acquire their

Advances in Stem Cells and their Niches, Volume 3
ISSN 2468-5097
https://doi.org/10.1016/bs.asn.2019.06.002

specialized status is unknown due to the complexity of the tissues and the lack of methods to isolate and characterize the cells.

The SC niche concept has been known for over 50 years now, when Schofield hypothesized that SC are influenced by their environment and that they would change their properties upon removal from the "niche" (Schofield, 1978). Experimentally, this was largely unstudied until the research conducted in the gonad systems of *Caenorhabditis elegans* (Kimble & Crittenden, 2007) and *Drosophila melanogaster* (Lin, 2002), where a distal tip cell (single cell niche) (Kimble & Crittenden, 2007) or hub and cap cells (multiple cell niche) (Kiger et al., 2001; Xie & Spradling, 2000) instruct the neighboring germ SCs, respectively. In the mammalian system where the SC microenvironments are more complex, similar niche concepts have been proposed for the support of SCs in the intestinal epithelium, the bone marrow, the brain, the mammary gland (LaBarge, Petersen, & Bissell, 2007), the lung (Liu & Engelhardt, 2008), kidney (Al-Awqati & Oliver, 2006) and the hair follicle/epidermis of the skin (Moore & Lemischka, 2006; Walker et al., 2009; Xie & Li, 2007). For instance, in the intestinal epithelium, SCs reside at the base of the crypts, tucked away from the cell surface, but are surrounded by less well-characterized mesenchymal cells from which instructive signals are believed to originate (Moore & Lemischka, 2006). In the bone marrow, hematopoietic SCs are among the best characterized SC populations, but their microenvironmental relationship with niche stromal cells and osteoblasts is just beginning to emerge (Moore & Lemischka, 2006; Raaijmakers & Scadden, 2008; Zhang & Li, 2008). Lastly, in the hair follicle the dynamic relationship between dermal papilla (DP) niche cells and the epithelial SCs has been appreciated since many years (Fuchs, 2007; Hardy, 1992; Millar, 2002; Schmidt-Ullrich & Paus, 2005). Recent developed tools to isolate dermal papilla (DP) cells in the hair follicle SC niche, which are thought to direct follicular SCs, has begun to dissect signaling pathways that could partially regulate the SC niche and fate in the hair follicle (Rendl, Lewis, & Fuchs, 2005; Rendl, Polak, & Fuchs, 2008).

2. Role of DP cells during skin embryogenesis

2.1 The hair follicle stem cell niche formation

Within the skin, there is a dynamic relationship between DP niche cells and epidermal SCs that has been proposed to be crucial for embryonic hair follicle formation and adult regeneration during the hair growth cycle

(Driskell, Clavel, Rendl, & Watt, 2011; Yang et al., 2017). In fact, hair follicle morphogenesis and regeneration is an excellent model system for SC activation and its niche fate specification. During mammalian skin development, a stratified epithelium called epidermis is formed, from which the epithelial component of the hair follicle develops. For the hair follicle formation to occur it requires a series of epithelial-mesenchymal interactions between dermal cells and epidermal stem cells that are separated by a basement membrane (Hardy, 1992; Millar, 2002). Hair follicle morphogenesis starts around embryonic day (E)13 when it is thought that a still unidentified signal from the dermis signals to the epidermal stem cells that switch from an epidermal to a hair follicle fate (Sengel & Mauger, 1976). Then at E14, the stem cells rearrange to form hair placodes, which in return send back a signal to the dermal compartment to form recognizable cell condensates of dermal papilla (DP) precursor cells (Hardy, 1992). Subsequent proliferation and migration of epithelial cells into the dermis then leads to hair germ formation. Then, reciprocal signaling between those cells leads to further epithelial proliferation and downgrowth of the hair peg, with the DP cells at the bottom (Millar, 2002). At that point, the epithelial cells engulf the DP cells, which now act as a signaling center to guide the surrounding epithelial cells, now called matrix cells, to proliferate and to turn around and move upward. During this process, the matrix cells differentiate into the concentric layers of the inner root sheath and the hair shaft, thus forming the outgrowing, visible hair (Saxena, Mok, & Rendl, 2019). During the initial downgrowth, hair follicle SCs from the placode are already set aside in the upper portion of the hair follicle (Nowak, Polak, Pasolli, & Fuchs, 2008), in a region that is later in the fully formed follicle recognizable as a protruding structure called the bulge, where the adult hair follicle SCs reside (Blanpain & Fuchs, 2009). During further downgrowth matrix cells, direct stem cell progeny at the base of follicle bulbs, further engulf DP precursor cells to form the mature DP of fully formed hair follicles. This basic morphogenetic sequence of hair follicle formation is repeated in three separate waves giving rise to different hair follicle types in mice. Hair forming from condensates at E14.5 will give rise to the longest and thicker hairs called guard hairs, this hair is evenly distributed along the coat and corresponds to only 1–5% or the total mice hair. The second wave of hair morphogenesis starts at E16.5, giving rise to awl and auchene hairs. These hairs comprise around 10% of the total coat. Finally, at E18.5 the third and final wave occurs giving rise to most common and shortest hair type, 80–90% of total coat, the zigzag hairs (Clavel et al., 2012; Dry, 1926). Notably, the model of dermal

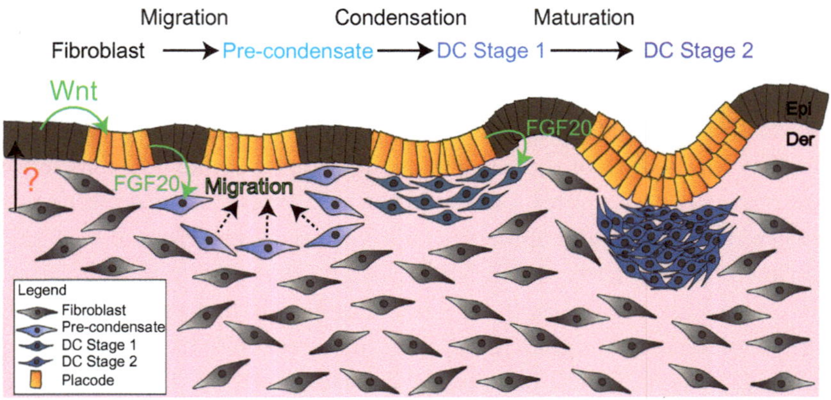

Fig. 1 Updated model for hair follicle fate specification and maturation of the dermal condensate. Unknown dermal signals are thought to induce a Wnt-dependent placode formation. In turn, FGF20 is upregulated in the placode and induces the fibroblast migration toward the placode. FGF20 signaling is required for fate specification and dermal condensate initiation and maturation.

condensate formation during the initiation of hair follicle morphogenesis has been recently fine-tuned (Biggs et al., 2018; Glover et al., 2017; Mok et al., 2019; Saxena et al., 2019). In this new updated model, the DC condensate is formed by non-proliferating dermal cells migrating in response to signaling from the just being formed epithelial placode (Biggs et al., 2018; Glover et al., 2017; Mok et al., 2019) (Fig. 1). This indicates that the fate of the dermal condensate may be already dictated prior any condensation has occurred (Biggs et al., 2018; Mok et al., 2019).

Interestingly, this dermal condensate formation during hair follicle morphogenesis resembles other mesenchymal condensations occurring in ectodermal appendages formations like tooth, feather, scales or mammary gland (Biggs & Mikkola, 2014; Widelitz et al., 1997). Therefore, a better understanding of the signaling involved in condensate formation will not be only key for hair growth, but also important to understand how condensates drive organs morphogenesis in the rest of the body (Hall & Miyake, 2000).

2.2 DP signaling during hair follicle morphogenesis

The hair induction capacity of the dermis and DP cells was first described in experiments done during the early 1970s (Dhouailly, 1973; Kollar, 1970; Oliver, 1970). In these studies, the researchers performed interspecific recombination experiments with epidermis and dermis, from hair-forming and non-hair forming sites, which were separated and recombined to assess hair induction. Results indicated that only hair-forming dermis was able to induce hair growth in early mouse embryos. Moreover, this hair induction occurred regardless of the origin of the epidermal tissues, as mouse dermis was able to induce feather-like appendages when it was recombined with feather-forming epidermis (Dhouailly, 1973). Then, the requirement of DP for hair growth induction was demonstrated in extirpation and re-engraftment experiments using follicles from rat vibrissae (Oliver, 1970). No hair follicle regeneration was observed after the lower third of the follicle was removed. However, hair growth was induced when an isolated DP was transplanted (Oliver, 1970). Later on, rat pelage DP was also shown to induce hair growth in the non-hair forming mouse paws (Reynolds & Jahoda, 1992).

As it has been described above, during the early stages of hair follicle morphogenesis there is a very tightly regulated temporal and spatial crosstalk between the epidermis and dermis. To prospectively identify the key crosstalk players, recent efforts have used different strategies to identify the molecular identity of the hair follicle morphogenesis, right when the hair fate is dictated (Biggs et al., 2018; Mok et al., 2019). And, during the last couple decades a battery of DP/DC and Pc signature genes, upregulated during the early stages of hair follicle morphogenesis, have been available (Higgins et al., 2013; Ohyama, 2012; Rendl et al., 2005; Rezza et al., 2014, 2016; Sennett et al., 2015). The canonical Wnt/β-catenin pathway members are the best characterized placode markers due to its absolute requirement for hair follicle morphogenesis (Alonso & Fuchs, 2003; Zhou, Byrne, Jacobs, & Fuchs, 1995). In addition, studies where β-catenin was ablated resulted in failure of placode formation (Huelsken et al., 2001; Rognoni et al., 2016; Zhang et al., 2009). Moreover, when β-catenin's expression is stabilized in the nucleus of epidermal or hair follicle epithelia cells, it will lead to *de novo* hair follicles and ectopic hair follicle formations, respectively (Gat, DasGupta, Degenstein, & Fuchs, 1998; Kratochwil et al., 1996; Lo Celso, Prowse, & Watt, 2004). Downstream of Wnt/β-catenin, we can find two of the early placode markers ectodysplasin (Eda) receptor (Edar) and Dickkopf WNT Signaling Pathway Inhibitor (Dkk) 4 (Ahtiainen et al., 2014; Bazzi et al., 2007). Dkk4 expression has been linked to BMP ligands' expression as key inhibitor and regulator of the placode spacing along the epidermis. A member of the TNF family, Eda signals through NF-κB after its receptor, Edar, binds. During skin embryogenesis Eda signaling is active across all the epidermis, whereas Edar expression is restricted to the placode. This pathway is essential for placode morphogenesis as primary hair follicles will fail to fully develop after its ablation (Headon & Overbeek, 1999; Schmidt-Ullrich et al., 2006). Notably, downstream of Wnt and Eda signaling you can find one of the earliest key factors needed for placode maturation and dermal condensate formation, Fgf20 (Huh et al., 2013; Lefebvre, Fliniaux, Schneider, & Mikkola, 2012). Moreover, two recent studies have also shown how Fgf20 dictates dermal cells migration toward condensation and fate even before cellular clustering is appreciable (Biggs et al., 2018; Mok et al., 2019).

As it happens with the placode, during the first stages of hair follicle fate specification, several dermal condensate/DP signatures genes upregulate their expression and signaling right before the dermal condensate (DC) formation (Mok et al., 2019; Rendl et al., 2005; Sennett et al., 2015). During hair follicle initiation, BMP signaling has an inhibitory effect with BMP ligands upregulated in the DC and BMP receptors in the epidermis (Bitgood & McMahon, 1995). Whereas Noggin, a BMP signaling inhibitor, is also selectively expressed by the dermal condensate (Botchkarev et al., 1999). In this way, the combination of BMPs and Noggin fine-tunes the signals reaching the epidermis in order to stimulate hair follicle induction. Indeed, ablation of Noggin impairs hair follicle induction and its overexpression resulted in an excessive numbers of placodes (Botchkarev et al., 1999, 2002). Sonic hedgehog (Shh) signaling is another important factor for DC formation and maturation. Shh is expressed by the placode and when absent, the hair follicle morphogenesis gets arrested (Chiang et al., 1999; Karlsson, Bondjers, & Betsholtz, 1999; St-Jacques et al., 1998). However, the Shh receptor Smoothened (Smo) is expressed by the dermal condensate. When Smo is ablated in the condensate, the hair follicle morphogenesis also gets arrested after its initiation, indicating its role for proper DC end-development and maturation (Woo, Zhen, & Oro, 2012). Another epithelial-mesenchymal interaction playing a role after the hair follicle initiation is PDGF signaling. During early skin morphogenesis, Pdgfa is expressed along the epidermis but its expression is then specifically shifted toward the placodes. Pdgfa knockout mice develop fully grown hair follicles, however, their coats are sparse and their skin are thin with diminished white adipose tissue. Then, Pdgfa receptor alpha (Pdgfra) is broadly expressed by the upper dermis during morphogenesis but, as the hair follicle matures, its expression gets restricted to the DP and dermal sheath cells (Karlsson et al., 1999). Of note, Pdgfra knockout mice die during embryogenesis but when Pdgfra was conditionally knocked out in the DP, hair follicles developed and mature without any abnormal hair phenotype (Rezza et al., 2015).

Tbx18 is another specific marker of early DC formation that even though its expression is not required for DC formation, it has been characterized as novel DP-specific genetic driver to target the DC during embryogenesis. In this way, Tbx18Cre mice has been recently used to conditionally knock out several DP signature genes including Sox2, Pdgfra, Enpp2 and Cxcr4 (Clavel et al., 2012; Grisanti et al., 2012; Rezza et al., 2015; Sennett et al., 2014). However, their ablation did not impede the development of DC or complete hair morphogenesis. The fact that there was not effect on hair morphogenesis, after knocking down DP signature genes specifically at the DC, may be an indication of how rapidly and tightly regulated is the niche signaling between the epidermis and dermis. For this reason, recent studies have been shifting their efforts to characterize the molecular identity during the precise moment of dermal condensate fate-acquisition even if the physical condensate or placode has not yet been formed (Biggs et al., 2018; Glover et al., 2017; Mok et al., 2019). For example, Sox2 and Foxd1, two DP signature transcriptions factors expressed by early condensates, are actually being already expressed even before DC clustering (Clavel et al., 2012; Mok et al., 2019). Now, we have the molecular identity right during the fate-acquisition of the placode and dermal condensate (Mok et al., 2019). Thanks to the new molecular signatures and improved culture strategies, we are in position to tackle the loss of dermal papilla hair induction *in vitro* and improve the current therapeutic approaches against hair growth disorders.

3. Dermal papilla role and signaling during the hair follicle cycle

Remarkably, after initial morphogenesis and after a period of continuous hair growth (anagen), hair follicles in the adult undergo cyclical periods of regression (catagen) and rest (telogen) followed by regeneration of a newly formed growing anagen hair follicle (Dry, 1926). During the destruction period, around two-thirds of the epithelial portion within the follicle is destroyed through programmed cell death. Interestingly enough, it has been recently shown that both terminal differentiation and apoptosis take place (Mesa et al., 2015). The DP niche cells, however, remain intact and are pulled upward in the process, through their attachment to the diminishing basement membrane, until they come to rest just beneath the permanent SC reservoir of the bulge. This period of rest can last from a few days to weeks, depending on the hair cycle. Signals from the DP niche then activates bulge/germ SCs to proliferate and generate a new growth down to

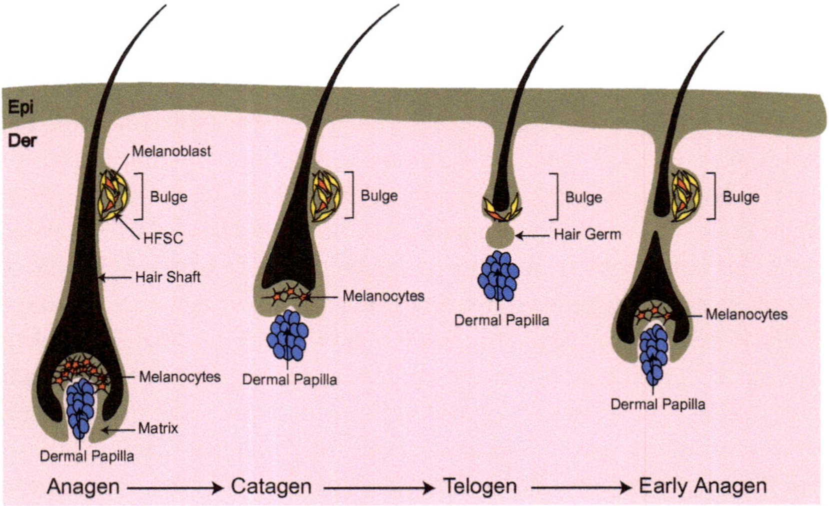

Fig. 2 Schematic of the hair growth cycle. After continuous HF growth (Anagen) the epithelial strand below the bulge stem cells degenerates by apoptosis (Catagen). The DP niche remains intact and gets pulled upward toward the bulge/hair germ. After a period of rest (Telogen), signals from the DP reactivate the stem cells in the bulge/hair germ to proliferate, grow down and regenerate a new HF.

regenerate new hair follicles (anagen), thereby closing the cycle (Cotsarelis, Sun, & Lavker, 1990; Wilson et al., 1994) (Fig. 2). Throughout life, the hair follicles repeat this precisely orchestrated sequence of events. In this way, DP cells are able to sustain their hair inducing fate from initial hair morphogenesis throughout adulthood. Proliferation of SC progeny, their migration downward together with the DP niche cells into the dermis, and subsequent continued proliferation and differentiation highly resemble initial follicle morphogenesis. At all three stages of initial hair formation, hair growth and hair regeneration, the DP niche cells act as a signaling center that closely interacts with and instructs the SCs and their progeny, yet how the DP cells become programmed to fulfil their specialized niche function is not entirely known.

3.1 DP signaling during the growth phase

By the end of morphogenesis, the DP compartment is situated at the lower end bulb of the hair follicle and separated from the SC niche situated at the bulge (Ohyama, 2007). At this stage the distance between the DP and the SC makes highly improbable that DP signaling will have any role in the maintenance or activation of hair follicle stem cells. However, DP signaling is thought to instruct epithelial matrix cells, the hair follicle stem cell progeny, to proliferate, migrate and differentiate into the different hair shaft compartments. Wnt signaling is needed for the keratins to mature and differentiate in the hair shaft (DasGupta & Fuchs, 1999; Merrill, Gat, DasGupta, & Fuchs, 2001). The DP is an important source of Wnt signaling during the sustained growth phase. When β-catenin is conditionally ablated in the DP using a postnatal Cre driver, it will alter hair growth (Enshell-Seijffers, Lindon, Kashiwagi, & Morgan, 2010). Also, DP *in vitro* co-cultures supplemented with Wnt ligands prolonged their hair induction capacity (Kishimoto, Burgeson, & Morgan, 2000). As described above, the DP is an important regulator of BMP signaling and it has been also shown to be important for retaining the DP hair induction capacity of DP cells. When DP cells are isolated or Bmp1ra is knocked out *in vitro*, DP cells fail to form hair follicles in hair reconstitution assays like the chamber graft assay (Rendl et al., 2008). In addition, the DP-specific ablation of Sox2 directly upregulated BMP signaling in the surrounding matrix cells. This excessive BMP signaling was responsible for slowing down the matrix migration toward their hair shaft differentiation compartment and as a consequence, the Sox2 knocked out mice grew shorter hair shafts (Clavel et al., 2012).

3.2 DP signaling during hair follicle regeneration

The hair growth induction capacity during the hair cycle has been well documented during the last decade. This hair induction is the result of the activation of the stem cells by the dermal papilla and it is facilitated by the close proximity of both populations during the telogen step. In an effort to show how SC gets activated, an study showed how the keratinocytes were activated by the DP as part of a two-step mechanism for hair follicle stem cell activation (Greco et al., 2009). In this last study, DP-derived Fgf7 was pointed out as a factor capable to promote telogen to anagen transition by inducing the proliferation of the bulge cells (Greco et al., 2009). Also, other factors have been shown to have an inhibitory effect like Fgf18 or BMP signaling. Ffg18 is expressed by both, the bulge and the DP compartments. When Fgf18 was ablated in the SC compartment it induced early anagen, and when Fgf18 was injected *in vivo* it induced hair growth arrest (Blanpain et al., 2004; Kimura-Ueki et al., 2012). Another key player is BMP signaling, since it has been shown to maintain quiescence of the SC compartment and the telogen stage when expressed by the DP (Botchkarev et al., 2001; Kobielak et al., 2007).

As commented before, during the telogen to anagen stages of the hair follicle, a BMP/Wnt signaling balance is key to either maintain quiescence or induce activation/proliferation. The DP-derived and Bmp-inhibitor Noggin plays a pivotal role in this balance, as its upregulation at the end of telogen allows the activation of Wnt and the initiation of anagen (Blanpain & Fuchs, 2009; Botchkarev et al., 2001). Moreover, when β-catenin is DP-ablated during the initial hair growth, the follicles are not able to transit from telogen to anagen and downregulate anagen activator genes (Enshell-Seijffers et al., 2010). All these DP factors regulating the telogen to anagen transition are an indication of the importance of the DP for the regulation of hair follicle regeneration. Indeed, a recent study using two-photon confocal microscopy has shown the absolute need of the dermal papilla for hair regeneration. In this study, the whole DP compartment was laser-ablated *in vivo* during the telogen phase. Follicles where the DP was ablated were tracked during the telogen to anagen transition. Remarkably, DP-depleted follicles failed to enter anagen whereas the non DP-depleted follicle transited to anagen as expected (Rompolas et al., 2012). A simultaneous study corroborated these results by using toxin to selectively destroy the DP compartment (Chi, Eleanor, & Morgan, 2013). In this last study, even though the authors were not able to achieve 100% of DP cells death within the follicle, the anagen entry was significantly delayed. Interestingly, it was found that within the DP compartment there was a minimum cell number threshold needed for inducing the telogen to anagen transition. Hair follicles coupled with DP compartments, in which the DP cell number threshold did not reach the minimum, failed to transit into anagen (Chi et al., 2013). Moreover, when there was sufficient DP cells to advance to anagen, the number of DP cells dictated the hair size and/or type in the mice. For example, awl and auchene hair types, with a reduced number of DP cells moving into anagen, instead gave rise to the smaller zigzag hair type in the new hair follicles when they entered into anagen (Chi et al., 2013). These results and the fact that a hair type switch happens during normal hair cycles, raised the question of how DP numbers are maintained by the DP. This aspect is particularly intriguing, since it is widely acknowledged that DP cells are quiescent as no cell proliferation is detectable within the DP compartment (Tobin et al., 2003). One possible scenario is the contribution of Dermal sheath (DS) cells, which have been proposed as an alternative cell population to replenish the DP compartment (Chi, Enshell-Seijffers, & Morgan, 2010; Jahoda, 2003; McElwee et al., 2003; Rahmani et al., 2014; Reynolds et al., 1999). The strongest evidence in favor of DS cells as a reservoir for new

DP cells comes from *in vivo* fate mapping experiments (Chi et al., 2010; McElwee et al., 2003; Rahmani et al., 2014). These results exposed that DS cells are capable of self-renewal and of being able to incorporate into both DS and DP compartments. Moreover, when these DS cells were ablated, there was a hair type switch and a delayed entry into anagen (Rahmani et al., 2014). Overall, by using novel genetic approaches, these results have been instrumental to fully demonstrate the DP requirement for hair follicle growth, and have shed light into the relevance of the DP population numbers in the regulation of hair type/size and cycle.

4. DP heterogenicity and the new microniches

As it has been reviewed above, the DP size and cell numbers dictate hair type specification and size (Chi et al., 2013). During early morphogenesis in the mouse, the dermal condensate formation leads to three consecutive signaling waves during hair follicle morphogenesis. These waves will generate four different types of hair follicles including guard, awl/auchene and zigzag which are initiated at embryonic day E14.5, 16.5 and 18.5, respectively. Recent findings have indicated that wave-specific DP condensates don't share the same molecular identity, thereby DP waves are characterized by a distinct DP gene expression (Clavel et al., 2012; Driskell et al., 2009; Sennett et al., 2015). For example, the transcription factor Sox2 is expressed only by DP condensates associated to guard and awl/achene hair types, whereas in zigzag hairs DP condensates are Sox2 negative (Clavel et al., 2012; Driskell et al., 2009; Morgan, 2014). In addition, DP cells located in different skin regions of our bodies respond differently to the surrounding tissue environment. For example, in androgenetic alopecia, androgen receptors and 5α-OH-reductase will induce miniaturization of scalp DPs but they are capable to stimulate growth in beard DPs (Rutberg et al., 2006).

Remarkably, recent lineage tracing studies in mice have shown that even within a single hair follicle we can find seven different microniches (Yang et al., 2017). These seven hair follicle microniches corresponded to the lineage identification of seven hair follicle stem cell progenitors that were spatially organized along the DP compartment. Each one of these seven lineage progenitors differentiated into different epithelial layers within the hair follicle (Yang et al., 2017). These findings raised the possibility that DP signals arising from the DP compartment are capable of dictating all seven differentiation lineages; therefore, it is plausible that cell heterogeneity also exists within a DP compartment. Indeed, single-cell RNA-seq analysis of anagen DPs revealed the presence of four different cell populations within the DP compartment. Along these four populations it was found that known DP signature genes and signaling pathways were differentially distributed. For example, BMP signals were distributed in a gradient fashion from the upper to the lower part of the DP, and the expression of Noggin, a BMP inhibitor, was higher in the lower DP in contrast to the upper region (Yang et al., 2017).

These findings are adding new dimensions to the potential functional implications of the dermal papilla to hair follicle growth based on its heterogenicity. Future research will provide insight to the relevance of specific DP subsets in orchestrating the crosstalk between the stem cell niche or microniches to generate different hair types and keratinocytes lineages.

5. DP regulation of hair follicle pigmentation

As stated above, hair follicles constantly undertake phases of cyclic regeneration and regression to regrow new hair. This phenomenon is driven by the activation of hair follicle stem cells (HFSCs) located in the bulge region of individual hair follicles, induced by the DP (Hsu, Li, & Fuchs, 2014; Nishimura, 2011). The bulge region is also home to another pivotal population of cells; the melanocyte stem cells (MSCs). MSCs originate from the neural crest cells and migrate toward the epithelia to predominantly populate the basal layer of the skin epidermis and bulge region in the hair follicle (Cichorek, Wachulska, Stasiewicz, & Tymińska, 2013; D'Mello, Finlay, Baguley, & Askarian-Amiri, 2016; Mort, Jackson, & Patton, 2015). Upon activation through inductive signals, MSCs in the bulge give rise to melanin-producing melanocytes that migrate downward to the hair matrix area, where they synthesize and transfer melanin to pigment the outgrowing hair shaft (Hsu et al., 2014; Slominski et al., 2005). Hair follicle pigmentation is therefore intricately linked to the hair cycle, specifically to the anagen phase when it gets activated. Collectively, melanotropic melanocytes present on the hair bulb form the pigmentary unit with neighboring keratinocytes, which received the melanized granules to pigment the outgrowing hair shaft. Additionally, a crosstalk between the bulbar melanocytes and DP is evident by their anatomical location whereby the matrix and DP are in close proximity and is separated by a very thin and permeable basal lamina membrane (Slominski et al., 2005). Interestingly, the fates of melanocytes are synchronized with hair cycle progression, with phases of proliferation and differentiation of the melanocyte pool during anagen, and apoptosis of melanocytes in catagen. This coincidence hence suggests that follicular melanocytes are tightly regulated by hair follicle cycling (Nishimura, 2011; Slominski et al., 1994, 2005; Tobin, Slominski, Botchkarev, & Paus, 1999).

Melanocytes contain lysosome-like organelles called melanosomes where melanin is synthesized via a complex biochemical pathway called melanogenesis (Marks & Seabra, 2001). There are two main types of melanin—eumelanin and pheomelanin which confer black and red/yellow colors, respectively (D'Mello et al., 2016; Lamoreux, Wakamatsu, & Ito, 2001). In both cases, the distinct melanin utilize the same precursor, L-tyrosine, and derive from a common tyrosine-dependent pathway. The first step involves the hydroxylation of L-tyrosine to L-DOPA (L-3,4-dihydroxyphenylalanine) catalyzed by tyrosinase and the subsequent oxidation to dopaquinone. Dopaquinone is a common substrate from which both pathways diverge (Slominski et al., 2005). The two crucial enzymes to eumelanogenesis are the tyrosinase-related proteins TRP1 and TRP2, also known as dopachrome tautomerase (DCT). On the other hand, pheomelanogenesis occurs through the conjugation of dopaquinone with L-cysteine (Videira, Moura, & Magina, 2013). The main signaling that directs melanogenesis comes from the melanocortin 1 receptor, MC1R. Apart from the tyrosinase-dependent pathway, eumelanogenesis and pheomelanogenesis can also be stimulated by the MC1R agonist α-melanocyte stimulating hormone 1 (α-MSH 1) and the agouti signaling protein (ASP), respectively (Lin & Fisher, 2007). As melanocytes typically produce both pigments, their relative concentration ratio ultimately depends on the availability of substrates, regulators and enzymes present (Cichorek et al., 2013).

Interestingly, in addition to the well-known ability of the DP to induce growth, DP cells are also a major source of hair follicle pigmentation regulators. For example, DP cells express the Stem Cell factor and when coupled with its melanocyte receptor, c-KIT, it activates a pathway crucial for the regulation of melanogenesis. Kit signaling is needed for melanocyte migration, proliferation and differentiation (Alexeev & Yoon, 2006; Botchkareva et al., 2001). Endothelin 3 is a another DP signature gene that has been shown to induce the differentiation and proliferation of melanocytes (Baynash et al., 1994; Lahav, Ziller, Dupin, & Le Douarin, 1996). Also, Endothelin 3 signaling upregulates skin pigmentation when induced in transgenic mice (Garcia et al., 2008). In addition, *in vitro* studies have demonstrated that human DP cells secrete different melanocyte chemoattractants, and it is speculated that these signaling molecules lead to observed melanocyte migration to hair follicles *in vivo* (Ideta, Soma, Tsunenaga, & Ifuku, 2002).

Regarding melanin synthesis, one of the most representative ways by which the DP cells contribute to hair follicle pigmentation is through the DP factor agouti signaling protein (ASP), mentioned above. The Mc1r receptor is usually found on the surface of melanocytes during the growth phase. Interestingly, ASP expression occurs in DP cells and is highly expressed in the early growth phase of the hair cycle (Enshell-Seijffers, Lindon, & Morgan, 2007, 2008; Vrieling et al., 1994). In turn, this allows the binding of ASP to suppress Mc1r activity during the window period when both melanocytes and dermal papilla are in close proximity as described previously. The resultant switch to pheomelanin will lead the generation of a subapical yellow band in the hair tip, which will create an overall appearance of a mottled brown hair coat (Hoekstra, 2006). Variations in the length of the pheomelanin band can alter the coat appearance and allow adaptive coloration changes to take place (Enshell-Seijffers et al., 2008; Linnen, Kingsley, Jensen, & Hoekstra, 2009; Nachman, Hoekstra, & D'Agostino, 2003). The interaction between Mc1r and ASP is known to be modulated by several other genes. For example, *proopiomelanocortin* (*Pomc*) is a gene that encodes for the precursor of α-MSH which competitively inhibits the ASP binding and its binding will upregulate the activity of Mc1r. In addition, β-defensin is another known antagonist of ASP, which functions by competitively inhibiting its binding to Mc1r.

In conjunction with ASP, the protein Corin, a transmembrane protease, is also highly expressed in the DP cells. Corin suppress the interaction of Mc1r and ASP by altering the effective period of ASP activity. When the DP-specific expression of Corin is ablated, it leads to a prolonged ASP activity, resulting in the production of more pheomelanin bands (Enshell-Seijffers et al., 2008). Interestingly, this process seems to be more complex, since the expression of ASP occurs in a transient short period during the early growth phase of the hair cycle, suggesting the occurrence of additional transcriptional regulatory events within that period. These regulatory mechanisms play an important role in contributing to pigment type switching, and the DP, therefore, plays key roles in controlling melanocyte behavior and pigmentation through the dynamic expression of ASP and Corin. Notably, β-catenin activity is also involved in regulating pigment-type switching. β-Catenin in the DP inhibits Agouti expression but activates Corin.

The ablation of β-catenin leads to a yellow coat color, whereas an increase in β-catenin results in the promotion of black pigment production. Interestingly, this study also reveals the presence of an additional level of regulation for the expression of eumelanogenic genes in the melanocytes that is dependent on the β-catenin activity in the DP (Enshell-Seijffers et al., 2010).

Even though still largely unexplored, all these recently acquired knowledge opens a new scenario where the DP and melanocyte cross-signaling could be regulated to re-induce pigmentation in different hair pigmentation disorders like hair graying.

6. DP perspectives as a therapeutic tool against hair loss

Hair de-pigmentation and thinning are probably two of the more common visual signs of human aging. In fact, in our current society, the visual appearance of our hair can be linked to social status, chances of employment or even our quality of life in general. Specially, hair loss or alopecia is frequently causes serious psychological burden in men and women. Therefore, there is little doubt about the need of finding novel therapeutic approaches to prevent or revert hair loss. In certain types of hair alopecia, like androgenetic alopecia, the link between DP losing their cell numbers and hair thinning or even hair loss has been documented (Elliott, Messenger, & Stephenson, 1999). Although the molecular mechanisms are not yet understood, and with the ability of DP cells to induce hair follicle morphogenesis and regenerate hair growth during the adult phase, it seems logic to agree with the general public and scientific community hopes that DP cells are an ideal therapeutic tool against hair loss.

With certain limitations, hair transplantation from the growing to the balding area of the scalp is the only current treatment where almost permanent full hair follicle recovery is achieved. In most cases, especially in AGA or senescence alopecia, the donor or unaffected area is the temporal/occipital area of the scalp and the recipient or affected are is the parietal area. Interestingly, temporal/occipital and parietal areas of the scalp have different embryological origins (Yoshida et al., 2008). Besides the high cost, significant time and the laborious technique required for the transplantation, the main limitation is the insufficient number of follicles available from the donor area. However, not only transplantation of entire follicles induce hair growth but implanted DP can also induce new hair follicles in animal experimental models (Jahoda, Horne, & Oliver, 1984). Therefore, as an alternative to the whole hair follicle transplantation technique, there has been a lot of effort on implanting *in vitro* expanded DP to induce hair growth. However, human DP cells quickly lose their hair induction capacity once they are isolated from the scalp and expanded *in vitro* mainly due to the loss of their molecular identity (Higgins et al., 2013; Ohyama, 2012). Thus, different methods and strategies have been published regarding efforts to maintain the DP hair induction capacity or even regain hair induction *in vitro* (Higgins et al., 2013; Kishimoto et al., 2000; Ohyama, 2012; Osada et al., 2007; Rendl et al., 2005; Thangapazham et al., 2014). However, in most of these reports, human DP were able to induce hair growth on murine keratinocytes but not in human keratinocytes. One potential explanation for the failure to induce human hair growth *in vitro* may be the change in their microenvironment, from a three-dimensional configuration within the hair follicle *in vivo* to a 2D culture system *in vitro*. So, in order to maintain their *in vivo* microenvironment, encouraging results have been obtained by creating 3D spheroids that partially restored the expression of the DP gene signature. Even more encouraging, it was found that 3D spheroids where able to induce *de novo* hair growth in human skin xenografts (Higgins et al., 2013). In addition, other ways to mimic the *in vivo* microenvironment is the enrichment of the culture media with keratinocytes and/or secreted substrates (Driskell et al., 2012; Havlickova et al., 2009; Higgins et al., 2013). Other strategies involve the media supplementation with drugs or factors developed against hair loss like minoxidil. Minoxidil is commonly used to treat androgenetic alopecia as it suppresses the androgen receptor-mediated actions. Minoxidil has shown to stimulate proliferation and survival of DP cells when added to the media of DP cell *in vitro* (Madaan, Verma, Singh, & Jaggi, 2018; Messenger & Rundegren, 2004).

Thanks to all these advances in the understanding of how DP can induce and maintain their hair induction activity, we can hold the promise of their therapeutic potential against hair loss.

References

Ahtiainen, L., et al. (2014). Directional cell migration, but not proliferation, drives hair placode morphogenesis. *Developmental Cell*, *28*(5), 588–602. http://www.ncbi.nlm.nih.gov/pubmed/24636260 [February 25, 2019].

Al-Awqati, Q., & Oliver, J. A. (2006). The kidney papilla is a stem cells niche. *Stem Cell Reviews*, *2*, 181–184.

Alexeev, V., & Yoon, K. (2006). Distinctive role of the CKit receptor tyrosine kinase signaling in mammalian melanocytes. *Journal of Investigative Dermatology*, *126*, 1102–1110.

Alonso, L., & Fuchs, E. (2003). Stem cells in the skin: Waste not, Wnt not. *Genes & Development*, *17*(10), 1189–1200. http://www.ncbi.nlm.nih.gov/pubmed/12756224 [February 25, 2019].

Baynash, A. G., et al. (1994). Interaction of endothelin-3 with endothelin-B receptor is essential for development of epidermal melanocytes and enteric neurons. *Cell, 79*(7), 1277–1285. http://www.ncbi.nlm.nih.gov/pubmed/8001160 [February 25, 2019].

Bazzi, H., et al. (2007). The Wnt inhibitor, Dickkopf 4, is induced by canonical Wnt signaling during ectodermal appendage morphogenesis. *Developmental Biology, 305*(2), 498–507. http://www.ncbi.nlm.nih.gov/pubmed/17397822 [February 25, 2019].

Biggs, L. C., & Mikkola, M. L. (2014). Early inductive events in ectodermal appendage morphogenesis. *Seminars in Cell and Developmental Biology, 25–26*, 11–21.

Biggs, L. C., et al. (2018). Hair follicle dermal condensation forms via FGF20 primed cell cycle exit, cell motility, and aggregation. *eLife, 7*, e36468.

Bitgood, M. J., & McMahon, A. P. (1995). Hedgehog and Bmp genes are coexpressed at many diverse sites of cell–cell interaction in the mouse embryo. *Developmental Biology, 172*(1), 126–138. http://linkinghub.elsevier.com/retrieve/pii/S0012160685700103 [February 25, 2019].

Blanpain, C., & Fuchs, E. (2009). Epidermal homeostasis: A balancing act of stem cells in the skin. *Nature Reviews Molecular Cell Biology, 10*(3), 207–217. http://www.pubmedcentral.nih.gov/articlerender.fcgi?artid=2760218&tool=pmcentrez&rendertype=abstract [October 15, 2015].

Blanpain, C., et al. (2004). Self-renewal, multipotency, and the existence of two cell populations within an epithelial stem cell niche. *Cell, 118*(5), 635–648. http://www.ncbi.nlm.nih.gov/pubmed/15339667 [February 25, 2019].

Botchkarev, V. A., et al. (1999). Noggin is a mesenchymally derived stimulator of hair-follicle induction. *Nature Cell Biology, 1*(3), 158–164. http://www.ncbi.nlm.nih.gov/pubmed/10559902 [February 25, 2019].

Botchkarev, V. A., et al. (2001). Noggin is required for induction of the hair follicle growth phase in postnatal skin. *FASEB Journal: Official Publication of the Federation of American Societies for Experimental Biology, 15*(12), 2205–2214. http://www.ncbi.nlm.nih.gov/pubmed/11641247 [February 28, 2017].

Botchkarev, V. A., et al. (2002). Modulation of BMP signaling by noggin is required for induction of the secondary (nontylotrich) hair follicles. *Journal of Investigative Dermatology, 118*, 3–10.

Botchkareva, N. V., et al. (2001). SCF/c-kit signaling is required for cyclic regeneration of the hair pigmentation unit. *The FASEB Journal, 15*(3), 645–658. http://www.ncbi.nlm.nih.gov/pubmed/11259383 [February 25, 2019].

Chi, W., Eleanor, W., & Morgan, B. A. (2013). Dermal papilla cell number specifies hair size, shape and cycling and its reduction causes follicular decline. *Development (Cambridge, England), 140*(8), 1676–1683. http://www.ncbi.nlm.nih.gov/pubmed/23487317 [January 27, 2017].

Chi, W. Y., Enshell-Seijffers, D., & Morgan, B. A. (2010). De novo production of dermal papilla cells during the anagen phase of the hair cycle. *The Journal of Investigative Dermatology, 130*(11), 2664–2666. https://linkinghub.elsevier.com/retrieve/pii/S0022202X15345978 [February 25, 2019].

Chiang, C., et al. (1999). Essential role for Sonic hedgehog during hair follicle morphogenesis. *Developmental Biology, 205*(1), 1–9. http://www.ncbi.nlm.nih.gov/pubmed/9882493 [February 25, 2019].

Cichorek, M., Wachulska, M., Stasiewicz, A., & Tymińska, A. (2013). Skin melanocytes: Biology and development. *Postepy Dermatologii i Alergologii, 30*(1), 30–41. http://www.termedia.pl/doi/10.5114/pdia.2013.33376 [February 25, 2019].

Clavel, C., et al. (2012). Sox2 in the dermal papilla niche controls hair growth by fine-tuning BMP signaling in differentiating hair shaft progenitors. *Developmental Cell, 23*(5), 981–994. http://www.ncbi.nlm.nih.gov/pubmed/23153495 [December 29, 2016].

Cotsarelis, G., Sun, T. T., & Lavker, R. M. (1990). Label-retaining cells reside in the bulge area of pilosebaceous unit: Implications for follicular stem cells, hair cycle, and skin carcinogenesis. *Cell*, *61*(7), 1329–1337. http://www.ncbi.nlm.nih.gov/pubmed/2364430 [February 25, 2019].

D'Mello, S., Finlay, G., Baguley, B., & Askarian-Amiri, M. (2016). Signaling pathways in melanogenesis. *International Journal of Molecular Sciences*, *17*(7), 1144. http://www.mdpi.com/1422-0067/17/7/1144 [February 25, 2019].

DasGupta, R., & Fuchs, E. (1999). Multiple roles for activated LEF/TCF transcription complexes during hair follicle development and differentiation. *Development (Cambridge, England)*, *126*(20), 4557–4568. http://www.ncbi.nlm.nih.gov/pubmed/10498690 [February 25, 2019].

Dhouailly, D. (1973). Dermo-epidermal interactions between birds and mammals: Differentiation of cutaneous appendages. *Development*, *30*(3), 587–603.

Driskell, R. R., Clavel, C., Rendl, M., & Watt, F. M. (2011). Hair follicle dermal papilla cells at a glance. *Journal of Cell Science*, *124*, 1179–1182.

Driskell, R. R., et al. (2009). Sox2-positive dermal papilla cells specify hair follicle type in mammalian epidermis. *Development*, *136*(16), 2815–2823. http://www.ncbi.nlm.nih.gov/pubmed/19605494 [February 25, 2019].

Driskell, R. R., et al. (2012). Clonal growth of dermal papilla cells in hydrogels reveals intrinsic differences between Sox2-positive and -negative cells in vitro and in vivo. *Journal of Investigative Dermatology*, *132*(4), 1084–1093. http://www.ncbi.nlm.nih.gov/pubmed/22189784 [February 25, 2019].

Dry, F. W. (1926). The coat of the mouse (*Mus musculus*). *Journal of Genetics*, *16*, 287–340.

Elliott, K., Messenger, A. G., & Stephenson, T. J. (1999). Differences in hair follicle dermal papilla volume are due to extracellular matrix volume and cell number: Implications for the control of hair follicle size and androgen responses. *The Journal of Investigative Dermatology*, *113*(6), 873–877. http://www.ncbi.nlm.nih.gov/pubmed/10594724 [February 25, 2019].

Enshell-Seijffers, D., Lindon, C., Kashiwagi, M., & Morgan, B. A. (2010). β-Catenin activity in the dermal papilla regulates morphogenesis and regeneration of hair. *Developmental Cell*, *18*(4), 633–642. https://linkinghub.elsevier.com/retrieve/pii/S1534580710001140 [February 25, 2019].

Enshell-Seijffers, D., Lindon, C., & Morgan, B. A. (2007). The serine protease Corin is a novel modifier of the agouti pathway. *Development*, *135*(2), 217–225.

Enshell-Seijffers, D., Lindon, C., & Morgan, B. A. (2008). The serine protease Corin is a novel modifier of the agouti pathway. *Development (Cambridge, England)*, *135*(2), 217–225. http://dev.biologists.org/cgi/doi/10.1242/dev.011031 [February 25, 2019].

Fuchs, E. (2007). Scratching the surface of skin development. *Nature*, *445*, 834–842.

Garcia, R. J., et al. (2008). Endothelin 3 induces skin pigmentation in a keratin-driven inducible mouse model. *The Journal of Investigative Dermatology*, *128*(1), 131–142. https://linkinghub.elsevier.com/retrieve/pii/S0022202X15336022 [February 25, 2019].

Gat, U., DasGupta, R., Degenstein, L., & Fuchs, E. (1998). De novo hair follicle morphogenesis and hair tumors in mice expressing a truncated beta-catenin in skin. *Cell*, *95*(5), 605–614. http://www.ncbi.nlm.nih.gov/pubmed/9845363 [September 20, 2015].

Glover, J. D., et al. (2017). Hierarchical patterning modes orchestrate hair follicle morphogenesis. *PLoS Biology*, *15*(7), e2002117. http://www.ncbi.nlm.nih.gov/pubmed/28700594 [February 9, 2019].

Greco, V., et al. (2009). A two-step mechanism for stem cell activation during hair regeneration. *Cell Stem Cell*, *4*(2), 155–169. http://www.pubmedcentral.nih.gov/articlerender.fcgi?artid=2668200&tool=pmcentrez&rendertype=abstract [November 8, 2015].

Grisanti, L., et al. (2012). Tbx18 targets dermal condensates for labeling, isolation, and gene ablation during embryonic hair follicle formation. *The Journal of Investigative Dermatology*, *133*, 344–353.

Hall, B. K., & Miyake, T. (2000). All for one and one for all: Condensations and the initiation of skeletal development. *BioEssays*, *22*, 138–147.
Hardy, M. H. (1992). The secret life of the hair follicle. *Trends in Genetics*, *8*(2), 55–61. http://www.ncbi.nlm.nih.gov/pubmed/1566372 [November 8, 2015].
Havlickova, B., et al. (2009). A human folliculoid microsphere assay for exploring epithelial-mesenchymal interactions in the human hair follicle. *The Journal of Investigative Dermatology*, *129*(4), 972–983. https://linkinghub.elsevier.com/retrieve/pii/S0022202X15342767 [February 25, 2019].
Headon, D. J., & Overbeek, P. A. (1999). Involvement of a novel Tnf receptor homologue in hair follicle induction. *Nature Genetics*, *22*(4), 370–374. http://www.ncbi.nlm.nih.gov/pubmed/10431242 [February 25, 2019].
Higgins, C. A., et al. (2013). Microenvironmental reprogramming by three-dimensional culture enables dermal papilla cells to induce de novo human hair-follicle growth. *Proceedings of the National Academy of Sciences of the United States of America*, *110*(49), 19679–19688. http://www.pnas.org/cgi/doi/10.1073/pnas.1309970110 [February 25, 2019].
Hoekstra, H. E. (2006). Genetics, development and evolution of adaptive pigmentation in vertebrates. *Heredity*, *97*(3), 222–234. http://www.ncbi.nlm.nih.gov/pubmed/16823403 [February 25, 2019].
Hsu, Y.-C., Li, L., & Fuchs, E. (2014). Emerging interactions between skin stem cells and their niches. *Nature Medicine*, *20*(8), 847–856. http://www.nature.com/articles/nm.3643 [February 25, 2019].
Huelsken, J., et al. (2001). Beta-catenin controls hair follicle morphogenesis and stem cell differentiation in the skin. *Cell*, *105*(4), 533–545. http://www.ncbi.nlm.nih.gov/pubmed/11371349 [February 25, 2019].
Huh, S.-H., et al. (2013). Fgf20 governs formation of primary and secondary dermal condensations in developing hair follicles. *Genes & Development*, *27*(4), 450–458. http://www.pubmedcentral.nih.gov/articlerender.fcgi?artid=3589561&tool=pmcentrez&rendertype=abstract [November 10, 2015].
Ideta, R., Soma, T., Tsunenaga, M., & Ifuku, O. (2002). Cultured human dermal papilla cells secrete a chemotactic factor for melanocytes. *Journal of Dermatological Science*, *28*(1), 48–59. http://www.ncbi.nlm.nih.gov/pubmed/11916130 [February 25, 2019].
Jahoda, C. A. B. (2003). Cell movement in the hair follicle dermis—More than a two-way street? *Journal of Investigative Dermatology*, *121*(6), ix–xi. http://www.ncbi.nlm.nih.gov/pubmed/14675219 [February 25, 2019].
Jahoda, C. A. B., Horne, K. A., & Oliver, R. F. (1984). Induction of hair growth by implantation of cultured dermal papilla cells. *Nature*, *311*, 560–562.
Karlsson, L., Bondjers, C., & Betsholtz, C. (1999). Roles for PDGF-A and sonic hedgehog in development of mesenchymal components of the hair follicle. *Development (Cambridge, England)*, *126*(12), 2611–2621. http://www.ncbi.nlm.nih.gov/pubmed/10331973 [February 25, 2019].
Kiger, A. A., et al. (2001). Stem cell self-renewal specified by JAK-STAT activation in response to a support cell cue. *Science*, *294*, 2542–2545.
Kimble, J., & Crittenden, S. L. (2007). Controls of germline stem cells, entry into meiosis, and the sperm/oocyte decision in *Caenorhabditis elegans*. *Annual Review of Cell and Developmental Biology*, *23*(1), 405–433. http://www.annualreviews.org/doi/10.1146/annurev.cellbio.23.090506.123326 [February 25, 2019].
Kimura-Ueki, M., et al. (2012). Hair cycle resting phase is regulated by cyclic epithelial FGF18 signaling. *The Journal of Investigative Dermatology*, *132*(5), 1338–1345. https://linkinghub.elsevier.com/retrieve/pii/S0022202X15357390 [February 25, 2019].
Kishimoto, J., Burgeson, R. E., & Morgan, B. A. (2000). Wnt signaling maintains the hair-inducing activity of the dermal papilla. *Genes & Development*, *14*(10), 1181–1185. http://www.ncbi.nlm.nih.gov/pubmed/10817753 [February 25, 2019].

Kobielak, K., et al. (2007). Loss of a quiescent niche but not follicle stem cells in the absence of bone morphogenetic protein signaling. *Proceedings of the National Academy of Sciences of the United States of America*, *104*(24), 10063–10068. http://www.pnas.org/cgi/doi/10.1073/pnas.0703004104 [February 25, 2019].

Kollar, E. J. (1970). The induction of hair follicles by embryonic dermal papillae. *The Journal of Investigative Dermatology*, *55*(6), 374–378. http://www.ncbi.nlm.nih.gov/pubmed/5489079 [February 25, 2019].

Kratochwil, K., et al. (1996). Lef1 expression is activated by BMP-4 and regulates inductive tissue interactions in tooth and hair development. *Genes & Development*, *10*(11), 1382–1394. http://www.ncbi.nlm.nih.gov/pubmed/8647435 [February 25, 2019].

LaBarge, M. A., Petersen, O. W., & Bissell, M. J. (2007). Of microenvironments and mammary stem cells. *Stem Cell Reviews*, *3*, 137–146.

Lahav, R., Ziller, C., Dupin, E., & Le Douarin, N. M. (1996). Endothelin 3 promotes neural crest cell proliferation and mediates a vast increase in melanocyte number in culture. *Proceedings of the National Academy of Sciences of the United States of America*, *93*(9), 3892–3897. http://www.ncbi.nlm.nih.gov/pubmed/8632985 [February 25, 2019].

Lamoreux, M. L., Wakamatsu, K., & Ito, S. (2001). Interaction of major coat color gene functions in mice as studied by chemical analysis of eumelanin and pheomelanin. *Pigment Cell Research*, *14*(1), 23–31. http://www.ncbi.nlm.nih.gov/pubmed/11277491 [February 25, 2019].

Lefebvre, S., Fliniaux, I., Schneider, P., & Mikkola, M. L. (2012). Identification of ectodysplasin target genes reveals the involvement of chemokines in hair development. *The Journal of Investigative Dermatology*, *132*(4), 1094–1102. https://linkinghub.elsevier.com/retrieve/pii/S0022202X15357110 [February 25, 2019].

Lin, H. (2002). The stem-cell niche theory: Lessons from flies. *Nature Reviews. Genetics*, *3*, 931–940.

Lin, J. Y., & Fisher, D. E. (2007). Melanocyte biology and skin pigmentation. *Nature*, *445*(7130), 843–850. http://www.nature.com/articles/nature05660 [February 25, 2019].

Linnen, C. R., Kingsley, E. P., Jensen, J. D., & Hoekstra, H. E. (2009). On the origin and spread of an adaptive allele in deer mice. *Science*, *325*(5944), 1095–1098. http://www.sciencemag.org/cgi/doi/10.1126/science.1175826 [February 25, 2019].

Liu, X., & Engelhardt, J. F. (2008). The glandular stem/progenitor cell niche in airway development and repair. *Proceedings of the American Thoracic Society*, *5*(6), 682–688. http://pats.atsjournals.org/cgi/doi/10.1513/pats.200801-003AW [February 25, 2019].

Lo Celso, C., Prowse, D. M., & Watt, F. M. (2004). Transient activation of beta-catenin signalling in adult mouse epidermis is sufficient to induce new hair follicles but continuous activation is required to maintain hair follicle tumours. *Development (Cambridge, England)*, *131*(8), 1787–1799. http://dev.biologists.org/cgi/doi/10.1242/dev.01052 [February 25, 2019].

Madaan, A., Verma, R., Singh, A. T., & Jaggi, M. (2018). Review of hair follicle dermal papilla cells as in vitro screening model for hair growth. *International Journal of Cosmetic Science*, *40*(5), 429–450. http://doi.wiley.com/10.1111/ics.12489 [February 25, 2019].

Marks, M. S., & Seabra, M. C. (2001). The melanosome: Membrane dynamics in black and white. *Nature Reviews. Molecular Cell Biology*, *2*(10), 738–748. http://www.nature.com/articles/35096009 [February 25, 2019].

McElwee, K. J., et al. (2003). Cultured peribulbar dermal sheath cells can induce hair follicle development and contribute to the dermal sheath and dermal papilla. *The Journal of Investigative Dermatology*, *121*(6), 1267–1275. http://www.ncbi.nlm.nih.gov/pubmed/14675169 [February 25, 2019].

Merrill, B. J., Gat, U., DasGupta, R., & Fuchs, E. (2001). Tcf3 and Lef1 regulate lineage differentiation of multipotent stem cells in skin. *Genes & Development*, *15*(13), 1688–1705. http://www.ncbi.nlm.nih.gov/pubmed/11445543 [February 25, 2019].

Mesa, K. R., et al. (2015). Niche-induced cell death and epithelial phagocytosis regulate hair follicle stem cell pool. *Nature, 522*(7554), 94–97. http://www.ncbi.nlm.nih.gov/pubmed/25849774 [June 10, 2019].

Messenger, A. G., & Rundegren, J. (2004). Minoxidil: Mechanisms of action on hair growth. *The British Journal of Dermatology, 150*(2), 186–194. http://www.ncbi.nlm.nih.gov/pubmed/14996087 [February 25, 2019].

Millar, S. E. (2002). Molecular mechanisms regulating hair follicle development. *The Journal of Investigative Dermatology, 118*, 216–225.

Mok, K.-W., et al. (2019). Dermal condensate niche fate specification occurs prior to formation and is placode progenitor dependent. *Developmental Cell, 48*(1) 32-48.e5. http://www.ncbi.nlm.nih.gov/pubmed/30595537 [February 20, 2019].

Moore, K. A., & Lemischka, I. R. (2006). Stem cells and their niches. *Science, 311*, 1880–1885.

Morgan, B. A. (2014). The dermal papilla: An instructive niche for epithelial stem and progenitor cells in development and regeneration of the hair follicle. *Cold Spring Harbor Perspectives in Medicine, 4*(7), a015180. http://perspectivesinmedicine.cshlp.org/lookup/doi/10.1101/cshperspect.a015180 [February 25, 2019].

Mort, R. L., Jackson, I. J., & Patton, E. E. (2015). The melanocyte lineage in development and disease. *Development, 142*(4), 620–632. http://www.ncbi.nlm.nih.gov/pubmed/25670789 [February 25, 2019].

Nachman, M. W., Hoekstra, H. E., & D'Agostino, S. L. (2003). The genetic basis of adaptive melanism in pocket mice. *Proceedings of the National Academy of Sciences of the United States of America, 100*(9), 5268–5273. http://www.ncbi.nlm.nih.gov/pubmed/12704245 [February 25, 2019].

Nishimura, E. K. (2011). Melanocyte stem cells: A melanocyte reservoir in hair follicles for hair and skin pigmentation. *Pigment Cell & Melanoma Research, 24*(3), 401–410. http://doi.wiley.com/10.1111/j.1755-148X.2011.00855.x [February 25, 2019].

Nowak, J. A., Polak, L., Pasolli, H. A., & Fuchs, E. (2008). Hair follicle stem cells are specified and function in early skin morphogenesis. *Cell Stem Cell, 3*, 33–43.

Ohyama, M. (2007). Hair follicle bulge: A fascinating reservoir of epithelial stem cells. *Journal of Dermatological Science, 46*, 81–89.

Ohyama, M. (2012). Restoration of the intrinsic properties of human dermal papilla in vitro. *Journal of Cell Science, 125*(17), 4114–4125. http://www.ncbi.nlm.nih.gov/pubmed/22623722 [February 25, 2019].

Oliver, R. F. (1970). The induction of hair follicle formation in the adult hooded rat by vibrissa dermal papillae. *Journal of Embryology and Experimental Morphology, 23*(1), 219–236. http://www.ncbi.nlm.nih.gov/pubmed/4926619 [February 25, 2019].

Osada, A., et al. (2007). Long-term culture of mouse vibrissal dermal papilla cells and *de novo* hair follicle induction. *Tissue Engineering, 13*(5), 975–982. http://www.ncbi.nlm.nih.gov/pubmed/17341162 [February 25, 2019].

Raaijmakers, M. H. G. P., & Scadden, D. T. (2008). Evolving concepts on the microenvironmental niche for hematopoietic stem cells. *Current Opinion in Hematology, 15*, 301–306.

Rahmani, W., et al. (2014). Hair follicle dermal stem cells regenerate the dermal sheath, repopulate the dermal papilla, and modulate hair type. *Developmental Cell, 31*(5), 543–558. http://www.ncbi.nlm.nih.gov/pubmed/25465495 [February 25, 2019].

Rendl, M., Lewis, L., & Fuchs, E. (2005). Molecular dissection of mesenchymal-epithelial interactions in the hair follicle. *PLoS Biology, 3*, e331.

Rendl, M., Polak, L., & Fuchs, E. (2008). BMP signaling in dermal papilla cells is required for their hair follicle-inductive properties. *Genes and Development, 22*(4), 543–557.

Reynolds, A. J., & Jahoda, C. A. (1992). Cultured dermal papilla cells induce follicle formation and hair growth by transdifferentiation of an adult epidermis. *Development*

(Cambridge, England), *115*(2), 587–593. http://www.ncbi.nlm.nih.gov/pubmed/1425341 [February 25, 2019].

Reynolds, A. J., et al. (1999). Trans-gender induction of hair follicles. *Nature*, *402*(6757), 33–34. http://www.nature.com/articles/46938 [February 25, 2019].

Rezza, A., Sennett, R., & Rendl, M. (2014). Adult stem cell niches: Cellular and molecular components. *Current Topics in Developmental Biology*, *107*, 333–372. https://linkinghub.elsevier.com/retrieve/pii/B9780124160224000123 [February 25, 2019].

Rezza, A., et al. (2015). PDGF signalling in the dermis and in dermal condensates is dispensable for hair follicle induction and formation. *Experimental Dermatology*, *24*(6), 468–470. http://doi.wiley.com/10.1111/exd.12672 [February 19, 2019].

Rezza, A., et al. (2016). Signaling networks among stem cell precursors, transit-amplifying-progenitors, and their niche in developing hair follicles. *Cell Reports*, *14*(12), 3001–3018. https://linkinghub.elsevier.com/retrieve/pii/S2211124716302133[February 25, 2019].

Rognoni, E., et al. (2016). Inhibition of β-catenin signalling in dermal fibroblasts enhances hair follicle regeneration during wound healing. *Development (Cambridge, England)*, *143*(14), 2522–2535. http://www.ncbi.nlm.nih.gov/pubmed/27287810 [August 31, 2016].

Rompolas, P., et al. (2012). Live imaging of stem cell and progeny behaviour in physiological hair-follicle regeneration. *Nature*, *487*(7408), 496–499. http://www.ncbi.nlm.nih.gov/pubmed/22763436 [February 25, 2019].

Rutberg, S. E., et al. (2006). Differences in expression of specific biomarkers distinguish human beard from scalp dermal papilla cells. *Journal of Investigative Dermatology*, *126*(12), 2583–2595. http://www.ncbi.nlm.nih.gov/pubmed/16810298 [February 25, 2019].

Saxena, N., Mok, K.-W., & Rendl, M. (2019). An updated classification of hair follicle morphogenesis. *Experimental Dermatology*, *28*(4), 332–344. http://www.ncbi.nlm.nih.gov/pubmed/30887615 [June 10, 2019].

Schmidt-Ullrich, R., & Paus, R. (2005). Molecular principles of hair follicle induction and morphogenesis. *BioEssays*, *27*, 247–261.

Schmidt-Ullrich, R., et al. (2006). NF- B transmits Eda A1/EdaR signalling to activate Shh and cyclin D1 expression, and controls post-initiation hair placode down growth. *Development*, *133*(6), 1045–1057. http://www.ncbi.nlm.nih.gov/pubmed/16481354 [February 25, 2019].

Schofield, R. (1978). The relationship between the spleen colony-forming cell and the haemopoietic stem cell. *Blood Cells*, *4*(1–2), 7–25. http://www.ncbi.nlm.nih.gov/pubmed/747780 [February 19, 2019].

Sengel, P., & Mauger, A. (1976). Peridermal cell patterning in the feather-forming skin of the chick embryo. *Developmental Biology*, *51*, 166–171.

Sennett, R., et al. (2014). Cxcr4 is transiently expressed in both epithelial and mesenchymal compartments of nascent hair follicles but is not required for follicle formation. *Experimental Dermatology*, *23*(10), 748–750. http://doi.wiley.com/10.1111/exd.12523 [February 25, 2019].

Sennett, R., et al. (2015). An integrated transcriptome atlas of embryonic hair follicle progenitors, their niche, and the developing skin. *Developmental Cell*, *34*(5), 577–591. https://linkinghub.elsevier.com/retrieve/pii/S153458071500430X [February 25, 2019].

Slominski, A., et al. (1994). Melanogenesis during the anagen-catagen-telogen transformation of the murine hair cycle. *The Journal of Investigative Dermatology*, *102*(6), 862–869. http://www.ncbi.nlm.nih.gov/pubmed/8006449 [February 25, 2019].

Slominski, A., et al. (2005). Hair follicle pigmentation. *The Journal of Investigative Dermatology, 124*(1), 13–21.
St-Jacques, B., et al. (1998). Sonic hedgehog signaling is essential for hair development. *Current Biology: CB, 8*(19), 1058–1068. http://www.ncbi.nlm.nih.gov/pubmed/9768360 [February 25, 2019].
Thangapazham, R. L., et al. (2014). Dissociated human dermal papilla cells induce hair follicle neogenesis in grafted dermal–epidermal composites. *Journal of Investigative Dermatology, 134*(2), 538–540. http://www.ncbi.nlm.nih.gov/pubmed/23924901 [February 25, 2019].
Tobin, D. J., Slominski, A., Botchkarev, V., & Paus, R. (1999). The fate of hair follicle melanocytes during the hair growth cycle. *The Journal of Investigative Dermatology Symposium Proceedings, 4*(3), 323–332. http://www.ncbi.nlm.nih.gov/pubmed/10674391 [February 25, 2019].
Tobin, D. J., et al. (2003). Plasticity and cytokinetic dynamics of the hair follicle mesenchyme: Implications for hair growth control. *Journal of Investigative Dermatology, 120*(6), 895–904. http://www.ncbi.nlm.nih.gov/pubmed/12787113 [February 25, 2019].
Videira, I. F. d. S., Moura, D. F. L., & Magina, S. B. L. M. V. (2013). Mechanisms regulating melanogenesis. *Anais Brasileiros de Dermatologia, 88*(1), 76–83. http://www.scielo.br/scielo.php?script=sci_arttext&pid=S0365-05962013000100076&lng=en&nrm=isc&tlng=en [February 26, 2016].
Vrieling, H., et al. (1994). Differences in dorsal and ventral pigmentation result from regional expression of the mouse agouti gene. *Proceedings of the National Academy of Sciences of the United States of America, 91*(12), 5667–5671. http://www.ncbi.nlm.nih.gov/pubmed/8202545 [February 25, 2019].
Walker, M. R., Patel, K. K., & Stappenbeck, T. S. (2009). The stem cell niche. *The Journal of Pathology, 217*, 169–180.
Widelitz, R. B., et al. (1997). Molecular histology in skin appendage morphogenesis. *Microscopy Research and Technique, 38*(4), 452–465. http://doi.wiley.com/10.1002/%28SICI%291097-0029%2819970815%2938%3A4%3C452%3A%3AAID-JEMT13%3E3.0.CO%3B2-I [February 25, 2019].
Wilson, C., et al. (1994). Cells within the bulge region of mouse hair follicle transiently proliferate during early anagen: Heterogeneity and functional differences of various hair cycles. *Differentiation; Research in Biological Diversity, 55*(2), 127–136. http://www.ncbi.nlm.nih.gov/pubmed/8143930 [February 25, 2019].
Woo, W.-M., Zhen, H. H., & Oro, A. E. (2012). Shh maintains dermal papilla identity and hair morphogenesis via a Noggin-Shh regulatory loop. *Genes & Development, 26*(11), 1235–1246. http://www.ncbi.nlm.nih.gov/pubmed/22661232 [February 25, 2019].
Xie, T., & Li, L. (2007). Stem cells and their niche: An inseparable relationship. *Development (Cambridge, England), 134*(11), 2001–2006. http://dev.biologists.org/cgi/doi/10.1242/dev.002022 [February 25, 2019].
Xie, T., & Spradling, A. C. (2000). A niche maintaining germ line stem cells in the Drosophila ovary. *Science, 290*, 328–330.
Yang, H., et al. (2017). Epithelial-mesenchymal micro-niches govern stem cell lineage choices. *Cell, 169*(3), 483-496.e13. http://www.ncbi.nlm.nih.gov/pubmed/28413068 [February 22, 2019].
Yoshida, T., et al. (2008). Cell lineage in mammalian craniofacial mesenchyme. *Mechanisms of Development, 125*(9–10), 797–808. http://www.ncbi.nlm.nih.gov/pubmed/18617001 [February 25, 2019].

Zhang, J., & Li, L. (2008). Stem cell niche: Microenvironment and beyond. *The Journal of Biological Chemistry*, *283*(15), 9499–9503. http://www.jbc.org/lookup/doi/10.1074/jbc.R700043200 [February 25, 2019].
Zhang, Y., et al. (2009). Reciprocal requirements for EDA/EDAR/NF-KB and Wnt/β-catenin signaling pathways in hair follicle induction. *Developmental Cell*, *17*(1), 49–61. http://www.ncbi.nlm.nih.gov/pubmed/19619491 [February 25, 2019].
Zhou, P., Byrne, C., Jacobs, J., & Fuchs, E. (1995). Lymphoid enhancer factor 1 directs hair follicle patterning and epithelial cell fate. *Genes & Development*, *9*(6), 700–713. http://www.ncbi.nlm.nih.gov/pubmed/7537238 [February 25, 2019].

Further reading

Mok, K.-W., et al. (2018). Fate before function: Specification of the hair follicle niche occurs prior to its formation and is progenitor dependent. *bioRxiv*. https://doi.org/10.1101/414839 [November 16, 2018].
Tsai, S. Y., et al. (2014). Wnt/β-catenin signaling in dermal condensates is required for hair follicle formation. *Developmental Biology*, *385*(2), 179–188.

CHAPTER FIVE

Krox20 in epithelial and glial stem cells and their niches

Chung-Ping Liao[a], Edem Tchegnon[a,b], Lu Q. Le[a,b,c,d,1]

[a]Department of Dermatology, University of Texas Southwestern Medical Center, Dallas, TX, United States
[b]Integrative Molecular and Biomedical Sciences Program, University of Texas Southwestern Medical Center, Dallas, TX, United States
[c]Hamon Center for Regenerative Science and Medicine, University of Texas Southwestern Medical Center, Dallas, TX, United States
[d]Simmons Comprehensive Cancer Center, University of Texas Southwestern Medical Center, Dallas, TX, United States
[1]Corresponding author: e-mail address: lu.le@utsouthwestern.edu

Contents

1. Introduction

Stem cells are a population of cells capable of giving rise to multiple cell types in an organism, known as "pluripotency." Stem cells maintain themselves through "self-renewal" which is to generate identical stem cells for the subsequent differentiations and regenerations. Stem cells give rise to new cell types through "differentiation" which allows the generation of multiple different cell types in the tissue. The differentiation process from stem cells to terminally differentiated cells involves multiple steps, and the in-between cell types are precursor or progenitor cells. For example, the hair follicle epithelial stem cells differentiate into a type of progenitor cells called transit-amplifying cells which rapidly divide for several rounds in the bulb

Advances in Stem Cells and their Niches, Volume 3
ISSN 2468-5097
https://doi.org/10.1016/bs.asn.2019.04.002

to amplify the cell numbers and then soon become terminally differentiated cells that produce the hair shaft and the inner root sheath (Hsu, Li, & Fuchs, 2014).

Stem cells commonly appear as a small group of cells, bordered by other non-stem cells in the surrounding microenvironment called "the niche" that play a very important role in the maintenance and activation of stem cells. Therefore, stem cells usually appear in defined histological areas. In addition, transplanted stem cells require an appropriate niche to maintain their survival and proliferation. The reciprocal interactions between stem cells and their niche have been shown to be critical during both stem cell quiescence and activation states (Fuchs, Tumbar, & Guasch, 2004; Morrison & Spradling, 2008).

In this chapter, we will review the role of Krox20 and Krox20-expressing cells in the stem cells and their niches. Krox20 (also known as early growth response 2, Egr2) is a zinc finger transcription factor (Chavrier et al., 1989) (Fig. 1), and it is recently discovered to also have an E3 ligase activity (Garcia-Gutierrez, Juarez-Vicente, Gallardo-Chamizo, Charnay, & Garcia-Dominguez, 2011). During development, Krox20 is initially found in the hindbrain rhombomere 3 (r3) and r5 (Wilkinson, Bhatt, Chavrier, Bravo, & Charnay, 1989). Krox20 is also expressed in the neural crest boundary cap cells which eventually differentiate into the glial cells in the peripheral nervous system (Maro et al., 2004). Furthermore, Krox20 is expressed in a subset of hair follicle cells primarily located in the upper hair follicle area (Gambardella, Schneider-Maunoury, Voiculescu, Charnay, & Barrandon, 2000). Recently, these hair follicle Krox20-expressing cells are found to differentiate into hair shaft progenitor cells in the hair matrix and eventually become the hair shaft structural cells (Liao, Booker, Morrison, & Le, 2017) (Fig. 2). Interestingly, Krox20 lineage hair shaft progenitor cells in the hair matrix provide an essential niche for hair follicle mature melanocytes through SCF-KIT signaling as an indispensable signaling for hair pigmentation (Liao et al., 2017).

2. The physiological function of Krox20 and its expression during development

Krox20 is a transcription factor containing three zinc finger DNA-binding sites (Chavrier et al., 1989) (Fig. 1). The expression of Krox20 was initially found to be transiently induced in quiescent NIH-3T3 fibroblast in response to serum stimulation (Chavrier et al., 1988); therefore,

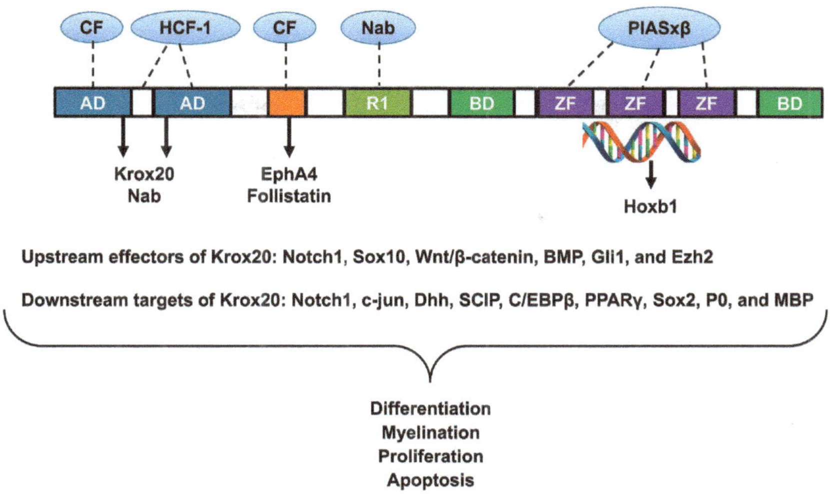

Fig. 1 Representation of Krox20 protein domains with known interactors, targets, and upstream effectors. Among known interactions, targets and regulators of Krox20, only a subset has been mapped to the various domains of the protein. For instance, HCF-1 interacts with the N-terminus of Krox20 which contains two activation domains (AD), in cooperation with unknown co-factors (CF) to regulate the transcription of Krox20 and Nab. Krox20 also influences EphA4 and follistatin expression through an interaction between an unknown co-factor (CF) and an uncharacterized domain. The repression domain (R1) is reported as the binding site of Nab, a negative regulator of Krox20 transcriptional activity. PIASxβ interacts with the DNA binding domain (DBD), made of up of three zinc fingers (ZF) to regulate the expression of Hoxb1. The DBD is flanked by two basic domains (BD). Although there are other effectors of Krox20, These are not mapped to a particular domain of the protein. Each or a combination of these krox20 interactions and effectors mediate various cellular processes, including differentiation, myelination, proliferation, or apoptosis.

Krox20 is also named early growth response 2 (Egr2). Krox20 can regulate the expression of hindbrain associated genes through either transcriptional activation (e.g., *Nab* and *EphA4*) or repression (e.g., *Follistatin* and *Hoxb1*) (Desmazieres, Charnay, & Gilardi-Hebenstreit, 2009). Krox20 is also necessary for the expression of myelin structural proteins in Schwann cells (Svaren & Meijer, 2008). The recent observation of Krox20 expression in potential hair follicle stem cells and their niche (Liao et al., 2017) warrants the investigation of Krox20 in hair follicle stem cell differentiation. However, the transcriptional regulation of Krox20 in epithelial cells is currently unclear.

Recently, Krox20 was shown to also possess the activity of E3 small ubiquitin-like modifier (SUMO)-protein ligase (Garcia-Gutierrez et al., 2011)

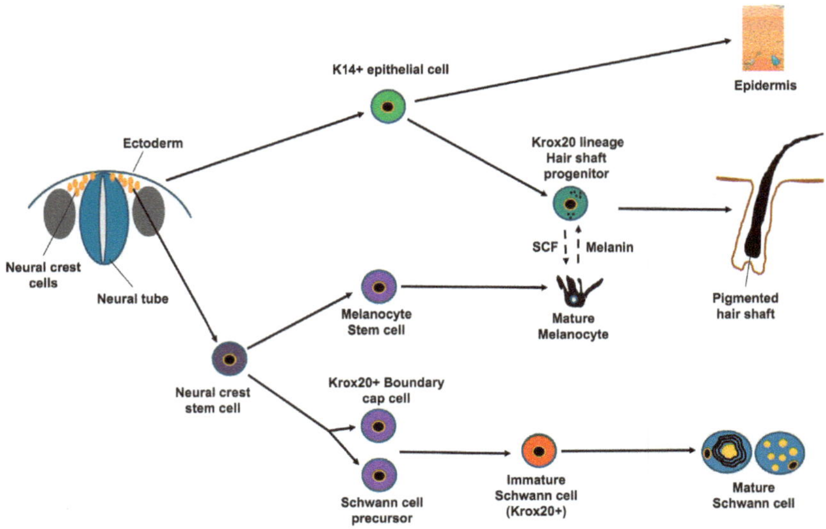

Fig. 2 Developmental scheme of Krox20 cells. Krox20 cells derive from two sources during development: the neural crest stem cells and K14 epithelial cells from the ectoderm. The Krox20 cells from the neural crest stem cells further differentiate to give rise to Schwann cells and melanocytes. On the other hand, K14-derived epithelial Krox20 lineage cells represent hair shaft progenitor cells which terminally differentiate to form the structural component of the hair shaft. When in the hair matrix, Krox20 lineage cells express SCF to provide a crucial niche for the differentiation of immature melanocytes into melanin-producing mature melanocytes. The melanin is then absorbed by the hair shaft progenitor cells to generate hair pigmentation.

to inactivate its transcription repressor NAB1 (Russo, Sevetson, & Milbrandt, 1995) and NAB2 (Svaren et al., 1996). Krox20 interacts with SUMO-conjugating enzyme Ubc9 (Knipscheer et al., 2008), through its zinc-finger domain (Garcia-Gutierrez et al., 2011). In this interaction, Krox20 was confirmed not to be the target of Ubc9 as Krox20 does not contain a consensus sequence for SUMOylation and SUMOylated form of Krox20 was not detected (Garcia-Dominguez, Gilardi-Hebenstreit, & Charnay, 2006). The sumoylation of Nab proteins repress Krox20 transcriptional activity (Garcia-Gutierrez et al., 2011).

Krox20 is an evolutionary conserved protein; human Krox20 shares high amino acid identities with that in chimpanzee (*P. troglodytes*, 100%) and in mouse (*M. musculus*, 89%) (NCBI Unigene), highlighting the essential role of Krox20 during organismal evolution. There is a total of four members in the same protein family with Krox20 (Egr1–4) (Beckmann & Wilce, 1997), and these proteins can have the same or distinct functions in context-dependent manners. For example, Krox20 and Krox24 (Egr1) both rapidly respond to skin wounding; the inductions of both *Krox20* and *Krox24* transcripts in the wounded skin can be detected as early as 30 min after injury (Grose, Harris, Cooper, Topilko, & Martin, 2002). However, in Schwann cells, Krox20 is a marker for myelinated Schwann cells, whereas Krox24 is restricted to unmyelinated Schwann cell precursors (Topilko et al., 1997). This is due to the role of Krox20 in activating myelin related gene expressions, while simultaneously suppressing the genes associated with immature Schwann cells (Jessen & Mirsky, 2005).

In mouse embryos, Krox20 is first detected at E7.75 (0-somite stage) as a thin stripe in the pre-r3 location. At 4-somite stage, the r3 band is broadened and a few isolated Krox20 cells can be detected in the pre-r5. At 6-somite stage, a thin band can be observed at pre-r5. At 8-somite stage, Krox20 clearly identified pre-r3 and pre-r5 in the hindbrain (Irving, Nieto, DasGupta, Charnay, & Wilkinson, 1996). Krox20 is essential for the identity of odd-numbered rhombomeres as overexpression or ablation of Krox20 interrupted the correct hindbrain segmentation during development (Giudicelli, Taillebourg, Charnay, & Gilardi-Hebenstreit, 2001; Schneider-Maunoury et al., 1993). In addition, Krox20 homozygous mice are perinatally lethal with the loss of hindbrain r3 and r5 and their derivatives (Swiatek & Gridley, 1993). These neurological defects affect the central neuronal networks and result in slow rhythms of jaw opening and respiration (Jacquin et al., 1996).

During embryonic development, Krox20 is also expressed in both the dorsal and ventral boundary cap cells starting E11 (Gresset et al., 2015). These Krox20 cells differentiate into the Schwann cell lineages in the dorsal root ganglia and the peripheral nerves (Maro et al., 2004). The expression of Krox20 is turned off at the Schwann cell precursor stage; whereas Krox20 resumes its expression in the immature Schwann cell stage and plays the role of a key regulator in the myelination process (Le et al., 2011). Prss56 is a similar boundary cap cell marker as Krox20; however, a clear distinction between these two boundary cap markers is that Prss56 cells give rise to glial cells in the skin epidermis region, whereas Krox20 cells do not (Gresset et al., 2015). This observation highlights the different origins of glial cells in the skin and their potential to give rise to cutaneous Schwann cell neoplasm (Chen et al., 2019).

In human, genetic mutations in Krox20 gene is associated with a genetic disorder called Charcot-Marie-Tooth (CMT) disease. CMT neuropathy is a group of hereditary disorders characterized by a chronic motor and sensory neuropathy. Currently, more than 80 of different genes have been identified to be associated with CMT (Stojkovic, 2016). The inheritance mode of Krox20 mutation in CMT can be either autosomal dominant (CMT1D) or autosomal recessive (CMT4E). Both forms of Krox20 mutation present demyelination in the peripheral nervous system as the primary clinical manifestation (Bird, 1993).

3. Molecular functions of Krox20 in key biological signaling pathways

Stem cell quiescence and activation is tightly regulated by both intrinsic and extrinsic factors through a complex interlacing signaling pathway network. For instance, Wnt/β-catenin and BMP signaling are central signaling pathways in embryonic development and stem cell activation. Overexpression of Wnt-1 or Wnt-3a in the neuroectoderm of *Xenopus* embryos causes an increase number of Krox-20-expressing cells; however, this does not affect the normal Krox20 expression at the hindbrain r3 and r5 (Saint-Jeannet, He, Varmus, & Dawid, 1997). Exogenous Wnt-3a induces Krox20 expression in osteoblast cells (Leclerc et al., 2008; Zaman et al., 2012). Although Wnt-3a alone is able to induce Krox20 expression through β-catenin; interestingly, this induction is synergized by the co-treatment of BMP antagonist Noggin (McGrew, Lai, & Moon, 1995; Saint-Jeannet et al., 1997). BMP4 can also induce the expression of Krox20 in the ectodermal cells in *Xenopus* embryos (Ishimura et al., 2000).

Notch is a negative regulator of Schwann cell myelination, and its expression is suppressed by Krox20 (Mirsky et al., 2008). Krox20 inhibits Notch signaling in myelinating Schwann cells as evidenced by the suppression of Notch intracellular domain (NICD) protein levels after Krox20 transfection in vitro, as well as the increased NICD in the nerves of Krox20 knockout mice (Woodhoo et al., 2009). Krox20 overexpression also impairs Notch1 expression in immature single-positive thymocytes (Li et al., 2011), affecting their lineage differentiation. Like Notch, c-Jun is a negative regulator of Schwann cell myelination. c-Jun is a downstream transcription factor of JNK signaling pathway, and its homeostasis is critical for Schwann cell differentiation, proliferation, and myelination. c-Jun conditional knockout mice showed increased Schwann cell myelination in the nerve (Svaren & Meijer, 2008) and c-Jun expression in Schwann cells is negatively regulated by Krox20, as overexpression of Krox-20 suppresses c-Jun protein expression and maintained Schwann cells in the myelinated state (Parkinson et al., 2004).

The hedgehog (hh) secreted proteins family plays critical roles in development, tissue homeostasis, and stem cell quiescence and differentiation (Briscoe & Therond, 2013). Dhh secretion by Schwann cells is very important for the formation of perineurium and the maintenance of myelinated nerve fibers (Parmantier et al., 1999). In Krox20 deficient mice, the expression levels of Dhh are significantly reduced (Jang, LeBlanc, Roopra, Wrabetz, & Svaren, 2006). ChIP analysis has been performed to determine whether Dhh is directly regulated by Krox20, and the results showed the binding of Krox20 to the first intron of the Dhh gene, but not its promoter region (Jang et al., 2006). Recently, the expression of Shh in the transit-amplifying cells in the hair follicle has been shown to be critical for the hair follicle stem cell activation (Hsu et al., 2014). Furthermore, this Shh expression can also regulate dermal adipose formation in a non-cell autonomous manner (Zhang et al., 2016). Gli1 is a downstream effector of Shh signaling and it is also a zinc finger transcription factor like Krox20. Gli1 can bind to the regulatory domain of Krox20 and directly suppresses Krox20 transcription in medulloblastoma cells (Yoon, Gilbertson, Iannaccone, Iannaccone, & Walterhouse, 2009). Thus, Krox20 interact with multiple signaling pathways to modulate stem cell activity and homeostasis.

4. Krox20 in stem/progenitor cell differentiation

Krox20 is important for multiple types of stem/precursor cell differentiation. In C3H10T1/2 pluripotent embryonic stem cells, when committed to adipocyte lineage, BMP signaling is activated to down-regulate Krox20 expression. In turn, overexpression of Krox20 in the C3H10T1/2 cells disrupted the BMP signaling mediated adipocyte lineage commitment and redirected the differentiation toward smooth muscle-like cells (Wang et al., 2013). Moreover, in 3T3-L1 preadipocyte cells, Krox20 overexpression induces adipogenesis through the transactivation of C/EBPβ promoter. Conversely, Krox20 knockdown reduces adipogenesis (Chen, Torrens, Anand, Spiegelman, & Friedman, 2005). Furthermore, in primary hematopoietic progenitor cells, Krox20 acts as a downstream target of PU.1 to activate the macrophage program and repress the neutrophil associated genes (Gabet et al., 2010; Laslo et al., 2006).

Krox20 is a key regulator of Schwann cell differentiation. The differentiation of Schwann cell is initiated by the activation of transcription factor Sox10, which then activates Krox20 expression (Topilko et al., 1997).

Sox10 and Krox20 then act in concert to upregulate the expression of myelin associated proteins, including protein zero (P0) and myelin basic protein (MBP) (Svaren & Meijer, 2008). Thus, Krox20 controls Schwann cell differentiation from non-myelinating to myelinating stage by activating the expressions of myelin structural proteins. In fact, Krox20 deficient mice showed severe demyelination phenotype in the peripheral nervous system (Topilko et al., 1994). In Krox20 knockout mice, nerve myelination fails completely, although Schwann cell and the larger axons are still able to form a normal 1:1 relationship; however, these Schwann cells can only make one and a half spiral wraps around the axons and thus the normal compact thickness of the myelin layer cannot fully develop. Krox20 deficient Schwann cells in vivo still express pan Schwann cell markers S100 and MAG, but not myelin-associated markers P0 and MBP (Topilko et al., 1994).

Krox20 is also a critical regulator in T cell differentiation and activation. In a set of microarray analysis, Krox20 was identified as a negative regulator of T cell activation. This finding was confirmed by a Krox20 overexpression experiment which showed an increase of E3 ubiquitin ligase Cbl-b and suppression of T cell activation (Safford et al., 2005). Interestingly, Krox20 can also serve as a positive regulator in T cell by promoting peripheral naïve T-cell differentiation. In Krox20 conditional knockout mice, naïve T cells showed delayed T-cell receptor-induced proliferation and decreased expressions of IFN-γ, IL-4, IL-9 and IL-17A that are important for T-helper differentiation and viral defense (Du et al., 2014). Krox20 ablation in T cells also results in impaired clonal expansion and increased differentiation. Conversely, Krox20 over-expression enhanced clonal expansion but inhibited T effector cell differentiation (Miao et al., 2017). Krox20 might also regulate T cells through T-bet, a master regulator of T cell differentiation; Krox20 physically interacts with the T-box DNA binding domain and inhibits the production of IFN-γ mediated by T-bet (Singh et al., 2017).

Krox20 is expressed in osteoblasts and chondrocytes, and the expression of Krox20 in these cells during embryogenesis is required for trabecular bone formation (Leclerc et al., 2008), as evidenced by the osteopenia phenotype in Krox20 knockout mice (Levi et al., 1996). Exogenous Wnt-3a induces Krox20 expression in osteoblast cells (Leclerc et al., 2008; Zaman et al., 2012) and treatment of glucocorticoid dexamethasone in osteoblasts rapidly abolishes the expression of Krox20 (Leclerc et al., 2004, 2008). This Krox20 inhibition is partly through Wnt signaling as treatment of Wnt antagonist Dkk1 can relieve this suppression (Leclerc et al., 2008).

Krox20 is also required for embryonic tendon differentiation by modulating collagen expression in tendon. Krox20 deficient mice showed reduced *Col1a1* transcripts and collagen fibrils in embryonic tendon. Transcription analysis revealed that Krox20 trans-activated *Col1a1* proximal promoter by binding to its regulatory regions (Lejard et al., 2011).

5. Krox20 in epidermal stem cells and their niches

Skin is the largest and the outermost organ of the human body and is responsible for the defense against microorganisms and injury (Gravitz, 2018). Epidermal homeostasis is maintained by skin epithelial adult stem cells which are a population of cells featured by their quiescent and slow-cycling characteristics (Morrison & Spradling, 2008). In the early 90', stem cells in the skin were successfully identified by their label retaining ability due to their slow mitotic rate (Cotsarelis, Sun, & Lavker, 1990). These labeled stem cells are predominantly located in the bulge region of hair follicle (Cotsarelis et al., 1990). A later experiment grafted LacZ+ bugle tissue to replace another normal bulge hair follicle resulting in the entire newly synthesized hair follicle with LacZ positive cells. This is direct evidence confirming the regenerative potential of bulge stem cells to generate the entire hair follicle (Oshima, Rochat, Kedzia, Kobayashi, & Barrandon, 2001).

However, the bulge stem cells appear to only give rise to the hair follicle but not the interfollicular epithelium surrounding the hair follicle. In the labeling experiment mentioned above, label retaining cells were not observed in the interfollicular epithelium (Cotsarelis et al., 1990). This finding revealed an open question about the location of stem cells for the interfollicular epithelium. A later experiment employed a dual labeling assay which showed that the label retaining cells were also found in the infundibulum zone above the bulge; and importantly, these infundibular stem cells are able to repopulate the interfollicular epidermis (Taylor, Lehrer, Jensen, Sun, & Lavker, 2000). However, these hair follicle infundibular stem cells do not seem to be the only stem cells for the interfollicular epidermis, because volar skin in the palm and sole, which represents a large surface of skin without any hair follicles, is still able to regenerate during homeostasis and wounding. A recent study showed that some epithelial progenitor cells residing in the basal layer of epidermis possess the potential to generate the skin epithelium (Clayton et al., 2007).

In one of our recent studies, by genetic lineage tracing of Krox20-expressing cells in the skin, we found that Krox20 lineage cells were initially detected in the infundibulum region of hair follicle at the postnatal day 0 (P0) mouse skin. As the hair follicle continue to develop, Krox20 cell population expanded downward to the junctional zone, isthmus, bulge and then the outer root sheath. At P6–P10, Krox20 lineage cells began to reside at the hair follicle matrix. At P12, Krox20 lineage cells terminally differentiate into the hair shaft producing cells, and shortly after, commit to cell death as the structural cells of hair shaft (Liao et al., 2017). The detection of Krox20 lineage cells in the hair follicle stem cells niches (infundibulum, junctional zone, isthmus, and bulge) suggests that Krox20 cells can be the stem cells or the niche cells interacting with stem cells. The fact that Krox20 cells can be traced to the hair shaft cells and their progenitors in the matrix and outer root sheath indicated that Krox20 cells at least represent a sub-lineage of hair follicle stem cells at the hair follicle morphogenetic phase (Liao et al., 2017) (Fig. 3).

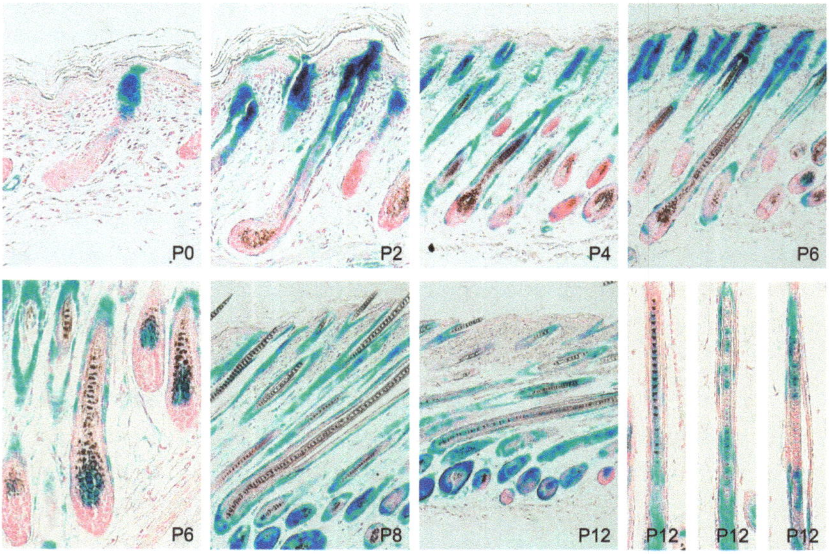

Fig. 3 Dynamic distribution of Krox20 lineage cells during hair follicle morphogenesis. Lineage tracing of Krox20 by R26-LacZ reporter shows that Krox20 lineage cells were detected in the infundibulum region of hair follicle at P0 in mouse skin. Along hair follicle development, Krox20 cell population expanded downward to the junctional zone, isthmus, bulge and then the outer root sheath. At P6-P10, Krox20 cells began to reside at the hair follicle matrix. At P12, LacZ was detected in the hair shaft showing that Krox20 marks a sub-lineage of hair follicle stem cells differentiating toward hair shaft.

6. Krox20 lineage cells provide an essential niche for mature melanocytes in the hair follicle

The bulge provides a niche not only for hair follicle epithelial stem cells but also melanocyte stem cells (Nishimura et al., 2002). The activation of melanocyte stem cell is coupled with the activation of hair follicle stem cells during anagen initiation (Rabbani et al., 2011). The maintenance of melanocyte stem cells is critical for hair pigmentation; for example, loss of Bcl2 causes melanocyte stem cell apoptosis and results in hair graying (Nishimura, Granter, & Fisher, 2005). Human aging-associated hair graying is also related to the loss of melanocyte stem cell in the bulge (Nishimura et al., 2005). While progressing toward their intrinsic fate of becoming the structural cells of the hair shaft, Krox20 lineage cells also act non-cell autonomously to influence melanocytes differentiation through the expression of stem cell factor (SCF) that can affect hair pigmentation and graying (Aoki, Hara, Motohashi, & Kunisada, 2013; Liao et al., 2017).

Similar to hair follicle stem cells, upon activation, melanocyte stem cells proliferate and differentiate into melanocyte precursor cells that migrate downward along the growing of hair follicle during anagen phase. The destination of unpigmented melanocyte precursor is in the hair follicle matrix below the Auber's line, the same niche where transit-amplifying hair follicle epithelial cells are also located (Fig. 4). Unpigmented melanocyte precursor cells differentiate into mature melanocytes and move upward to the niche above the Auber's line (Fig. 4). In this niche, mature melanocytes actively generate melanin pigments and then transfer them to the neighboring Krox20 lineage hair shaft progenitor cells. Pigmented hair shaft progenitor cells then terminally differentiate into hair shaft structural cells (Liao et al., 2017). This maturation of melanocytes is tightly controlled by the SCF signaling from the hair shaft progenitor cells, as loss of SCF in these cells results in failure of melanocyte maturation and lack of hair pigmentation (Liao et al., 2017). Interestingly, SCF-producing cells from the Krox20 lineage in the hair follicle only trace to the upper hair matrix above the Auber's line, where they co-localize with mature melanocytes. Thus, the regulation of hair pigmentation by SCF appears to occur at the final step of melanocytes differentiation as the KIT+ melanocyte precursor/transit-amplifying cells are still present in the hair follicle matrix below the Auber's line when *Scf* is ablated in the epithelial cells in vivo (Liao et al., 2017). This finding suggests that hair graying due to diminishing SCF expression in the hair follicle associated with aging could be reversible if Scf expression is restored.

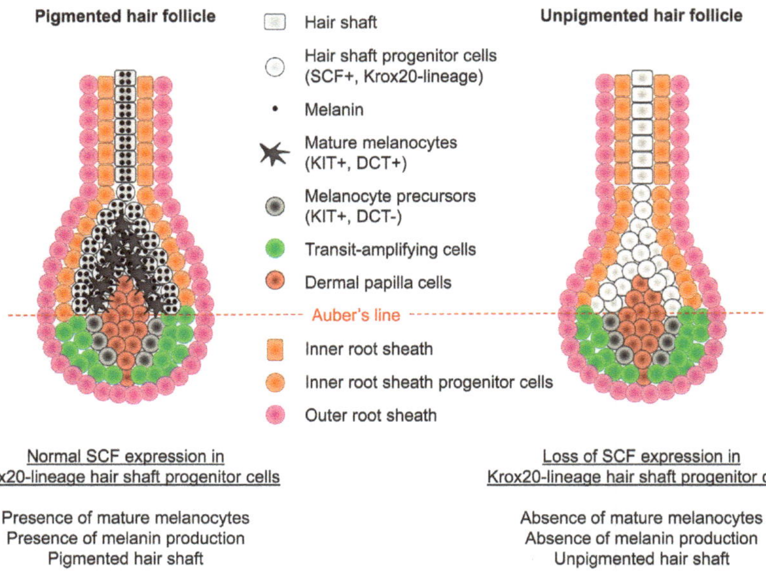

Fig. 4 SCF expression in hair shaft progenitor cells modulates hair pigmentation. In a normally pigmented hair follicle, SCF is expressed in hair shaft progenitor cells derived from Krox20 lineage hair follicle stem cells. This SCF expression is critical for the maturation of melanocytes. Mature melanocytes generate melanin pigments and transfer them to hair shaft progenitor cells to result in pigmentation in hair shaft. Conversely, loss of SCF expression in hair shaft progenitor cells completely abolishes the completion of melanocyte maturation, and thus hair shaft is unpigmented. Notably, in this hair depigmentation, melanocyte precursors are still viable below the Auber's line.

7. Conclusion

The transcription factor Krox20 is a critical regulator of hindbrain and glial cell differentiation. Recent observation of Krox20-expressing cells in the hair follicle epithelial stem cell niche (Liao et al., 2017) raised the important function of Krox20 in epithelial cell differentiation. Krox20 participates in the interlacing signaling pathways modulating stem cell quiescence and activation, including Wnt/β-catenin, BMP, and Notch singling. Future endeavors to delineate the contributions of Krox20 in epithelial cell differentiation will allow us to comprehend the role of Krox20 in skin and hair follicle homeostasis and disease.

Acknowledgments

CPL is a recipient of the Career Development Award from the Dermatology Foundation. L.Q.L. holds a Career Award for Medical Scientists from the Burroughs Wellcome Fund, the Thomas L. Shield, M.D. Professorship in Dermatology and supported by funding from the National Cancer Institute of the National Institutes of Health (Grant number R01 CA166593) and the US Department of Defense.

The authors have declared that no conflict of interest exists.

References

Aoki, H., Hara, A., Motohashi, T., & Kunisada, T. (2013). Keratinocyte stem cells but not melanocyte stem cells are the primary target for radiation-induced hair graying. *The Journal of Investigative Dermatology*, *133*(9), 2143–2151. https://doi.org/10.1038/jid.2013.155.

Beckmann, A. M., & Wilce, P. A. (1997). Egr transcription factors in the nervous system. *Neurochemistry International*, *31*(4), 477–510 [discussion 517-476].

Bird, T. D. (1993). Charcot-Marie-tooth (CMT) hereditary neuropathy overview. In M. P. Adam, H. H. Ardinger, R. A. Pagon, S. E. Wallace, L. J. H. Bean, K. Stephens, & A. Amemiya (Eds.), *GeneReviews((R))*. Seattle (WA).

Briscoe, J., & Therond, P. P. (2013). The mechanisms of Hedgehog signalling and its roles in development and disease. *Nature Reviews. Molecular Cell Biology*, *14*(7), 416–429. https://doi.org/10.1038/nrm3598.

Chavrier, P., Janssen-Timmen, U., Mattei, M. G., Zerial, M., Bravo, R., & Charnay, P. (1989). Structure, chromosome location, and expression of the mouse zinc finger gene Krox-20: Multiple gene products and coregulation with the proto-oncogene c-fos. *Molecular and Cellular Biology*, *9*(2), 787–797.

Chavrier, P., Zerial, M., Lemaire, P., Almendral, J., Bravo, R., & Charnay, P. (1988). A gene encoding a protein with zinc fingers is activated during G0/G1 transition in cultured cells. *The EMBO Journal*, *7*(1), 29–35.

Chen, Z., Mo, J., Brosseau, J. P., Shipman, T., Wang, Y., Liao, C. P., et al. (2019). Spatiotemporal loss of NF1 in Schwann cell lineage leads to different types of cutaneous neurofibroma susceptible to modification by the hippo pathway. *Cancer Discovery*, *9*(1), 114–129. https://doi.org/10.1158/2159-8290.CD-18-0151.

Chen, Z., Torrens, J. I., Anand, A., Spiegelman, B. M., & Friedman, J. M. (2005). Krox20 stimulates adipogenesis via C/EBPbeta-dependent and -independent mechanisms. *Cell Metabolism*, *1*(2), 93–106. https://doi.org/10.1016/j.cmet.2004.12.009.

Clayton, E., Doupe, D. P., Klein, A. M., Winton, D. J., Simons, B. D., & Jones, P. H. (2007). A single type of progenitor cell maintains normal epidermis. *Nature*, *446*(7132), 185–189. https://doi.org/10.1038/nature05574.

Cotsarelis, G., Sun, T. T., & Lavker, R. M. (1990). Label-retaining cells reside in the bulge area of pilosebaceous unit: Implications for follicular stem cells, hair cycle, and skin carcinogenesis. *Cell*, *61*(7), 1329–1337.

Desmazieres, A., Charnay, P., & Gilardi-Hebenstreit, P. (2009). Krox20 controls the transcription of its various targets in the developing hindbrain according to multiple modes. *The Journal of Biological Chemistry*, *284*(16), 10831–10840. https://doi.org/10.1074/jbc.M808683200.

Du, N., Kwon, H., Li, P., West, E. E., Oh, J., Liao, W., et al. (2014). EGR2 is critical for peripheral naive T-cell differentiation and the T-cell response to influenza. *Proceedings of the National Academy of Sciences of the United States of America*, *111*(46), 16484–16489. https://doi.org/10.1073/pnas.1417215111.

Fuchs, E., Tumbar, T., & Guasch, G. (2004). Socializing with the neighbors: Stem cells and their niche. *Cell*, *116*(6), 769–778.

Gabet, Y., Baniwal, S. K., Leclerc, N., Shi, Y., Kohn-Gabet, A. E., Cogan, J., et al. (2010). Krox20/EGR2 deficiency accelerates cell growth and differentiation in the monocytic lineage and decreases bone mass. *Blood*, *116*(19), 3964–3971. https://doi.org/10.1182/blood-2010-01-263830.

Gambardella, L., Schneider-Maunoury, S., Voiculescu, O., Charnay, P., & Barrandon, Y. (2000). Pattern of expression of the transcription factor Krox-20 in mouse hair follicle. *Mechanisms of Development, 96*(2), 215–218.

Garcia-Dominguez, M., Gilardi-Hebenstreit, P., & Charnay, P. (2006). PIASxbeta acts as an activator of Hoxb1 and is antagonized by Krox20 during hindbrain segmentation. *The EMBO Journal, 25*(11), 2432–2442. https://doi.org/10.1038/sj.emboj.7601122.

Garcia-Gutierrez, P., Juarez-Vicente, F., Gallardo-Chamizo, F., Charnay, P., & Garcia-Dominguez, M. (2011). The transcription factor Krox20 is an E3 ligase that sumoylates its Nab coregulators. *EMBO Reports, 12*(10), 1018–1023. https://doi.org/10.1038/embor.2011.152.

Giudicelli, F., Taillebourg, E., Charnay, P., & Gilardi-Hebenstreit, P. (2001). Krox-20 patterns the hindbrain through both cell-autonomous and non cell-autonomous mechanisms. *Genes & Development, 15*(5), 567–580. https://doi.org/10.1101/gad.189801.

Gravitz, L. (2018). Skin. *Nature, 563*(7732), S83. https://doi.org/10.1038/d41586-018-07428-4.

Gresset, A., Coulpier, F., Gerschenfeld, G., Jourdon, A., Matesic, G., Richard, L., et al. (2015). Boundary caps give rise to neurogenic stem cells and terminal glia in the skin. *Stem Cell Reports, 5*(2), 278–290. https://doi.org/10.1016/j.stemcr.2015.06.005.

Grose, R., Harris, B. S., Cooper, L., Topilko, P., & Martin, P. (2002). Immediate early genes krox-24 and krox-20 are rapidly up-regulated after wounding in the embryonic and adult mouse. *Developmental Dynamics, 223*(3), 371–378. https://doi.org/10.1002/dvdy.10064.

Hsu, Y. C., Li, L., & Fuchs, E. (2014). Transit-amplifying cells orchestrate stem cell activity and tissue regeneration. *Cell, 157*(4), 935–949. https://doi.org/10.1016/j.cell.2014.02.057.

Irving, C., Nieto, M. A., DasGupta, R., Charnay, P., & Wilkinson, D. G. (1996). Progressive spatial restriction of Sek-1 and Krox-20 gene expression during hindbrain segmentation. *Developmental Biology, 173*(1), 26–38. https://doi.org/10.1006/dbio.1996.0004.

Ishimura, A., Maeda, R., Takeda, M., Kikkawa, M., Daar, I. O., & Maeno, M. (2000). Involvement of BMP-4/msx-1 and FGF pathways in neural induction in the Xenopus embryo. *Development, Growth & Differentiation, 42*(4), 307–316.

Jacquin, T. D., Borday, V., Schneider-Maunoury, S., Topilko, P., Ghilini, G., Kato, F., et al. (1996). Reorganization of pontine rhythmogenic neuronal networks in Krox-20 knockout mice. *Neuron, 17*(4), 747–758.

Jang, S. W., LeBlanc, S. E., Roopra, A., Wrabetz, L., & Svaren, J. (2006). In vivo detection of Egr2 binding to target genes during peripheral nerve myelination. *Journal of Neurochemistry, 98*(5), 1678–1687. https://doi.org/10.1111/j.1471-4159.2006.04069.x.

Jessen, K. R., & Mirsky, R. (2005). The origin and development of glial cells in peripheral nerves. *Nature Reviews. Neuroscience, 6*(9), 671–682. https://doi.org/10.1038/nrn1746.

Knipscheer, P., Flotho, A., Klug, H., Olsen, J. V., van Dijk, W. J., Fish, A., et al. (2008). Ubc9 sumoylation regulates SUMO target discrimination. *Molecular Cell, 31*(3), 371–382. https://doi.org/10.1016/j.molcel.2008.05.022.

Laslo, P., Spooner, C. J., Warmflash, A., Lancki, D. W., Lee, H. J., Sciammas, R., et al. (2006). Multilineage transcriptional priming and determination of alternate hematopoietic cell fates. *Cell, 126*(4), 755–766. https://doi.org/10.1016/j.cell.2006.06.052.

Le, L. Q., Liu, C., Shipman, T., Chen, Z., Suter, U., & Parada, L. F. (2011). Susceptible stages in Schwann cells for NF1-associated plexiform neurofibroma development. *Cancer Research, 71*(13), 4686–4695. https://doi.org/10.1158/0008-5472.CAN-10-4577.

Leclerc, N., Luppen, C. A., Ho, V. V., Nagpal, S., Hacia, J. G., Smith, E., et al. (2004). Gene expression profiling of glucocorticoid-inhibited osteoblasts. *Journal of Molecular Endocrinology, 33*(1), 175–193.

Leclerc, N., Noh, T., Cogan, J., Samarawickrama, D. B., Smith, E., & Frenkel, B. (2008). Opposing effects of glucocorticoids and Wnt signaling on Krox20 and mineral deposition in osteoblast cultures. *Journal of Cellular Biochemistry, 103*(6), 1938–1951. https://doi.org/10.1002/jcb.21587.

Lejard, V., Blais, F., Guerquin, M. J., Bonnet, A., Bonnin, M. A., Havis, E., et al. (2011). EGR1 and EGR2 involvement in vertebrate tendon differentiation. *The Journal of Biological Chemistry, 286*(7), 5855–5867. https://doi.org/10.1074/jbc.M110.153106.

Levi, G., Topilko, P., Schneider-Maunoury, S., Lasagna, M., Mantero, S., Cancedda, R., et al. (1996). Defective bone formation in Krox-20 mutant mice. *Development, 122*(1), 113–120.

Li, S., Symonds, A. L., Zhu, B., Liu, M., Raymond, M. V., Miao, T., et al. (2011). Early growth response gene-2 (Egr-2) regulates the development of B and T cells. *PLoS One, 6*(4), e18498. https://doi.org/10.1371/journal.pone.0018498.

Liao, C. P., Booker, R. C., Morrison, S. J., & Le, L. Q. (2017). Identification of hair shaft progenitors that create a niche for hair pigmentation. *Genes & Development, 31*(8), 744–756. https://doi.org/10.1101/gad.298703.117.

Maro, G. S., Vermeren, M., Voiculescu, O., Melton, L., Cohen, J., Charnay, P., et al. (2004). Neural crest boundary cap cells constitute a source of neuronal and glial cells of the PNS. *Nature Neuroscience, 7*(9), 930–938. https://doi.org/10.1038/nn1299.

McGrew, L. L., Lai, C. J., & Moon, R. T. (1995). Specification of the anteroposterior neural axis through synergistic interaction of the Wnt signaling cascade with noggin and follistatin. *Developmental Biology, 172*(1), 337–342. https://doi.org/10.1006/dbio.1995.0027.

Miao, T., Symonds, A. L. J., Singh, R., Symonds, J. D., Ogbe, A., Omodho, B., et al. (2017). Egr2 and 3 control adaptive immune responses by temporally uncoupling expansion from T cell differentiation. *The Journal of Experimental Medicine, 214*(6), 1787–1808. https://doi.org/10.1084/jem.20160553.

Mirsky, R., Woodhoo, A., Parkinson, D. B., Arthur-Farraj, P., Bhaskaran, A., & Jessen, K. R. (2008). Novel signals controlling embryonic Schwann cell development, myelination and dedifferentiation. *Journal of the Peripheral Nervous System, 13*(2), 122–135. https://doi.org/10.1111/j.1529-8027.2008.00168.x.

Morrison, S. J., & Spradling, A. C. (2008). Stem cells and niches: Mechanisms that promote stem cell maintenance throughout life. *Cell, 132*(4), 598–611. https://doi.org/10.1016/j.cell.2008.01.038.

Nishimura, E. K., Granter, S. R., & Fisher, D. E. (2005). Mechanisms of hair graying: Incomplete melanocyte stem cell maintenance in the niche. *Science, 307*(5710), 720–724. https://doi.org/10.1126/science.1099593.

Nishimura, E. K., Jordan, S. A., Oshima, H., Yoshida, H., Osawa, M., Moriyama, M., et al. (2002). Dominant role of the niche in melanocyte stem-cell fate determination. *Nature, 416*(6883), 854–860. https://doi.org/10.1038/416854a.

Oshima, H., Rochat, A., Kedzia, C., Kobayashi, K., & Barrandon, Y. (2001). Morphogenesis and renewal of hair follicles from adult multipotent stem cells. *Cell, 104*(2), 233–245.

Parkinson, D. B., Bhaskaran, A., Droggiti, A., Dickinson, S., D'Antonio, M., Mirsky, R., et al. (2004). Krox-20 inhibits Jun-NH2-terminal kinase/c-Jun to control Schwann cell proliferation and death. *The Journal of Cell Biology, 164*(3), 385–394. https://doi.org/10.1083/jcb.200307132.

Parmantier, E., Lynn, B., Lawson, D., Turmaine, M., Namini, S. S., Chakrabarti, L., et al. (1999). Schwann cell-derived Desert hedgehog controls the development of peripheral nerve sheaths. *Neuron, 23*(4), 713–724.

Rabbani, P., Takeo, M., Chou, W., Myung, P., Bosenberg, M., Chin, L., et al. (2011). Coordinated activation of Wnt in epithelial and melanocyte stem cells initiates pigmented hair regeneration. *Cell*, *145*(6), 941–955. https://doi.org/10.1016/j.cell.2011.05.004.

Russo, M. W., Sevetson, B. R., & Milbrandt, J. (1995). Identification of NAB1, a repressor of NGFI-A- and Krox20-mediated transcription. *Proceedings of the National Academy of Sciences of the United States of America*, *92*(15), 6873–6877.

Safford, M., Collins, S., Lutz, M. A., Allen, A., Huang, C. T., Kowalski, J., et al. (2005). Egr-2 and Egr-3 are negative regulators of T cell activation. *Nature Immunology*, *6*(5), 472–480. https://doi.org/10.1038/ni1193.

Saint-Jeannet, J. P., He, X., Varmus, H. E., & Dawid, I. B. (1997). Regulation of dorsal fate in the neuraxis by Wnt-1 and Wnt-3a. *Proceedings of the National Academy of Sciences of the United States of America*, *94*(25), 13713–13718.

Schneider-Maunoury, S., Topilko, P., Seitandou, T., Levi, G., Cohen-Tannoudji, M., Pournin, S., et al. (1993). Disruption of Krox-20 results in alteration of rhombomeres 3 and 5 in the developing hindbrain. *Cell*, *75*(6), 1199–1214.

Singh, R., Miao, T., Symonds, A. L. J., Omodho, B., Li, S., & Wang, P. (2017). Egr2 and 3 inhibit T-bet-mediated IFN-gamma production in T cells. *Journal of Immunology*, *198*(11), 4394–4402. https://doi.org/10.4049/jimmunol.1602010.

Stojkovic, T. (2016). Hereditary neuropathies: An update. *Revue Neurologique (Paris)*, *172*(12), 775–778. https://doi.org/10.1016/j.neurol.2016.06.007.

Svaren, J., & Meijer, D. (2008). The molecular machinery of myelin gene transcription in Schwann cells. *Glia*, *56*(14), 1541–1551. https://doi.org/10.1002/glia.20767.

Svaren, J., Sevetson, B. R., Apel, E. D., Zimonjic, D. B., Popescu, N. C., & Milbrandt, J. (1996). NAB2, a corepressor of NGFI-A (Egr-1) and Krox20, is induced by proliferative and differentiative stimuli. *Molecular and Cellular Biology*, *16*(7), 3545–3553.

Swiatek, P. J., & Gridley, T. (1993). Perinatal lethality and defects in hindbrain development in mice homozygous for a targeted mutation of the zinc finger gene Krox20. *Genes & Development*, 7(11), 2071–2084.

Taylor, G., Lehrer, M. S., Jensen, P. J., Sun, T. T., & Lavker, R. M. (2000). Involvement of follicular stem cells in forming not only the follicle but also the epidermis. *Cell*, *102*(4), 451–461.

Topilko, P., Levi, G., Merlo, G., Mantero, S., Desmarquet, C., Mancardi, G., et al. (1997). Differential regulation of the zinc finger genes Krox-20 and Krox-24 (Egr-1) suggests antagonistic roles in Schwann cells. *Journal of Neuroscience Research*, *50*(5), 702–712. https://doi.org/10.1002/(SICI)1097-4547(19971201)50:5<702::AID-JNR7>3.0.CO;2-L.

Topilko, P., Schneider-Maunoury, S., Levi, G., Baron-Van Evercooren, A., Chennoufi, A. B., Seitanidou, T., et al. (1994). Krox-20 controls myelination in the peripheral nervous system. *Nature*, *371*(6500), 796–799. https://doi.org/10.1038/371796a0.

Wang, S. S., Huang, H. Y., Chen, S. Z., Li, X., Liu, Y., Zhang, W. T., et al. (2013). Early growth response 2 (Egr2) plays opposing roles in committing C3H10T1/2 stem cells to adipocytes and smooth muscle-like cells. *The International Journal of Biochemistry & Cell Biology*, *45*(8), 1825–1832. https://doi.org/10.1016/j.biocel.2013.06.003.

Wilkinson, D. G., Bhatt, S., Chavrier, P., Bravo, R., & Charnay, P. (1989). Segment-specific expression of a zinc-finger gene in the developing nervous system of the mouse. *Nature*, *337*(6206), 461–464. https://doi.org/10.1038/337461a0.

Woodhoo, A., Alonso, M. B., Droggiti, A., Turmaine, M., D'Antonio, M., Parkinson, D. B., et al. (2009). Notch controls embryonic Schwann cell differentiation, postnatal myelination and adult plasticity. *Nature Neuroscience*, *12*(7), 839–847. https://doi.org/10.1038/nn.2323.

Yoon, J. W., Gilbertson, R., Iannaccone, S., Iannaccone, P., & Walterhouse, D. (2009). Defining a role for Sonic hedgehog pathway activation in desmoplastic medulloblastoma by identifying GLI1 target genes. *International Journal of Cancer, 124*(1), 109–119. https://doi.org/10.1002/ijc.23929.
Zaman, G., Sunters, A., Galea, G. L., Javaheri, B., Saxon, L. K., Moustafa, A., et al. (2012). Loading-related regulation of transcription factor EGR2/Krox-20 in bone cells is ERK1/2 protein-mediated and prostaglandin, Wnt signaling pathway-, and insulin-like growth factor-I axis-dependent. *The Journal of Biological Chemistry, 287*(6), 3946–3962. https://doi.org/10.1074/jbc.M111.252742.
Zhang, B., Tsai, P. C., Gonzalez-Celeiro, M., Chung, O., Boumard, B., Perdigoto, C. N., et al. (2016). Hair follicles' transit-amplifying cells govern concurrent dermal adipocyte production through Sonic Hedgehog. *Genes & Development, 30*(20), 2325–2338. https://doi.org/10.1101/gad.285429.116.

CHAPTER SIX

Dermal niche signaling and epidermal stem cells

Sophie Frech*, Agnes Forsthuber*, Ana Korosec*, Beate M. Lichtenberger[1]
Skin & Endothelium Research Division, Department of Dermatology, Medical University of Vienna, Vienna, Austria
[1]Corresponding author: e-mail address: beate.lichtenberger@meduniwien.ac.at

Contents

1. Skin architecture and cellular heterogeneity

The skin is the largest organ in the body in surface area and weight, with many functions ranging from protection against physical or chemical harm, pathogens or excessive water loss to camouflage, insulation, temperature regulation and sensation in addition to important roles in immunity (Blanpain & Fuchs, 2006, 2009). Adult skin is composed of various cell types descending from different embryonic origins. It is comprised of two layers: the epidermis and the dermis with its dermal white adipose tissue (DWAT; Fig. 1). Beneath the dermis lies the subcutaneous white adipose tissue (SWAT) or hypodermis, which is eventually separated from the dermis

*Contributed equally.

Advances in Stem Cells and their Niches, Volume 3
ISSN 2468-5097
https://doi.org/10.1016/bs.asn.2019.04.004

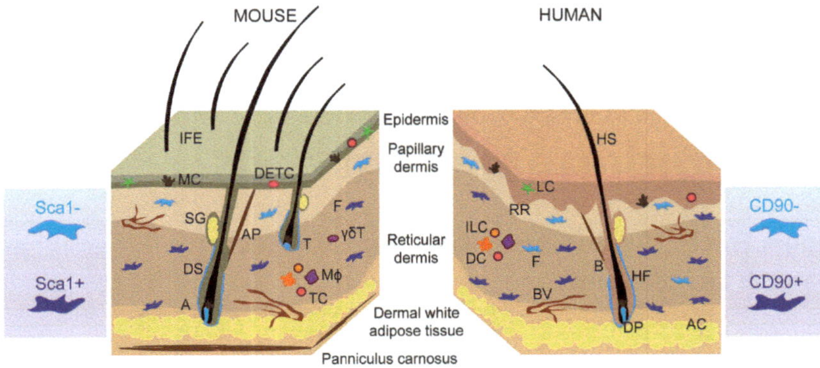

A	Anagen	HS	Hair shaft
AC	Adipocyte	IFE	Interfollicular epidermis
AP	Arrector pili	ILC	Innate lymphoid cell
B	Bulge	LC	Langerhans cell
BV	Blood vessel	MC	Melanocyte
DC	Dendritic cell	Mϕ	Macrophage
DETC	Dendritic epidermal T cell	RR	Rete ridge
DP	Dermal papilla	SG	Sebaceous gland
DS	Dermal sheath	T	Telogen
F	Fibroblast	TC	CD4/8+ T cell
HF	Hair follicle	γδT	γδ T cell

Fig. 1 Schematic cross-section of mouse and human skin depicting the epidermal/dermal structure, cellular composition and species differences. Mouse and human skin are divided into the epidermis, the papillary and reticular dermis and the dermal white adipose tissue. Mouse skin harbors a large number of hair follicles (HF), while human skin has only sparse hair but a much thicker interfollicular epidermis (IFE) with the characteristic rete ridges (RR). Dermal fibroblasts (F) are divided into papillary (light blue) and reticular fibroblasts (dark blue), which can be distinguished by the expression of Sca1 in mouse and CD90 in humans, respectively. Throughout life, HFs undergo cycles of degeneration and regeneration, thereby shifting from telogen (T) to anagen (A), which is characterized by a longer HF ranging into the dermal white adipose tissue and its pronounced dermal papilla (DP). Shown are also the appendages of the HF, sebaceous glands (SG) and the arrector pili muscle (AP). Various immune cell types are present in murine and human skin.

by the panniculus carnosus (developed in many lower mammals but not in humans) (Driskell, Jahoda, Chuong, Watt, & Horsley, 2014; Fuchs, 2008, 2016). In mammals, the epidermis develops from neuroectodermal cells remaining at the embryo's surface shortly after gastrulation and starts off as a single layer of unspecified progenitor cells, whereas mesoderm-derived cells contribute to fibroblasts and adipocytes of the underlying dermis, the vasculature, mesenchymal stem cells (MSCs) and immune cells that reside in the skin (Blanpain & Fuchs, 2006). Interestingly, while fibroblasts in the facial skin originate from the neural crest, dorsal skin fibroblasts derive from the dermato-myotome, and ventral skin comes from the lateral plate mesoderm (Sriram, Bigliardi, & Bigliardi-Qi, 2015).

Mature epidermis is a multilayered and stratified squamous epithelium that harbors several appendages: hair follicles (HF) with its sebaceous glands (SG) and associated arrector pili muscles (APM), and sweat glands in non-haired regions (Fuchs, 2007). Its remarkable ability to regenerate throughout life is ensured by tissue-resident epidermal stem cells (SCs), which serve as a reservoir of cells with multi-lineage potential that maintain the skin during homeostasis and ensure skin regeneration upon injury (Ge & Fuchs, 2018; Gonzales & Fuchs, 2017). These stem cells reside in the basal layer of the epidermis, which together with fibroblasts of the underlying dermis, synthesize, secrete and assemble the basement membrane which separates the epidermis from the dermis (Blanpain & Fuchs, 2006, 2009; Hsu, Li, & Fuchs, 2014). Only the basal layer of the epidermis is mitotically active. Importantly, the majority of cell divisions occur asymmetrically, thereby generating a basal progenitor cell and a differentiating cell positioned in the suprabasal layer (Kulukian & Fuchs, 2013; Mills, Szczerkowski, & Habib, 2017). As cells leave this layer and move toward the skin's surface, they withdraw from the cell cycle and undergo a defined program of terminal differentiation (Simpson, Patel, & Green, 2011).

Both, mouse and human dermis are divided into the papillary dermis, located directly beneath the epidermis, and the reticular dermis, lying between the papillary dermis and the DWAT, which can easily be distinguished histologically as the papillary dermis displays a high cell density and loose connective tissue, whereas reticular dermis harbors a low number of cells but is rich in collagens and other connective tissue proteins (Watt & Fujiwara, 2011). Mammalian dermis contains various cell types including endothelial, lymphatic and neuronal cells, immune cells, MSCs, adipocytes and fibroblasts, which are the predominant dermal cell type.

In vivo lineage tracing studies and various reporter constructs have characterized stem cell and progenitor populations in the interfollicular epidermis (IFE), the SG, and specific niches of the HFs (Alcolea & Jones, 2014; Kretzschmar & Watt, 2014; Niemann & Horsley, 2012; Petersson & Niemann, 2012). While several SC pools and their exact location within the HF have been defined, IFE SCs have not been identified yet, neither their determining markers nor their location (Cheng et al., 2018; Joost et al., 2016; Tan et al., 2013; Viswanathan et al., 2016). Single-cell transcriptomics in mice revealed 25 distinct populations of interfollicular and follicular epidermal cells, and their molecular relationship (Joost et al., 2016). Importantly, the different SC populations lack a "stemness" signature but rather share a common strong basal-cell gene signature and each of them has their own spatial expression pattern (Joost et al., 2016). Similarly, transcriptional analysis of human epidermis at single cell level uncovered 15 clusters of epidermal cells but no specific progenitor population (Cheng et al., 2018), indicating that long-term self-renewing cells may not require a "stemness" signature but the capacity to attach to the basement membrane and to occupy specific positions within a tissue (Clevers, 2015; Joost et al., 2016), and may thus be influenced by the local niche.

In addition to the cell heterogeneity found in the epidermis, recent studies revealed high cell plasticity and heterogeneity within the dermis as well. Fibroblasts, the principal cell type within the dermis, were considered an underwhelming cell type, which's main function is the production of structural and non-structural extracellular matrix (ECM) molecules important for skin architecture and functional integrity (Sorrell & Caplan, 2004). However, it is becoming evident that fibroblasts have important functions in tissue development, homeostasis and regeneration, inflammatory regulation, and also in pathological conditions (Kalluri & Zeisberg, 2006; Sorrell & Caplan, 2009) and aging (Tigges et al., 2014), and that functionally distinct fibroblast subsets exist.

2. Dermal fibroblast subsets

2.1 Fibroblast lineages in murine skin

Apart from being a scaffold rich in ECM that provides essential strength and elasticity to the skin, the dermis comprises various cells including highly dynamic fibroblasts, mesenchymal stem cells (MSCs), cells of the vasculature and sensory neurons as well as diverse immune cells, all of which influence the behavior of epidermal stem cells. Historically, heterogeneity among skin fibroblasts has only been described in regard to their localization within the dermis: papillary fibroblasts reside directly beneath the epidermis, while reticular fibroblasts are located deeper in the dermis, both of these subsets being distinct from fibroblasts associated with HFs (Sorrell & Caplan, 2004; Watt, 2014). Although the papillary and reticular dermis can easily be distinguished by their cellular and ECM content, how the dermal architecture develops and is maintained remained elusive until recently.

In vivo fate mapping revealed that two fibroblast lineages constitute the dermis in murine skin (Driskell et al., 2013; Rinkevich et al., 2015). These lineages originate from a common mesenchymal progenitor during embryogenesis and commit to the respective lineage around embryonic day 14.5 (E14.5), possibly earlier. One lineage gives rise to the papillary fibroblasts located primarily in the superficial dermis, including fibroblasts of the dermal papilla (DP), dermal sheath (DS) and arrector pili muscle (APM). These cells are essential for hair follicle (HF) formation, as discussed below. The other lineage forms the lower dermis, including the reticular fibroblasts that synthesize the majority of the extra-cellular matrix (ECM), and the preadipocytes and adipocytes of the DWAT (Driskell et al., 2013). While the papillary dermis is prominent during embryonic and early postnatal development, it seems to shrink to a thin layer beneath the IFE and surrounding HFs with age, and 70–90% of fibroblasts are of reticular origin in adult telogen (resting phase of the HF cycle) skin (Lichtenberger, Mastrogiannaki, & Watt, 2016; Rinkevich et al., 2015). However, histological analysis with antibodies targeting the two lineages (Dlk1 and later Sca1 expression for reticular fibroblasts as Dlk1 expression diminishes shortly after birth; Lrig1 for the papillary lineage; Fig. 2) and lineage tracing into adulthood with different promoters revealed that while in young skin papillary fibroblasts are located in the upper, and reticular fibroblasts in the lower dermis, they are intermingled in adult skin (Lichtenberger et al., 2016; Rognoni et al., 2016). Nevertheless, reticular fibroblasts are predominant in the deep dermis and hypodermis. Fate mapping with Dlk1CreER and En1Cre to target reticular fibroblasts and Lrig1CreER aiming at the papillary lineage demonstrated that these two lineages have unique functions in the development of the skin and its appendages, skin homeostasis and regeneration and in pathological conditions like fibrosis and cancer (Driskell et al., 2013; Jiang et al., 2018; Lichtenberger et al., 2016; Mastrogiannaki et al., 2016; Rinkevich et al., 2015; Rognoni et al., 2016; Walmsley et al., 2015). Of note, lineage labeling via LrigCreER mainly targets papillary fibroblasts in embryonic and early postnatal skin but also a small subset of Sca1 positive fibroblasts that have the capacity to undergo adipogenesis (Lichtenberger et al., 2016), a hallmark of reticular fibroblasts. Thus, careful interpretation of experimental data is suggested. Intriguingly, fibroblasts of the two lineages respond to different secreted signals in a unique and niche-dependent manner. For example, epidermis-derived Sonic Hedgehog (Shh) promotes proliferation of papillary fibroblasts (Lichtenberger et al., 2016) but also induces adipogenic differentiation of preadipocytes when secreted by transit-amplifying cells (TAC) during HF cycling (Zhang et al., 2016).

Fibroblasts associated with the HFs descend from papillary progenitor cells (Driskell et al., 2013). Cells of the (DS) surround the HFs, whereas DP cells are part of the HF. Reciprocal interactions between epidermal stem

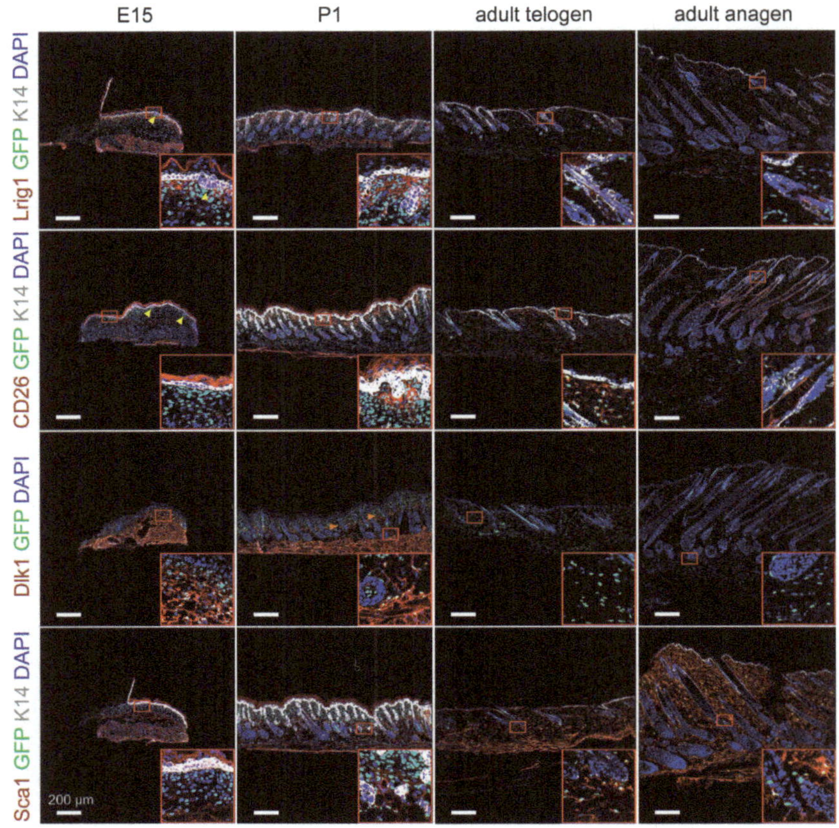

Fig. 2 Mouse fibroblast lineages during skin development. Fibroblast progenitors express Dlk1, which is lost during fibroblast commitment and marks reticular fibroblasts until after birth. Later, Sca1 expression identifies reticular fibroblasts. Papillary fibroblasts express Lrig1. CD26 is characteristic for papillary fibroblasts in embryonic and neonatal skin but is expressed by reticular fibroblasts and telogen DPs in adult skin. New hair follicle formation in embryonic skin (E15) is marked by yellow arrow heads. All fibroblasts are labeled green (PDGFRaH2BeGFP).

cells and DP precursor cells are essential for HF morphogenesis, and DP cells control HF growth during HF cycling. Intriguingly, DP cells rarely proliferate but activation of Wnt/β-catenin signaling, a key pathway in HF morphogenesis and cycling, enhances cell cycle entry and drives hair growth (Kaushal et al., 2015; Zhou, Xu, et al., 2016; Zhou, Yang, et al., 2016). Interestingly, DS cells were shown to migrate in and out of the DP, suggesting that DS cells replenish the DP with new cells and modulate DP function (Rahmani et al., 2014). Whether skin-resident and bone-marrow-derived MSCs, which share common fibroblast cell surface markers

(Halfon, Abramov, Grinblat, & Ginis, 2011; Vaculik et al., 2012), contribute to the dermal fibroblast population during homeostatic conditions and wound regeneration remains controversial (Caplan, 2015; Driskell et al., 2013).

2.2 Human dermal fibroblast subsets

Various studies implied that also human skin dermis comprises diverse fibroblast subpopulations, which harbor structural and functional differences *in vivo* and *in vitro* (Chen et al., 2007; Harper & Grove, 1979; Hiraoka et al., 2016; Sriram et al., 2015). Already in 1979, Harper and Groove showed that the papillary dermis harbors a higher fibroblast density and is metabolically more active, and that fibroblasts of the superficial dermis display a higher proliferation rate than those of the reticular layer (Harper & Grove, 1979). Other functional differences include tissue remodeling abilities and immune response modulation characterized by differential expression of cytokines and chemokines or matrix proteins and ECM modifying enzymes (Janson, Saintigny, van Adrichem, Mahe, & El Ghalbzouri, 2012; Mine, Fortunel, Pageon, & Asselineau, 2008; Nauroy et al., 2017), and the capability to facilitate epidermal SC growth in two- or three-dimensional *in vitro* culture systems (Higgins et al., 2017; Korosec & Lichtenberger, 2017; Lee & Cho, 2005), which works best with HF-associated fibroblasts. Interestingly, genomic profiling of human fibroblasts revealed that they are highly heterogeneous depending on their anatomical origins (Chang et al., 2002), with 267 transcripts discriminated between human fibroblast from distinct anatomic sites (Rinn, Bondre, Gladstone, Brown, & Chang, 2006). Roughly 30% of these transcripts were found to be differentially expressed between fibroblasts of the superficial or lower dermis, respectively (Haydont, Neiveyans, Fortunel, & Asselineau, 2019; Haydont, Neiveyans, Zucchi, Fortunel, & Asselineau, 2019). Differential expression of several markers has been suggested for fibroblasts located in the papillary or reticular dermis, respectively, dipeptidyl peptidase 4 (DPP4/CD26), Netrin 1 (NTN1) and Podoplanin (PDPN) for papillary and Smooth muscle actin alpha (ACTA2), Calponin 1 (CNN1), Collagen 11A1 (COL11A1) for reticular fibroblasts, to name but a few (Janson et al., 2012; Nauroy et al., 2017; Sriram et al., 2015). Since cell surface markers allowing the discrimination of these subsets had not been identified, the majority of the expression and functional analyses were generated with fibroblast explant cultures from (dermatomed) superficial and lower dermal layers or *in situ*.

Recent spatial and single cell transcriptomic analyses revealed additional proteins with differential expression in distinct fibroblast subsets (Haydont, Neiveyans, Fortunel, et al., 2019; Haydont, Neiveyans, Zucchi, et al., 2019; Philippeos et al., 2018; Tabib, Morse, Wang, Chen, & Lafyatis, 2018). In their screen, Tabib et al. identified two major fibroblast populations expressing either Secreted Frizzled Related Protein 2 (SFRP2) and DPP4/CD26 or Flavin Containing Monooxygenase 1 (FMO1) and lymphocyte specific protein 1 (LSP1), respectively. In addition, five minor subsets clustered separately each expressing discrete genes: COL11A1, Flavin Containing Monooxygenase 2 (FMO2), Proteoglycan 4 (PRG4), Mediator Of Cell Motility 1 (MEMO1/C2ORF4), or Cellular Retinoic Acid Binding Protein 1 (CRABP1), the last subset representing DP cells (Tabib et al., 2018). Differential gene expression of these subsets points to different functional roles in matrix deposition, inflammatory cell retention, and connective tissue cell differentiation. Philippeos et al. performed RNA sequencing on total tissue of microdissected upper and lower dermis to identify cell surface markers for distinct fibroblast subsets, and found CD39 expression upregulated in the papillary dermis, while FAT/CD36, which is involved in the fatty-acid metabolism and adipogenesis (Gao, Li, et al., 2017; Gao, Volat, et al., 2017; Siersbaek, Nielsen, & Mandrup, 2010), was strongly expressed in the lower dermis (Philippeos et al., 2018). Next they used thymocyte antigen 1 (THY-1)/CD90, a presumable pan-fibroblast marker (Jiang & Rinkevich, 2018), to isolate skin fibroblast subsets expressing either CD39 or CD36 or none of these markers. These subsets differed in morphology and their capability to support epidermal differentiation in skin equivalents, and thus CD39 was suggested as a marker for papillary fibroblasts, and CD36 for reticular fibroblasts (Philippeos et al., 2018). Single cell sequencing of human skin from only one donor however revealed additional fibroblast subsets that were negative for CD90 expression (Philippeos et al., 2018). Another study used fibroblast activation protein (FAP) and CD90 to isolate fibroblast subsets that displayed different proliferation rates and functional differences such as the capability to undergo adipogenic differentiation, which is a hallmark for reticular fibroblasts (Korosec et al., 2019). In addition to the functional analysis, expression profiling for previously established markers for papillary or reticular fibroblasts confirmed CD90 as a marker for reticular fibroblasts, and $FAP^{+}/CD90^{-}$ cells as papillary fibroblasts (Korosec et al., 2019). In accordance with the study by Philippeos et al. CD36 was strongly expressed in $CD90^{+}$ fibroblasts and defined preadipocytes, while CD39 was upregulated in the papillary fibroblast subset. Interestingly, the three fibroblast subsets ($FAP^+/CD90^-$; $FAP^-/CD90^+$; $FAP^+/CD90^+$) were also identified in a single cell transcriptomics study of oral squamous cell carcinoma (Puram et al., 2017), implying that these cell surface markers can also be used to study functional heterogeneity of fibroblasts in skin pathologies.

Intriguingly, the study by Korosec et al. uncovered that, similar to adult mouse skin, also human skin papillary and reticular fibroblasts are not spatially restricted to the superficial or deep dermis, respectively, but harbors an opposing gradient of the two subsets from the skin surface to the deep dermis (Korosec et al., 2019). Furthermore, the specific niche within the dermis in which the fibroblast subsets are located (superficial or deep dermis) affects the expression of various enzymes, and possibly their function (Korosec et al., 2019).

3. Dermal signaling in embryonic skin development and hair follicle morphogenesis

Like in other tissues, cell fate transitions are essential for epidermal SC specification, and these SC fate decisions are determined by cell intrinsic mechanisms but also by the special microenvironments in which they reside and through direct interactions with neighboring niche cells, as well as by long-range hormonal and neuronal signals. Such signals specify whether SCs remain quiescent over a long period and regulate SC self-renewal or their differentiation into specific lineages. It is established that fibroblasts of the underlying dermis regulate these fate decisions and that reciprocal signaling between cells of the epidermis and the mesenchyme are indispensable for the development of the skin and its appendages (Rendl, Lewis, & Fuchs, 2005). Many of the key signaling pathways involved in niche modulation in the skin have been identified, and include Wingless/Integrated (WNT)/β-catenin, transforming growth factor (TGF) β, Bone Morphogenic Protein (BMP), fibroblast growth factors (FGFs), Sonic Hedgehog (SHH) and Notch, which can act both in an autocrine and paracrine manner (Greco et al., 2009; Rendl et al., 2005; Rezza, Sennett, Tanguy, Clavel, & Rendl, 2015; Sennett & Rendl, 2012; Sennett et al., 2015). Furthermore, besides unraveling the morphogenetic events associated with the developmental processes during embryonic HF development, various studies have investigated SC niche compartments and the appearance of SC (Sennett & Rendl, 2012). Nevertheless, our understanding of how niche fate specification is controlled and the precise timing of epidermal SC and niche cell appearance and their interaction during embryonic skin development remains limited.

Skin development begins shortly after gastrulation with a single layer of unspecified epidermal progenitors (Blanpain & Fuchs, 2006). Around E12.5, epidermis-derived Wnt proteins broadly activate dermal Wnt/β-catenin signaling (Chen, Jarrell, Guo, Lang, & Atit, 2012). Then at around E13.5 yet unidentified dermal signals induce fate specification of HF progenitors, which form the pre-placode (Zhang et al., 2009). Elongated keratinocytes in the epidermal layer organize into clusters to form the epidermal placode (EP), while placode progenitors signal back to the dermis and cells of mesenchymal origin simultaneously form the dermal condensate (DC) directly underneath (Sennett & Rendl, 2012; Sennett et al., 2015). These specialized fibroblasts serve as signaling niches for the placode precursors and regulate further HF development. Continued reciprocal signaling between placode and DC cells induce progenitor proliferation and downgrowth, and cells of the epithelial matrix then engulf the DC (Alonso & Fuchs, 2006). DC cells were previously believed to be of unique mesenchymal origin. However, embryonic lineage tracing in murine dermis revealed their provenance from the upper dermal papillary lineage (Driskell et al., 2013). The morphological detection of the DC after formation of the placode suggested that niche fate specification occurs in response to signals from the placode when the DC forms. Recent studies revealed that this is not the case. Strikingly, even without epidermal pre-patterning fibroblasts have the capacity to self-organize (Glover et al., 2017), and DC cells develop from fibroblasts of the utmost vicinity of the developing hair follicle through directed migration (Biggs et al., 2018). Single-cell transcriptomics of upper dermal fibroblasts during murine HF morphogenesis showed that DC cell specification occurs before DC clustering in response to placode-derived FGF20, and that pre-DC cells derive from fibroblasts and follow a transcriptional trajectory of distinct stages of differentiation ultimately resulting in the development of the mature DC (Gupta et al., 2019; Mok et al., 2019). Differentiated DC cells undergo morphological changes, migrate in clusters toward the epidermal placode, exit the cell cycle and remain predominantly in a quiescent state (Biggs et al., 2018). Cell cycle exit is thus a feature of pre-DC fate specification. Like pre-DC cells, aggregated DCs (Grisanti et al., 2013) and also mature DP niche cells are quiescent and rarely divide (Kaushal et al., 2015; Tobin, Gunin, Magerl, Handijski, & Paus, 2003). Ultimately, DC cells give rise to the DP, which is the main driver of HF cycling, and the DS (Chi, Wu, & Morgan, 2013). Importantly, adult fibroblasts retain the capacity to induce DP and HF formation in skin reconstitution assays even though HF neogenesis does not spontaneously occur in adult skin (Collins, Jensen, MacRae, Mansfield, & Watt, 2012). Data on human fibroblasts and HF neogenesis are more conflicting (Korosec & Lichtenberger, 2017): while Thangapazham et al. demonstrated that DP cells in skin constructs grafted on mice formed *de novo* hair follicles (Thangapazham et al., 2014), Higgins et al. showed human DP cells maintained the epidermis in bioengineered skin but failed to induce hair follicle neogenesis *in vitro* (Higgins et al., 2017). Intriguingly, DP cells change their gene signature instantaneously in two-dimensional cultures and thus lose their hair-inducting properties (Higgins, Chen, Cerise, Jahoda, & Christiano, 2013). However, three-dimensional hanging drop cultures of human scalp DP cells resulted in DP-like structures, which maintained a DP-gene signature and induced human hair *in vivo* when placed between separated epidermis and dermis and transplanted onto immunocompromised mice (Higgins et al., 2013).

As previously stated, multiple reciprocal signals between the epidermal and the mesenchymal compartment are required to induce placode and DC formation and subsequent HF development. For instance, epidermal pre-patterning and multiple regulatory interactions between WNT, FGF and BMP signaling pathways are required for the initiation of DC formation (Andl, Reddy, Gaddapara, & Millar, 2002; Glover et al., 2017; Mok et al., 2019). Moreover, HF spacing and epidermal appendage density is dictated by inversely correlating expression of WNT and its antagonist Dickkopf-related protein 1 (Dkk1) (Gupta et al., 2019; Sick, Reinker, Timmer, & Schlake, 2006) and upregulation of Dkk2 in non-hair-bearing skin is responsible for the inherent lack of HF morphogenesis such as in the plantar dermis of mice (Song et al., 2018). Sonic Hedgehog (Shh), main effector molecule of the Hedgehog signaling pathway, is required for the advancement of hair follicle morphogenesis beyond the hair germ stage (Chiang et al., 1999; Oro & Higgins, 2003). Interestingly, suppression of Shh by upregulation of mesenchyme-derived BMP and FGFs that signal to epithelial buds confines fate determination into hair rather than sweat glands in developing skin (Lu, Polak, Keyes, & Fuchs, 2016).

In adult skin, HF neogenesis only occurs in large wounds in mice (Ito et al., 2007) and humans (Kligman, 1959) or upon activation of epidermal Wnt/β-catenin signaling (Collins, Kretzschmar, & Watt, 2011; Lichtenberger et al., 2016), which redirects stem cells of the IFE and SG into the HF lineage, and induces paracrine Shh signaling in the mesenchyme (Lichtenberger et al., 2016), thereby inducing fate switching toward DP identity shown by Blimp1 expression in fibroblasts in proximity to ectopic placodes (Telerman et al., 2017). Strikingly, while epidermis-specific overexpression of the Wnt inhibitor Dkk1 impaired *de novo* HF formation (Ito et al., 2007) and overexpression of the Wnt7a ligand (Ito et al., 2007) in the epidermis or genetic deletion of the negative regulator of Wnt signaling CXXC-type zinc finger protein 5 (CXXC5) (Lee et al., 2017) lead to increased wounding-induced HF neogenesis, activation of Wnt/β-catenin in the dermis, which is differentially regulated in papillary and reticular fibroblasts (Mastrogiannaki et al., 2016), inhibited *de novo* HF formation (Rognoni et al., 2016). In contrast, when β-catenin was inhibited in dermal neonatal and adult fibroblasts, the number of *de novo* hair follicles in the wound bed was significantly increased, supposedly because the inhibition of dermal β-catenin entailed an increase of papillary, hair-forming fibroblasts in the wound bed (Rognoni et al., 2016). Accordingly, increasing the density of papillary fibroblasts in adult skin prior to wounding resulted in the formation of new HFs within the scar tissue (Driskell et al., 2013). Moreover, epidermal Shh overexpression or constitutive Smoothened activation in fibroblasts (Lim et al., 2018) as well as overexpression of FGF9 in γδT-cells (Gay et al., 2013) lead to increased HF neogenesis in large wounds. Artificial manipulation of the above-mentioned signaling pathways therefore harbors novel therapeutic implications for the improvement of dermatological diseases such as alopecia or keloids that involve hair loss and scarring.

4. Epidermal-dermal crosstalk in the hair cycle

Murine skin comprises different types of HFs with varying length, thickness and hair shaft kinking: guard, auchene, awl and zigzag (Schlake, 2007). The formation of these different hair types is dictated by differential expression of genes and markers in DP cells: $CD133^+$ $Sox2^+$ DP cells expressing Wnt, FGF and BMP give rise to the more common guard and auchene hair shafts while $CD133^+$ $Sox2^-$ DP cells present with upregulated signaling activity of Hedgehog-, Notch- and Integrin-pathways, and induce formation of zigzag hair (Driskell, Giangreco, Jensen, Mulder, & Watt, 2009). Correspondingly, genetic ablation of Sox2 in the DP leads to a shift of awl and auchene hair shafts toward zigzag hairs

(Lesko, Driskell, Kretzschmar, Goldie, & Watt, 2013). Interestingly, expression analysis revealed that hair type-specific DP cells are highly similar and express only a small number of signature genes (Rezza et al., 2016). Additionally, overexpression of the BMP-antagonist Noggin in Keratin 5-expressing epidermal cells entails an increased proliferation in the hair bulb which causes the development of larger HFs and the replacement of zigzag and auchene hairs by awl-like hairs, while overexpression of BMP produced an opposite effect (Sharov et al., 2006). Moreover, the number of DP cells orchestrates the size and type of the produced hair and one single HF can produce different types of hair shafts in consecutive hair cycles (Chi et al., 2013). Nevertheless, Wnt/β-catenin stabilization in DP cells increased proliferation in the DP and thus its size but did not trigger HF type switching (Kaushal et al., 2015; Zhou, Xu, et al., 2016; Zhou, Yang, et al., 2016).

Adult HFs perpetually undergo different stages of growing (anagen), involution (catagen) and resting (telogen) during their hair cycle (Muller-Rover et al., 2001). Intravital imaging revealed that during the regression phase the majority of the epithelial cells are eliminated either by terminal differentiation of suprabasal cells or apoptosis of basal cells, thereby also reducing the SC pool (Mesa et al., 2015). Dying epithelial cells are phagocytosed by neighboring basal epithelial cells. Interestingly, apart from extrinsic TGFβ signaling, mesenchymal crosstalk is also essential for catagen induction as depletion of the DP prevents HFs from entering the hair cycle (Mesa et al., 2015). In mice, the first hair cycle is initiated at approximately 3 weeks of postembryonic development when morphogenesis is completed and HFs transition into catagen synchronously, followed by a short period of telogen and subsequent induction of anagen (Fuchs, 2007; Sundberg, Peters, & Paus, 2005). While the first hair cycle proceeds in a synchronized manner, it becomes continually desynchronized in adult mouse skin, leading to an alternation of anagen, catagen and telogen areas (Sundberg et al., 2005). Both epidermal and dermal signaling have been shown to control the hair cycle. Selective *in vivo* laser ablation of the dermal papilla (Rompolas et al., 2012) or inhibition of β-catenin signaling in Keratin14-expressing basal epidermal cells abrogated HF cycling (Huelsken, Vogel, Erdmann, Cotsarelis, & Birchmeier, 2001). Strikingly, induction of β-catenin in HF stem cells and their progeny was sufficient to initiate the recruitment of wild-type epithelial cells and HF growth when DP cells were ablated concomitantly (Deschene et al., 2014), which indicates a superior role of epidermal Wnt/β-catenin signaling over the DP. Transient epidermal activation of Wnt/β-catenin signaling entailed the induction of anagen, while sustained activation resulted in the formation of ectopic HFs and HF tumors such as trichofolliculomas and pilomatricomas (Lichtenberger et al., 2016; Lo Celso, Prowse, & Watt, 2004; Silva-Vargas et al., 2005). Upregulation of the BMP-antagonist Noggin and downregulation of BMP4 in the DP mediates transition from telogen to anagen, while converse regulation initiates telogen (Botchkarev et al., 2001). Interestingly, Noggin is a strong inductor of Shh (Botchkarev et al., 2001), which has been implicated in the induction of the growing phase of the HF as well (Chiang et al., 1999; Oro & Higgins, 2003). Additionally, transition of early to late telogen is mediated by FGF7/10 signaling from the DP to the hair germ (Greco et al., 2009). New evidence suggests that signaling from adipocytes (Festa et al., 2011) or regulatory T-cells (Ali et al., 2017) to follicular SCs may contribute to the initiation of the hair cycle as well.

Importantly, the cellular composition of the dermis changes throughout the HF stages. Reticular fibroblasts dominate the dermal compartment during telogen, while the dermis is predominantly populated by papillary fibroblasts during postembryonic completion of hair follicle morphogenesis and anagen (Lichtenberger et al., 2016), indicating a differential responsiveness of the fibroblast subpopulations to signals from epidermal SCs. In fact, sustained epidermal Wnt/β-catenin signaling leads to the expansion of both the upper and lower dermal compartments (Driskell et al., 2013; Lichtenberger et al., 2016) and to remodeling of the entire dermis (Collins et al., 2011). However, different epidermal effectors affect the distinct fibroblast lineages. For example, increased proliferation of papillary fibroblasts and subsequent ectopic HF induction was mediated by epidermal Shh upon sustained Wnt/β-catenin signaling, while ECM remodeling by reticular fibroblasts was conveyed by paracrine TGFβ signaling (Lichtenberger et al., 2016). Fibroblast-specific β-catenin stabilization results in an extension of the anagen stage in the mouse, while selective stabilization of β-catenin in reticular fibroblasts had no effect on HF cycling, indicating a role for the papillary lineage (likely the DP fibroblasts) in the regulation of the hair cycle (Mastrogiannaki et al., 2016). However, reticular fibroblasts have the ability to differentiate into adipocytes (Driskell et al., 2013; Mastrogiannaki et al., 2016), and stabilization of β-catenin in Platelet-derived Growth Factorα (PDGFRα)-positive fibroblasts as well as stabilization restricted to the $Dlk1^+$ reticular compartment lead to reduced adipogenesis and an increasingly fibrotic phenotype, possibly because of inhibited adipogenic differentiation of reticular fibroblasts (Mastrogiannaki et al., 2016). Importantly, both, dermal fibroblasts and adipocytes support epidermal integrity and functions (Driskell & Watt, 2015; Sorrell & Caplan, 2004). Until recently, adipocytes were considered as lipid reservoirs for energy storage, but adipocytes secrete a plethora of growth factors and cytokines known to control processes beyond metabolism. Their role in regulating hair follicle cycling in the skin is undisputed (Rivera-Gonzalez, Shook, & Horsley, 2014). Concomitantly with HF down-growth, the adipocyte layer within the dermis expands. Intradermal adipose tissue regeneration during hair follicle cycling occurs, at least in part, by hypertrophy of mature adipocytes but also because of regeneration through a proliferative adipocyte precursor cell during anagen induction (Festa et al., 2011). Studies in mice with mutations in fatty acid metabolism, lacking the transcription factor Ebf1 or with pharmacological PPARγ inhibition, which display defects both in dermal adipogenesis and hair growth (Chen, Smith, Tow, Elias, & Farese Jr., 2002; Festa et al., 2011; Herrmann et al., 2003; Jong et al., 1998; Karlsson, Bondjers, & Betsholtz, 1999; Stone et al., 2004; Weger & Schlake, 2005), have revealed a role for preadipocytes in HF cycling. Furthermore, both dermal preadipocytes and mature adipocytes have been shown to express BMPs and PDGF (Festa et al., 2011; Plikus et al., 2008; Rendl et al., 2005), which control hair cycling. *Vice versa*, epidermal cells have been shown to express adipogenic factors and to control adipocyte differentiation in a Wnt-dependent manner (Donati et al., 2014; Lichtenberger et al., 2016). Epidermal Wnt/β-catenin signaling seems to be a key player in regulating the hair cycle-synchronous oscillation of adipose layer thickness. Sustained epidermal Wnt/β-catenin signaling leads to the development of ectopic hair follicles, thickening of the adipose layer, increased number of adipocytes in the lower dermis and the appearance of adipocytes in the upper dermis, where they are usually absent (Donati et al., 2014). Importantly, the effects on the adipose tissue were visible in embryonic mouse skin before HF morphogenesis as well, thus indicating that Wnt/β-catenin signaling rather than mature hair follicles are required for adipocyte generation (Donati et al., 2014). Intriguingly, transit-amplifying cells (TACs) of the growing HF have been shown to orchestrate adipogenic differentiation of dermal preadipocytes via secretion of Shh, which directly affects preadipocytes by inducing their proliferation and expression of key adipogenic genes (Zhang et al., 2016). Thus, these TACs not only generate their own progeny but also a neighboring lineage, thereby assuring concomitant tissue synthesis across lineages. Plikus et al. demonstrate a necessity for HF-derived BMP-signaling for the reprogramming of α smooth muscle actin (αSMA)-expressing (myo)fibroblasts into adipocytes during skin development in mice and *in vitro* for the transformation of human keloid fibroblasts into adipocytes, suggesting that fibroblasts may be manipulated to prevent scarring *in vivo* (Plikus et al., 2017). Conversely, it has been suggested that the intradermal adipocyte lineage is necessary for follicular SC activation and the initiation of HF cycling, as anagen induction is inhibited upon genetic blockage of adipocyte precursor formation (Festa et al., 2011).

5. Interplay of the epidermis and the mesenchyme in wound regeneration

Tissue repair is critical for survival. The skin epidermis is particularly exposed to injuries, which necessitate rapid wound healing in order to prevent infection and to restore disrupted barrier function. This is a dynamic process of strong intercellular communication and a well-concerted action of distinct epidermal SC subsets together with other cell types, including fibroblasts and immune cells. A complex crosstalk ensures the activation, migration and plasticity of these cells during wound repair, and a coordinated termination of the healing process.

Mammalian wounds normally heal by fibrotic repair resulting in a scar without HF regeneration. Scarless wound healing only occurs during fetal development in mice and is restricted to the first 24 weeks of human gestation; only oral mucosa heals without scarring in adults (Leavitt et al., 2016; Longaker et al., 1990; Lorenz & Adzick, 1993; Lorenz, Lin, Longaker, Whitby, & Adzick, 1995; Rowlatt, 1979). Fibroblasts are pivotal in controlling complete tissue regeneration (Gurtner, Werner, Barrandon, & Longaker, 2008; Werner, Krieg, & Smola, 2007).

Cutaneous tissue regeneration after injury follows partly overlapping but consecutive steps: blood clot formation and immune cell recruitment, wound contraction, re-epithelialization and granulation tissue build-up, matrix construction and tissue remodeling (Clark, 1985; Greaves, Ashcroft, Baguneid, & Bayat, 2013; Gurtner et al., 2008; Singer & Clark, 1999). Fibroblasts and epidermal SCs play an essential role in synthesis, deposition and remodeling of the ECM and re-epithelialization, respectively. The preceding inflammatory phase is important in initiating these processes by releasing a wide range of cytokines and growth factors that further stimulate neovascularization and the formation of the granulation tissue (Singer & Clark, 1999). At first neutrophils, later macrophages, fibroblasts and endothelial cells jointly move into the wound space and transit the wound healing process from the inflammatory phase to the formation of granulation tissue (Clark, 1985). Proliferation and active migration of fibroblasts from upper and lower dermis reaches a maximum in the wound bed 7 days post injury. This is paralleled by an upregulation of alpha-smooth muscle actin (α-SMA) in fibroblasts associated with the wound, also referred to as myofibroblasts (Jiang et al., 2018; Rognoni et al., 2018). Although these so-called myofibroblasts are established as terminally differentiated cells, they seem to belong to the reticular fibroblast lineage and are thus capable of adipogenic differentiation (Plikus et al., 2017). Fibroblasts secrete MMPs and simultaneously synthesize fresh ECM compounds that gradually replace the provisional matrix. In early phases, fibroblasts produce type III collagen, which is majorly replaced by type I collagen in later stages of wound healing (Clark, 1985; Clore, Cohen, & Diegelmann, 1979). Collagen synthesis and Ki67-positive proliferating fibroblasts show an inverse correlation. Thus an ECM negative feedback loop stops fibroblast proliferation and restores homeostasis (Rognoni et al., 2018). Concomitantly with fibroblast activation and migration, dermal preadipocytes undergo adipogenic differentiation, and adiponectin- and perilipin-expressing adipocytes repopulate the wound area (Schmidt & Horsley, 2013). This repopulation of adipocytes within the wound bed is associated with proliferation of preadipocytes and can be blocked by pharmacological inhibition of PPARγ, indicating that adipogenesis occurs during wound repair (Rivera-Gonzalez et al., 2014; Schmidt & Horsley, 2013).

Re-epithelialization begins within few hours after wounding and is supported by recruitment of epidermal SCs from skin appendages (Clark, 1985; Singer & Clark, 1999). Keratinocyte-fibroblast crosstalk guides keratinocyte migration and proliferation by paracrine signaling and direct cell-cell contact. Stromal cells in the dermis secrete keratinocyte growth factor (KGF) within 24 h after injury, which signals via the keratinocyte growth factor receptor KGFR2IIIb, and is essential for proper re-epithelialization (Pierce et al., 1994; Werner et al., 1992, 1994). In turn, keratinocytes secrete IL1, IL6 and TNFα further stimulating KGF production by fibroblasts (Pierce et al., 1994; Tang & Gilchrest, 1996). Direct cell-cell contact compared to transwell-co-cultures of fibroblasts and keratinocytes showed enhanced migration and growth of keratinocytes, in line with an increase of HB-EGF, IL1α and TGFβ1 (Wang, Wang, Farhangfar, Zimmer, & Zhang, 2012). Keratin 14 (Krt14)-positive IFE basal cells are strongly recruited from the periphery to the center of the wound, showing long-term persistence and basal attachment, whereas only smaller clones from the

Involucrin (Inv)-positive suprabasal IFE cell pool were involved in wound repair that displayed reduced survival and had a minimal contribution to the basal layer (Mascre et al., 2012). Interestingly, it was recently shown that differentiated Gata6^{+} cells located in the sebaceous duct migrate into the IFE and dedifferentiate, thereby acquiring the ability to undergo long-term self-renewal and to differentiate into different epidermal lineages (Donati et al., 2017). Under homeostatic conditions epidermal SCs remain only in their specific regions within the HFs or IFE and are maintained separately by independent stem cell pools, but upon injury, signals from the damaged tissue unleash a plasticity within these SCs, which subsequently convert their behavior, and thus SC from both the HF and the IFE migrate into the damaged site and contribute to re-epithelialization and tissue repair (Blanpain & Fuchs, 2009; Fuchs, 2018; Ge & Fuchs, 2018; Gonzales & Fuchs, 2017; Hsu et al., 2014; Ito et al., 2005; Levy, Lindon, Zheng, Harfe, & Morgan, 2007; Mascre et al., 2012; Snippert et al., 2010). Thus, it is not surprising that mice lacking either HFs or their SCs display delayed wound healing (Langton, Herrick, & Headon, 2008). *In vivo* lineage tracing via the Sonic hedgehog (Shh) promoter in mice, in which all keratinocytes of the pilosebaceous unit are labeled, showed that these cells contribute to epithelialization of the wounds with long-term persistence in the IFE, but do not contribute to the IFE under homeostatic conditions (Levy et al., 2007). Lineage tracing under the keratin 1–15 promoter (Krt1–15) revealed that bulge cells migrate to areas of epithelial cell depletion and acquire an epidermal phenotype. Contribution to IFE after tape stripping or incisional wounds is limited to the superficial epidermis, whereas after excisional wounds, labeled cells were also found in the basal layers of the IFE. However, the percentage of labeled cells in the regenerated IFE of excisional wounds between post wounding day (PW) 8 and PW50 is reduced from around 26% to 2.1% (Ito et al., 2005). Further evidence for the contribution of follicular SCs to IFE regeneration comes from fate mapping experiments with Lgr6^{+} progenitor cells that mark a distinct cell cluster in the central isthmus of the HF. Lgr6^{+} cells are actively recruited during wound repair of full thickness wounds and contribute to the basal layer of the newly formed IFE for more than 3 months as well as to up to 10% of neogenetic HFs in the wound (Snippert et al., 2010). However, only few descendants originating from HF SCs persist in the re-epithelialized epidermis. *In vivo* fate mapping of Lrig1^{+} SCs of the junctional zone and Gli1^{+} epidermal cells located above the bulge showed that their progeny remains longer in the repaired epithelium (Brownell, Guevara, Bai, Loomis, & Joyner, 2011; Page, Lombard, Ng, Gottgens, & Jensen, 2013), but also these are eventually replaced, indicating that SCs of the IFE play a major role in wound re-epithelialization (Mascre et al., 2012).

Similar to the developing epidermis in the embryo, also the developing epithelium during wound repair is guided by reciprocal communication between epidermal cells and fibroblasts. Importantly, the first wave of dermal regeneration in full thickness wounds in adult mouse back skin is characterized by fibroblasts from the reticular fibroblast lineage producing large amounts of ECM and thus a scar, whereas papillary fibroblasts only appear when re-epithelialization is already completed (Driskell et al., 2013; Jiang et al., 2018; Fig. 3). Furthermore, with age, the percentage of papillary fibroblasts in the dermis declines (Gilchrest, 1982, 1996; Korosec et al., 2019; Lichtenberger et al., 2016). This is likely why adult wound repair does not include *de novo* HF morphogenesis. As mentioned above, increasing the number of papillary fibroblasts to a density similar to neonatal skin results in *de novo* HF formation in full thickness wounds (Driskell et al., 2013), confirming that signals from papillary fibroblasts are essential for HF formation. Transplantation of engrailed1-history-naïve fibroblasts (corresponding to the papillary lineage) did not induce *de novo* HF formation in wounds but reduced scarring (Jiang et al., 2018). Likewise, modulating canonical Wnt or Hh signaling in dermal fibroblasts directs the

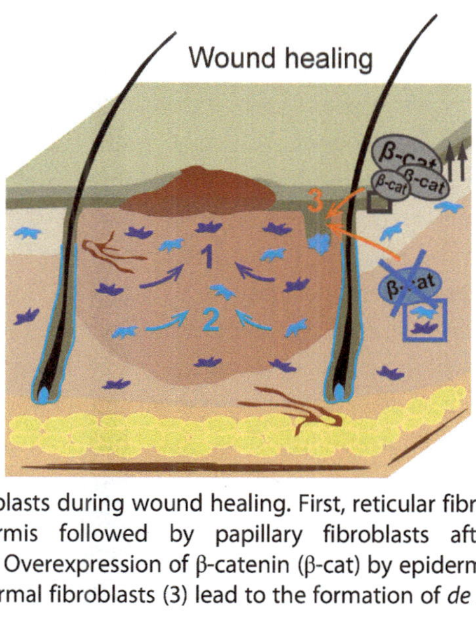

Fig. 3 Role of fibroblasts during wound healing. First, reticular fibroblasts (1) populate the wounded dermis followed by papillary fibroblasts after the wound is re-epithelialized (2). Overexpression of β-catenin (β-cat) by epidermal cells and/or inhibition of β-cat in dermal fibroblasts (3) lead to the formation of *de novo* hair follicles in wounds.

regenerating epithelium to form new hair (Lim et al., 2018; Rognoni et al., 2016), indicating that modulating dermal signaling may be a therapeutic approach to redirect wound healing toward more scarless skin regeneration.

Intriguingly, Shook et al. showed that during tissue repair, a larger number of reticular fibroblasts/preadipocytes express profibrotic cell surface proteins (CD26 and CD9), have reduced adipogenic potential, and are evenly distributed in the wound bed, while another fibroblast subset expressing CD29 at high level is primarily found in superficial areas of the wound (Shook et al., 2018). In aged mice, the abundance of preadipocytes decreased while CD29^{high} fibroblasts were more abundant, and the difference in gene expression between these subsets diminished. Furthermore, they showed that CD301b^{+} macrophages specifically promoted proliferation of the preadipocyte subset via PDGFC and IGF1 secretion, thereby promoting fibroblast heterogeneity within the wound and also its repair (Shook et al., 2018; Shook, Xiao, Kumamoto, Iwasaki, & Horsley, 2016). Although multiple factors likely influence fibroblast heterogeneity and plasticity during wound regeneration, targeting macrophage subsets might prove a valuable approach to improve wound repair or to treat keloid scars apart from directly modulating the abundance of fibroblast subsets or their signaling.

6. Dermal fibroblasts in the aging skin

As the outermost part of our body, the skin is subjected to intrinsic (genetic factors, somatic capacity) and extrinsic (life style, nutrition, environmental influences) factors that influence the aging process (Tigges et al., 2014). With age, skin undergoes a continuous degeneration of multiple structural components as well as in its function (Fig. 4). Interactions between the ECM and fibroblasts are essential for maintaining dermal integrity, and are subject to the aging process.

Intrinsic, chronological aging results in reduced epidermal and dermal thickness especially of the papillary dermis, reduced elasticity and loss of rete ridges, consequently leading to increased skin fragility, reduced skin turgor, wrinkling, declined barrier function and inefficient wound healing (Gilchrest, 1982, 1996; Harper & Grove, 1979; Varani et al., 2001). Furthermore, the function and growth of HFs is also compromised with age, and hair loss is associated with accelerated aging phenotypes in mouse models and chrono-aging in humans (Courtois, Loussouarn, Hourseau, & Grollier, 1995; Geyfman & Andersen, 2010; Harrison & Archer, 1988).

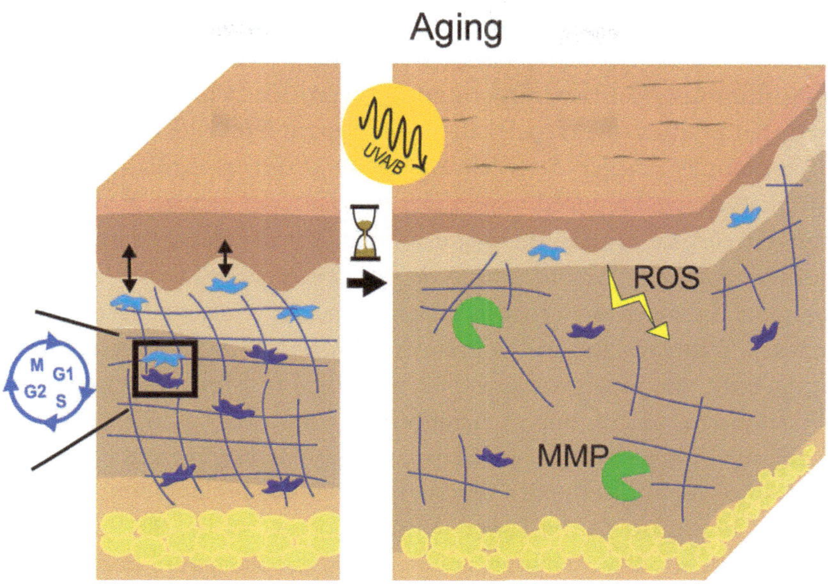

Fig. 4 Epidermal and dermal changes during chronological and/or photo-aging. On the left side a cross-section through young skin is shown, on the right side the aging processes are depicted. Chronologically aged skin is characterized by the reduction of skin elasticity, turgor and rete ridges leading to the formation of wrinkles. The number of fibroblasts and cell proliferation are reduced compared to young skin leading to a greater distance between fibroblasts. Furthermore, the collagen bundles in aged skin are disorganized and fragmented compared to young skin. Matrix metalloproteinases (MMP) are induced in chronologically aged skin leading to degradation of collagen fibers. UV exposure accelerates skin aging processes. Reactive oxygen species (ROS) induce stress and are also responsible for skin aging.

A pioneering study comparing skin histology of human autopsy specimen from infants to seniors (80 + years) reported an age-related dermal decrease in cellularity (Andrew, Behnke, & Sato, 1964). In particular, dermal fibroblast-like cell density accounts 7000 cells/mm^2 in fetuses at week 20–30 of pregnancy. Cell density drops to 4000 cells/mm^2 at week 30–40 of gestation until the end of the first decade of life and further reduces to 2000 cells/mm^2 from the second decade of life onward (Gunin, Kornilova, Vasilieva, & Petrov, 2011). Accordingly, in mice the dermal volume increases proportionally to fibroblast proliferation during prenatal development. However, postnatally, fibroblast proliferation declines and the dermis shows an inverse proportional relationship between Ki67-positive proliferating fibroblasts and collagen density (Rognoni et al., 2018). A recent study showed that fibroblasts in the aging dermis are lost as cell clusters and therefore the gaps between the remaining fibroblasts increase (Marsh, Gonzalez, Lathrop, Boucher, & Greco, 2018). During aging, fibroblasts lose their responsiveness to growth factors and hence show a reduced proliferative capacity in *in vitro* experiments (Plisko & Gilchrest, 1983; Stanulis-Praeger & Gilchrest, 1986). Primary cardiac ventricular fibroblasts showed reduced responsiveness to platelet derived growth factor (PDGF) receptor ligands, the major mitogen for fibroblasts, with no alterations in receptor number during *in vitro* senescence (Diez et al., 2001). Varani et al. have shown that in human skin, the contact surface between papillary fibroblasts and the ECM reduces with age (Varani et al., 2006). Intriguingly, the emerging gaps due to dermal cell loss are compensated by membrane protrusions of neighboring fibroblasts, whose cell bodies remain stable on their positions (Marsh et al., 2018). Human and mouse papillary fibroblasts in the upper dermis show higher proliferation rates than reticular fibroblasts in the lower dermis (Korosec et al., 2019; Rognoni et al., 2018). Short-term cultures of papillary fibroblasts from human mammary skin exhibit reduced growth rates when isolated from over 57-year-old individuals compared to 19–26 year olds, whereas growth rates of reticular fibroblasts show no significant change. Intriguingly, in skin reconstitution assays, only young papillary fibroblasts, but not reticular or old papillary fibroblasts had the potential to promote morphogenesis and differentiation of epidermal layers as well as rete ridges in reconstructed skin grafts *in vivo* (Mine, Fortunel, Pageon, & Asselineau, 2008). Accordingly, mouse fibroblasts displayed higher proliferation rates when transplanted onto decellularized dermis derived from neonatal compared to adult skin (Lichtenberger et al., 2016).

Apart from their higher proliferative and migratory activity that decreases with age, fibroblasts isolated from young and old individuals differ in their capability to synthesize and degrade the ECM (Schor, Schor, Rushton, & Smith, 1985; Sriram et al., 2015). ECM, mainly composed of collagen, is produced by fibroblasts (Hashimoto, 1978) and also regulates fibroblast function (Cole, Quan, Voorhees, & Fisher, 2018). Fibroblasts attach to collagen bundles via integrins, which leads to mechanical tensions that determine their cell shape and govern the balance between collagen synthesis and degradation. Dermal fibroblasts from younger (18–29 years) compared to over 80-year-old individuals show a spread cell shape with focal adhesions to collagen fibers as well as a higher proliferation rate and type I procollagen synthesis *in vitro* (Varani et al., 2006). In chronologically aged skin, collagen fibers are disorganized and fragmented in comparison to well-structured and tightly packed collagen bundles in young skin (Quan, Qin, Robichaud, Voorhees, & Fisher, 2011). Aged fibroblasts upregulate the expression of cysteine-rich 61 (CYR61/CCN1), which inhibits type I and III collagen synthesis and induces matrix metalloproteinase 1, 3 and 9 (MMP1/3/9) (Quan et al., 2006, 2011). Further key regulators of MMP1, α2β1 integrins and AP-1 were shown to be upregulated in aged fibroblasts, enforcing collagen degradation (Fisher et al., 2009). TGFβ stimulates fibroblast collagen synthesis and downregulates MMP1 expression. Elevated levels of CYR61 downregulate TGFβ type II receptor (TβRII) expression and interfere with TGFβ-dependent gene transcription (Quan et al., 2006).

Whole-transcriptome analysis of young and old murine skin dermal fibroblasts revealed increased expression of genes associated to pathways involving stress fiber formation, complement, innate immunity and *Staphylococcus aureus* infection in aged fibroblasts (Salzer et al., 2018). Moreover, increased fatty acid metabolism and brown fat cell differentiation signatures were found in aged dermal cells. On the other hand gene expression involved in collagen and glycosaminoglycan biosynthesis, protein folding and protein export were decreased in old fibroblasts. This change of gene expression pattern in the course of aging suggests that fibroblasts acquire pro-adipogenic traits instead of fibroblast features (Salzer et al., 2018). Transcriptomic profiling of human papillary and reticular fibroblasts from adult skin revealed alterations of the "tissue skeleton" gene network as a component of skin chrono-aging, especially in reticular fibroblasts, while analysis pinpointed to reduced contractility in papillary fibroblasts (Haydont, Neiveyans, Fortunel, et al., 2019; Haydont, Neiveyans, Zucchi, et al., 2019).

The most important extrinsic factor for skin aging as well as major risk factor for skin cancer is frequent and cumulative exposure to UV light. UV exposure expedites skin aging and can lead to extreme characteristics of aged skin. Photoaged skin shows areas of atrophy and hypertrophy in the epidermis, irregular hyperpigmentation, telangiectasia, deep wrinkles and laxity as well as sallowness and roughness (Gilchrest, 1996). Broken collagen fibers aggregate as twisted clumps leading to reduced mechanical tension and following reduced fibroblast collagen synthesis, similar to skin of over 80-year-olds (Fligiel et al., 2003). In sun-protected skin, fibroblasts show an extended cytoplasm adjacent to collagen bundles. Similar to young versus old skin, fibroblasts in photodamaged skin reveal a collapsed cytoplasm with lack of collagen fibers in close proximity (Cole et al., 2018). UV radiation impairs TGFβ/Smad signaling directly by downregulation of TβRII

(Quan, He, Kang, Voorhees, & Fisher, 2004) and further induces Smad7 upregulation, which interferes with TGFβ receptor Smads (Smad2, 3 and 4). Fibroblasts directly irradiated *in vitro* showed induced MMP1 and MMP3 expression, in a dose-, time- and wavelength-dependent manner, but marginally changes in expression of the tissue inhibitor of metalloproteinases TIMP1 (Brenneisen et al., 1996). In parallel to collagen breakdown, hyaluronan, also regulated by TGFβ1 and TβRII signaling, is decreased by UV-induced downregulation of hyaluronic acid synthases (HAS) 1, 2 and 3 in dermal fibroblasts (Dai et al., 2007). Reactive oxygen species (ROS), derived from mitochondria or NADP-oxidases, induces oxidative stress and is increased in aged dermal fibroblasts (Balaban, Nemoto, & Finkel, 2005; Kozieł, Greussing, Maier, Declercq, & Jansen-Dürr, 2011). Hydrogen peroxide increases MMP1 expression in human dermal fibroblasts *in vitro* (Brenneisen, Briviba, Wlaschek, Wenk, & Scharffetter-Kochanek, 1997). In chronologically as well as photoaged skin, oxidative stress may also be a result of attenuated anti-oxidative systems (Harman, 1981; Rhie et al., 2001).

Senescence exhibits characteristic cellular traits and can be induced by replicative exhaustion (Hayflick, 1980; Hayflick & Moorhead, 1961), oncogene overexpression (Guney & Sedivy, 2006; Serrano, Lin, McCurrach, Beach, & Lowe, 1997; Zhu, Woods, McMahon, & Bishop, 1998) or oxidative stress in cultured fibroblasts (Balin, Fisher, & Carter, 1984; Michiels, Toussaint, & Remacle, 1990). Most data about fibroblast senescence were generated in *in vitro* experiments. In culture, fibroblasts retain their replicative potential, but this does not correspond to donor age (Cristofalo, Allen, Pignolo, Martin, & Beck, 1998). Although fibroblasts in tissues proliferate only at low rates (Ruchti, Haller, Nuber, & Cottier, 1983), they also exhibit characteristic traits of cellular senescence *in vivo*: Increased expression of the cyclin-dependent kinase inhibitor protein $p16^{INK4a}$ in aged human dermis (Ressler et al., 2006) or increased markers for double strand breaks (DSB) (53BP1 and γ-H2AX) and colocalization to telomeres in fibroblasts of baboon skin biopsies (Herbig, Ferreira, Condel, Carey, & Sedivy, 2006).

With age, also the dermal adipose layer shrinks (Rivera-Gonzalez et al., 2014), indicating that adipocyte precursors may diminish or lose their activity. Since preadipocytes induce hair growth (Festa et al., 2011), the decline in adipogenesis may contribute to age-related hair loss. Another major alteration in aged skin is the immune cell density in the dermis with some subsets increased in number whereas other subsets decline (Giangreco, Qin, Pintar, & Watt, 2008; Khavkin & Ellis, 2011;

Zouboulis & Makrantonaki, 2011). These alterations may not only alter HF SC functions (Doles, Storer, Cozzuto, Roma, & Keyes, 2012) but may also contribute to the decline of dermal adipocytes (Kanneganti & Dixit, 2012; Schipper, Prakken, Kalkhoven, & Boes, 2012).

7. Concluding remarks

Skin homeostasis and repair is governed by epidermal SCs. These reside in special microenvironments, their niches, within the IFE and hair follicles, which influence SC fate decisions. Stress or signals from wounded tissue can redirect SCs of the different pools to change their normal behavior. This SC plasticity normally resolves as injuries heal and the tissue returns to normal homeostasis. In all these processes, reciprocal signaling between epithelial SCs and their microenvironment are crucial. Recent single cell transcriptomic analyses suggest that different epidermal SC pools share only a small number of SC signature genes including keratins and integrins allowing them to reside on the basement membrane. Corresponding to this model the local environment molds SC behavior, and thus SC plasticity during wound healing is a response to the altered microenvironment. Nevertheless, some findings are inconsistent with this notion such as the fact that each of these SCs has their site-specific gene signature.

Fibroblasts have been shown to be key players in the reciprocal niche signaling during epidermal development and injury-induced stress responses, as well as in molding the homeostatic niche. Like epidermal SCs, also fibroblasts display an immense heterogeneity and cell plasticity that seems to be controlled both cell intrinsically and by its neighboring cells and ECM. Various studies point toward a so-called cell memory within fibroblasts. When placed into new environments they seem to keep some of their properties related to their tissue or site of origin (Rinn et al., 2006). For example, Sox2-positive DP cells maintain their identity *in vitro* (Driskell et al., 2009), and in transplantation assays fibroblasts from the oral cavity or the skin maintained their specific site identity in migratory behavior and ECM synthesis (Rinkevich et al., 2015). Otherwise, the proliferative potential of fibroblasts changed when placed into ECM derived from tissue of old or young mice (Lichtenberger et al., 2016), and the local microenvironment affects the expression and function of human papillary and reticular fibroblasts as well (Korosec et al., 2019).

We are just starting to understand the heterogeneity and plasticity of SCs and cells in their niche. The discoveries yet to come hold great promise for regenerative medicine.

References

Alcolea, M. P., & Jones, P. H. (2014). Lineage analysis of epidermal stem cells. *Cold Spring Harbor Perspectives in Medicine*, *4*, a015206.

Ali, N., Zirak, B., Rodriguez, R. S., Pauli, M. L., Truong, H. A., Lai, K., et al. (2017). Regulatory T cells in skin facilitate epithelial stem cell differentiation. *Cell*, *169*, 1119-1129. e1111.

Alonso, L., & Fuchs, E. (2006). The hair cycle. *Journal of Cell Science*, *119*, 391–393.

Andl, T., Reddy, S. T., Gaddapara, T., & Millar, S. E. (2002). WNT signals are required for the initiation of hair follicle development. *Developmental Cell*, *2*, 643–653.

Andrew, W., Behnke, R. H., & Sato, T. (1964). Changes with advancing age in the cell population of human dermis. *Gerontologia*, *10*, 1–19.

Balaban, R. S., Nemoto, S., & Finkel, T. (2005). Mitochondria, oxidants, and aging. *Cell*, *120*, 483–495.

Balin, A. K., Fisher, A. J., & Carter, D. M. (1984). Oxygen modulates growth of human cells at physiologic partial pressures. *The Journal of Experimental Medicine*, *160*, 152–166.

Biggs, L. C., Makela, O. J., Myllymaki, S. M., Das Roy, R., Narhi, K., Pispa, J., et al. (2018). Hair follicle dermal condensation forms via Fgf20 primed cell cycle exit, cell motility, and aggregation. *eLife* 7. pii: e36468. https://doi.org/10.7554/eLife.36468.

Blanpain, C., & Fuchs, E. (2006). Epidermal stem cells of the skin. *Annual Review of Cell and Developmental Biology*, *22*, 339–373.

Blanpain, C., & Fuchs, E. (2009). Epidermal homeostasis: A balancing act of stem cells in the skin. *Nature Reviews. Molecular Cell Biology*, *10*, 207–217.

Botchkarev, V. A., Botchkareva, N. V., Nakamura, M., Huber, O., Funa, K., Lauster, R., et al. (2001). Noggin is required for induction of the hair follicle growth phase in postnatal skin. *FASEB Journal: Official Publication of the Federation of American Societies for Experimental Biology*, *15*, 2205–2214.

Brenneisen, P., Briviba, K., Wlaschek, M., Wenk, J., & Scharffetter-Kochanek, K. (1997). Hydrogen peroxide (H2O2) increases the steady-state mRNA levels of collagenase/ MMP-1 in human dermal fibroblasts. *Free Radical Biology & Medicine*, *22*, 515–524.

Brenneisen, P., Oh, J., Wlaschek, M., Wenk, J., Briviba, K., Hommel, C., et al. (1996). Ultraviolet B wavelength dependence for the regulation of two major matrix-metalloproteinases and their inhibitor TIMP-1 in human dermal fibroblasts. *Photochemistry and Photobiology*, *64*, 649–657.

Brownell, I., Guevara, E., Bai, C. B., Loomis, C. A., & Joyner, A. L. (2011). Nerve-derived sonic hedgehog defines a niche for hair follicle stem cells capable of becoming epidermal stem cells. *Cell Stem Cell*, *8*, 552–565.

Caplan, A. I. (2015). Adult mesenchymal stem cells: When, where, and how. *Stem Cells International*, *2015*, 628767.

Chang, H. Y., Chi, J. T., Dudoit, S., Bondre, C., van de Rijn, M., Botstein, D., et al. (2002). Diversity, topographic differentiation, and positional memory in human fibroblasts. *Proceedings of the National Academy of Sciences of the United States of America*, *99*, 12877–12882.

Chen, D., Jarrell, A., Guo, C., Lang, R., & Atit, R. (2012). Dermal beta-catenin activity in response to epidermal Wnt ligands is required for fibroblast proliferation and hair follicle initiation. *Development (Cambridge, England)*, *139*, 1522–1533.

Chen, H. C., Smith, S. J., Tow, B., Elias, P. M., & Farese, R. V., Jr. (2002). Leptin modulates the effects of acyl CoA: Diacylglycerol acyltransferase deficiency on murine fur and sebaceous glands. *The Journal of Clinical Investigation*, *109*, 175–181.

Chen, F. G., Zhang, W. J., Bi, D., Liu, W., Wei, X., Chen, F. F., et al. (2007). Clonal analysis of nestin(-) vimentin(+) multipotent fibroblasts isolated from human dermis. *Journal of Cell Science*, *120*, 2875–2883.

Cheng, J. B., Sedgewick, A. J., Finnegan, A. I., Harirchian, P., Lee, J., Kwon, S., et al. (2018). Transcriptional programming of normal and inflamed human epidermis at single-cell resolution. *Cell Reports, 25*, 871–883.

Chi, W., Wu, E., & Morgan, B. A. (2013). Dermal papilla cell number specifies hair size, shape and cycling and its reduction causes follicular decline. *Development (Cambridge, England), 140*, 1676–1683.

Chiang, C., Swan, R. Z., Grachtchouk, M., Bolinger, M., Litingtung, Y., Robertson, E. K., et al. (1999). Essential role for Sonic hedgehog during hair follicle morphogenesis. *Developmental Biology, 205*, 1–9.

Clark, R. A. F. (1985). Cutaneous tissue repair: Basic biologic considerations. I. *Journal of the American Academy of Dermatology, 13*, 701–725.

Clevers, H. (2015). STEM CELLS. What is an adult stem cell? *Science (New York, N.Y.), 350*, 1319–1320.

Clore, J. N., Cohen, I. K., & Diegelmann, R. F. (1979). Quantitation of collagen types I and III during wound healing in rat skin. *Proceedings of the Society for Experimental Biology and Medicine, 161*, 337–340.

Cole, M. A., Quan, T., Voorhees, J. J., & Fisher, G. J. (2018). Extracellular matrix regulation of fibroblast function: Redefining our perspective on skin aging. *Journal of Cell Communication and Signaling, 12*, 35–43.

Collins, C. A., Jensen, K. B., MacRae, E. J., Mansfield, W., & Watt, F. M. (2012). Polyclonal origin and hair induction ability of dermal papillae in neonatal and adult mouse back skin. *Developmental Biology, 366*, 290–297.

Collins, C. A., Kretzschmar, K., & Watt, F. M. (2011). Reprogramming adult dermis to a neonatal state through epidermal activation of beta-catenin. *Development (Cambridge, England), 138*, 5189–5199.

Courtois, M., Loussouarn, G., Hourseau, C., & Grollier, J. F. (1995). Ageing and hair cycles. *The British Journal of Dermatology, 132*, 86–93.

Cristofalo, V. J., Allen, R. G., Pignolo, R. J., Martin, B. G., & Beck, J. C. (1998). Relationship between donor age and the replicative lifespan of human cells in culture: A reevaluation. *Proceedings of the National Academy of Sciences of the United States of America, 95*, 10614–10619.

Dai, G., Freudenberger, T., Zipper, P., Melchior, A., Grether-Beck, S., Rabausch, B., et al. (2007). Chronic ultraviolet B irradiation causes loss of hyaluronic acid from mouse dermis because of down-regulation of hyaluronic acid synthases. *The American Journal of Pathology, 171*, 1451–1461.

Deschene, E. R., Myung, P., Rompolas, P., Zito, G., Sun, T. Y., Taketo, M. M., et al. (2014). beta-Catenin activation regulates tissue growth non-cell autonomously in the hair stem cell niche. *Science (New York, N.Y.), 343*, 1353–1356.

Diez, C., Nestler, M., Friedrich, U., Vieth, M., Stolte, M., Hu, K., et al. (2001). Down-regulation of Akt/PKB in senescent cardiac fibroblasts impairs PDGF-induced cell proliferation. *Cardiovascular Research, 49*, 731–740.

Doles, J., Storer, M., Cozzuto, L., Roma, G., & Keyes, W. M. (2012). Age-associated inflammation inhibits epidermal stem cell function. *Genes & Development, 26*, 2144–2153.

Donati, G., Proserpio, V., Lichtenberger, B. M., Natsuga, K., Sinclair, R., Fujiwara, H., et al. (2014). Epidermal Wnt/beta-catenin signaling regulates adipocyte differentiation via secretion of adipogenic factors. *Proceedings of the National Academy of Sciences of the United States of America, 111*, E1501–E1509.

Donati, G., Rognoni, E., Hiratsuka, T., Liakath-Ali, K., Hoste, E., Kar, G., et al. (2017). Wounding induces dedifferentiation of epidermal Gata6(+) cells and acquisition of stem cell properties. *Nature Cell Biology, 19*, 603–613.

Driskell, R. R., Giangreco, A., Jensen, K. B., Mulder, K. W., & Watt, F. M. (2009). Sox2-positive dermal papilla cells specify hair follicle type in mammalian epidermis. *Development (Cambridge, England)*, *136*, 2815–2823.
Driskell, R. R., Jahoda, C. A., Chuong, C. M., Watt, F. M., & Horsley, V. (2014). Defining dermal adipose tissue. *Experimental Dermatology*, *23*, 629–631.
Driskell, R. R., Lichtenberger, B. M., Hoste, E., Kretzschmar, K., Simons, B. D., Charalambous, M., et al. (2013). Distinct fibroblast lineages determine dermal architecture in skin development and repair. *Nature*, *504*, 277–281.
Driskell, R. R., & Watt, F. M. (2015). Understanding fibroblast heterogeneity in the skin. *Trends in Cell Biology*, *25*, 92–99.
Festa, E., Fretz, J., Berry, R., Schmidt, B., Rodeheffer, M., Horowitz, M., et al. (2011). Adipocyte lineage cells contribute to the skin stem cell niche to drive hair cycling. *Cell*, *146*, 761–771.
Fisher, G. J., Quan, T., Purohit, T., Shao, Y., Cho, M. K., He, T., et al. (2009). Collagen fragmentation promotes oxidative stress and elevates matrix metalloproteinase-1 in fibroblasts in aged human skin. *The American Journal of Pathology*, *174*, 101–114.
Fligiel, S. E., Varani, J., Datta, S. C., Kang, S., Fisher, G. J., & Voorhees, J. J. (2003). Collagen degradation in aged/photodamaged skin in vivo and after exposure to matrix metalloproteinase-1 in vitro. *The Journal of Investigative Dermatology*, *120*, 842–848.
Fuchs, E. (2007). Scratching the surface of skin development. *Nature*, *445*, 834–842.
Fuchs, E. (2008). Skin stem cells: Rising to the surface. *The Journal of Cell Biology*, *180*, 273–284.
Fuchs, E. (2016). Epithelial skin biology: Three decades of developmental biology, a hundred questions answered and a thousand new ones to address. *Current Topics in Developmental Biology*, *116*, 357–374.
Fuchs, E. (2018). Skin stem cells in silence, action, and cancer. *Stem Cell Reports*, *10*, 1432–1438.
Gao, H., Li, D., Yang, P., Zhao, L., Wei, L., Chen, Y., et al. (2017). Suppression of CD36 attenuates adipogenesis with a reduction of P2X7 expression in 3T3-L1 cells. *Biochemical and Biophysical Research Communications*, *491*, 204–208.
Gao, H., Volat, F., Sandhow, L., Galitzky, J., Nguyen, T., Esteve, D., et al. (2017). CD36 Is a marker of human adipocyte progenitors with pronounced adipogenic and triglyceride accumulation potential. *Stem cells (Dayton, Ohio)*, *35*, 1799–1814.
Gay, D., Kwon, O., Zhang, Z., Spata, M., Plikus, M. V., Holler, P. D., et al. (2013). Fgf9 from dermal gammadelta T cells induces hair follicle neogenesis after wounding. *Nature Medicine*, *19*, 916–923.
Ge, Y., & Fuchs, E. (2018). Stretching the limits: From homeostasis to stem cell plasticity in wound healing and cancer. *Nature Reviews. Genetics*, *19*, 311–325.
Geyfman, M., & Andersen, B. (2010). Clock genes, hair growth and aging. *Aging (Albany NY)*, *2*, 122–128.
Giangreco, A., Qin, M., Pintar, J. E., & Watt, F. M. (2008). Epidermal stem cells are retained in vivo throughout skin aging. *Aging Cell*, *7*, 250–259.
Gilchrest, B. A. (1982). Age-associated changes in the skin. *Journal of the American Geriatrics Society*, *30*, 139–143.
Gilchrest, B. A. (1996). A review of skin ageing and its medical therapy. *The British Journal of Dermatology*, *135*, 867–875.
Glover, J. D., Wells, K. L., Matthaus, F., Painter, K. J., Ho, W., Riddell, J., et al. (2017). Hierarchical patterning modes orchestrate hair follicle morphogenesis. *PLoS Biology*, *15*, e2002117.
Gonzales, K. A. U., & Fuchs, E. (2017). Skin and its regenerative powers: An alliance between stem cells and their niche. *Developmental Cell*, *43*, 387–401.

Greaves, N. S., Ashcroft, K. J., Baguneid, M., & Bayat, A. (2013). Current understanding of molecular and cellular mechanisms in fibroplasia and angiogenesis during acute wound healing. *Journal of Dermatological Science*, *72*, 206–217.

Greco, V., Chen, T., Rendl, M., Schober, M., Pasolli, H. A., Stokes, N., et al. (2009). A two-step mechanism for stem cell activation during hair regeneration. *Cell Stem Cell*, *4*, 155–169.

Grisanti, L., Clavel, C., Cai, X., Rezza, A., Tsai, S. Y., Sennett, R., et al. (2013). Tbx18 targets dermal condensates for labeling, isolation, and gene ablation during embryonic hair follicle formation. *The Journal of Investigative Dermatology*, *133*, 344–353.

Guney, I., & Sedivy, J. M. (2006). Cellular senescence, epigenetic switches and c-Myc. *Cell Cycle (Georgetown, Texas)*, *5*, 2319–2323.

Gunin, A. G., Kornilova, N. K., Vasilieva, O. V., & Petrov, V. V. (2011). Age-related changes in proliferation, the numbers of mast cells, eosinophils, and cd45-positive cells in human dermis. *The Journals of Gerontology. Series A, Biological Sciences and Medical Sciences*, *66*, 385–392.

Gupta, K., Levinsohn, J., Linderman, G., Chen, D., Sun, T. Y., Dong, D., et al. (2019). Single-cell analysis reveals a hair follicle dermal niche molecular differentiation trajectory that begins prior to morphogenesis. *Developmental Cell*, *48*. 17-31.e16.

Gurtner, G. C., Werner, S., Barrandon, Y., & Longaker, M. T. (2008). Wound repair and regeneration. *Nature*, *453*, 314–321.

Halfon, S., Abramov, N., Grinblat, B., & Ginis, I. (2011). Markers distinguishing mesenchymal stem cells from fibroblasts are downregulated with passaging. *Stem Cells and Development*, *20*, 53–66.

Harman, D. (1981). The aging process. *Proceedings of the National Academy of Sciences of the United States of America*, *78*, 7124–7128.

Harper, R. A., & Grove, G. (1979). Human skin fibroblasts derived from papillary and reticular dermis: Differences in growth potential in vitro. *Science (New York, N.Y.)*, *204*, 526–527.

Harrison, D. E., & Archer, J. R. (1988). Biomarkers of aging: Tissue markers. Future research needs, strategies, directions and priorities. *Experimental Gerontology*, *23*, 309–325.

Hashimoto, K. (1978). Normal and abnormal connective tissue of the human skin I. Fibroblast and collagen. *International Journal of Dermatology*, *17*, 459–471.

Haydont, V., Neiveyans, V., Fortunel, N. O., & Asselineau, D. (2019). Transcriptome profiling of human papillary and reticular fibroblasts from adult interfollicular dermis pinpoints the 'tissue skeleton' gene network as a component of skin chrono-ageing. *Mechanisms of Ageing and Development*, *179*, 60–77.

Haydont, V., Neiveyans, V., Zucchi, H., Fortunel, N. O., & Asselineau, D. (2019). Genome-wide profiling of adult human papillary and reticular fibroblasts identifies ACAN, Col XI alpha1, and PSG1 as general biomarkers of dermis ageing, and KANK4 as an exemplary effector of papillary fibroblast ageing, related to contractility. *Mechanisms of Ageing and Development*, *177*, 157–181.

Hayflick, L. (1980). Recent advances in the cell biology of aging. *Mechanisms of Ageing and Development*, *14*, 59–79.

Hayflick, L., & Moorhead, P. S. (1961). The serial cultivation of human diploid cell strains. *Experimental Cell Research*, *25*, 585–621.

Herbig, U., Ferreira, M., Condel, L., Carey, D., & Sedivy, J. M. (2006). Cellular senescence in aging primates. *Science (New York, N.Y.)*, *311*, 1257.

Herrmann, T., van der Hoeven, F., Grone, H. J., Stewart, A. F., Langbein, L., Kaiser, I., et al. (2003). Mice with targeted disruption of the fatty acid transport protein 4 (Fatp 4, Slc27a4) gene show features of lethal restrictive dermopathy. *The Journal of Cell Biology*, *161*, 1105–1115.

Higgins, C. A., Chen, J. C., Cerise, J. E., Jahoda, C. A., & Christiano, A. M. (2013). Microenvironmental reprogramming by three-dimensional culture enables dermal papilla cells to induce de novo human hair-follicle growth. *Proceedings of the National Academy of Sciences of the United States of America, 110*, 19679–19688.

Higgins, C. A., Roger, M. F., Hill, R. P., Ali-Khan, A. S., Garlick, J. A., Christiano, A. M., et al. (2017). Multifaceted role of hair follicle dermal cells in bioengineered skins. *The British Journal of Dermatology, 176*, 1259–1269.

Hiraoka, C., Toki, F., Shiraishi, K., Sayama, K., Nishimura, E. K., Miura, H., et al. (2016). Two clonal types of human skin fibroblasts with different potentials for proliferation and tissue remodeling ability. *Journal of Dermatological Science, 82*, 84–94.

Hsu, Y. C., Li, L., & Fuchs, E. (2014). Emerging interactions between skin stem cells and their niches. *Nature Medicine, 20*, 847–856.

Huelsken, J., Vogel, R., Erdmann, B., Cotsarelis, G., & Birchmeier, W. (2001). beta-catenin controls hair follicle morphogenesis and stem cell differentiation in the skin. *Cell, 105*, 533–545.

Ito, M., Liu, Y., Yang, Z., Nguyen, J., Liang, F., Morris, R. J., et al. (2005). Stem cells in the hair follicle bulge contribute to wound repair but not to homeostasis of the epidermis. *Nature Medicine, 11*, 1351–1354.

Ito, M., Yang, Z., Andl, T., Cui, C., Kim, N., Millar, S. E., et al. (2007). Wnt-dependent de novo hair follicle regeneration in adult mouse skin after wounding. *Nature, 447*, 316–320.

Janson, D. G., Saintigny, G., van Adrichem, A., Mahe, C., & El Ghalbzouri, A. (2012). Different gene expression patterns in human papillary and reticular fibroblasts. *The Journal of Investigative Dermatology, 132*, 2565–2572.

Jiang, D., Correa-Gallegos, D., Christ, S., Stefanska, A., Liu, J., Ramesh, P., et al. (2018). Two succeeding fibroblastic lineages drive dermal development and the transition from regeneration to scarring. *Nature Cell Biology, 20*, 422–431.

Jiang, D., & Rinkevich, Y. (2018). Defining skin fibroblastic cell types beyond CD90. *Frontiers in Cell and Development Biology, 6*, 133.

Jong, M. C., Gijbels, M. J., Dahlmans, V. E., Gorp, P. J., Koopman, S. J., Ponec, M., et al. (1998). Hyperlipidemia and cutaneous abnormalities in transgenic mice overexpressing human apolipoprotein C1. *The Journal of Clinical Investigation, 101*, 145–152.

Joost, S., Zeisel, A., Jacob, T., Sun, X., La Manno, G., Lonnerberg, P., et al. (2016). Single-cell transcriptomics reveals that differentiation and spatial signatures shape epidermal and hair follicle heterogeneity. *Cell Systems, 3*. 221-237.e229.

Kalluri, R., & Zeisberg, M. (2006). Fibroblasts in cancer. *Nature Reviews. Cancer, 6*, 392–401.

Kanneganti, T. D., & Dixit, V. D. (2012). Immunological complications of obesity. *Nature Immunology, 13*, 707–712.

Karlsson, L., Bondjers, C., & Betsholtz, C. (1999). Roles for PDGF-A and sonic hedgehog in development of mesenchymal components of the hair follicle. *Development (Cambridge, England), 126*, 2611–2621.

Kaushal, G. S., Rognoni, E., Lichtenberger, B. M., Driskell, R. R., Kretzschmar, K., Hoste, E., et al. (2015). Fate of prominin-1 expressing dermal papilla cells during homeostasis, wound healing and Wnt activation. *The Journal of Investigative Dermatology, 135*, 2926–2934.

Khavkin, J., & Ellis, D. A. (2011). Aging skin: Histology, physiology, and pathology. *Facial Plastic Surgery Clinics of North America, 19*, 229–234.

Kligman, A. M. (1959). Neogenesis of human hair follicles. *Annals of the New York Academy of Sciences, 83*, 507–511.

Korosec, A., Frech, S., Gesslbauer, B., Vierhapper, M., Radtke, C., Petzelbauer, P., et al. (2019). Lineage identity and location within the dermis determine the function of papillary and reticular fibroblasts in human skin. *The Journal of Investigative Dermatology, 139*, 342–351.

Korosec, A., & Lichtenberger, B. M. (2017). In vitro models to study hair follicle generation. In A. P. Marques, R. P. Pirraco, M. Cerqueira, & R. L. Reis (Eds.), *Skin tissue models* (pp. 279–301). Elsevier.

Kozieł, R., Greussing, R., Maier, A. B., Declercq, L., & Jansen-Dürr, P. (2011). Functional interplay between mitochondrial and proteasome activity in skin aging. *Journal of Investigative Dermatology, 131*, 594–603.

Kretzschmar, K., & Watt, F. M. (2014). Markers of epidermal stem cell subpopulations in adult mammalian skin. *Cold Spring Harbor Perspectives in Medicine, 4*, a01363.

Kulukian, A., & Fuchs, E. (2013). Spindle orientation and epidermal morphogenesis. *Philosophical Transactions of the Royal Society of London. Series B, Biological Sciences, 368*, 20130016.

Langton, A. K., Herrick, S. E., & Headon, D. J. (2008). An extended epidermal response heals cutaneous wounds in the absence of a hair follicle stem cell contribution. *The Journal of Investigative Dermatology, 128*, 1311–1318.

Leavitt, T., Hu, M. S., Marshall, C. D., Barnes, L. A., Lorenz, H. P., & Longaker, M. T. (2016). Scarless wound healing: Finding the right cells and signals. *Cell and Tissue Research, 365*, 483–493.

Lee, D. Y., & Cho, K. H. (2005). The effects of epidermal keratinocytes and dermal fibroblasts on the formation of cutaneous basement membrane in three-dimensional culture systems. *Archives of Dermatological Research, 296*, 296–302.

Lee, S. H., Seo, S. H., Lee, D. H., Pi, L. Q., Lee, W. S., & Choi, K. Y. (2017). Targeting of CXXC5 by a competing peptide stimulates hair regrowth and wound-induced hair neogenesis. *The Journal of Investigative Dermatology, 137*, 2260–2269.

Lesko, M. H., Driskell, R. R., Kretzschmar, K., Goldie, S. J., & Watt, F. M. (2013). Sox2 modulates the function of two distinct cell lineages in mouse skin. *Developmental Biology, 382*, 15–26.

Levy, V., Lindon, C., Zheng, Y., Harfe, B. D., & Morgan, B. A. (2007). Epidermal stem cells arise from the hair follicle after wounding. *FASEB Journal: Official Publication of the Federation of American Societies for Experimental Biology, 21*, 1358–1366.

Lichtenberger, B. M., Mastrogiannaki, M., & Watt, F. M. (2016). Epidermal beta-catenin activation remodels the dermis via paracrine signalling to distinct fibroblast lineages. *Nature Communications*, 7, 10537.

Lim, C. H., Sun, Q., Ratti, K., Lee, S. H., Zheng, Y., Takeo, M., et al. (2018). Hedgehog stimulates hair follicle neogenesis by creating inductive dermis during murine skin wound healing. *Nature Communications, 9*, 4903.

Lo Celso, C., Prowse, D. M., & Watt, F. M. (2004). Transient activation of beta-catenin signalling in adult mouse epidermis is sufficient to induce new hair follicles but continuous activation is required to maintain hair follicle tumours. *Development (Cambridge, England), 131*, 1787–1799.

Longaker, M. T., Whitby, D. J., Adzick, N. S., Crombleholme, T. M., Langer, J. C., Duncan, B. W., et al. (1990). Studies in fetal wound healing, VI. Second and early third trimester fetal wounds demonstrate rapid collagen deposition without scar formation. *Journal of Pediatric Surgery, 25*, 63–68. discussion 68–69.

Lorenz, H. P., & Adzick, N. S. (1993). Scarless skin wound repair in the fetus. *The Western Journal of Medicine, 159*, 350–355.

Lorenz, H. P., Lin, R. Y., Longaker, M. T., Whitby, D. J., & Adzick, N. S. (1995). The fetal fibroblast: The effector cell of scarless fetal skin repair. *Plastic and Reconstructive Surgery, 96*, 1251–1259. discussion 1260–1251.

Lu, C. P., Polak, L., Keyes, B. E., & Fuchs, E. (2016). Spatiotemporal antagonism in mesenchymal-epithelial signaling in sweat versus hair fate decision. *Science (New York, N.Y.), 354*, aah6102-1–aah6102-12.

Marsh, E., Gonzalez, D. G., Lathrop, E. A., Boucher, J., & Greco, V. (2018). Positional stability and membrane occupancy define skin fibroblast homeostasis in vivo. *Cell, 175*(6), 1620.e13–1633.e13.

Mascre, G., Dekoninck, S., Drogat, B., Youssef, K. K., Brohee, S., Sotiropoulou, P. A., et al. (2012). Distinct contribution of stem and progenitor cells to epidermal maintenance. *Nature, 489*, 257–262.

Mastrogiannaki, M., Lichtenberger, B. M., Reimer, A., Collins, C. A., Driskell, R. R., & Watt, F. M. (2016). beta-catenin stabilization in skin fibroblasts causes fibrotic lesions by preventing adipocyte differentiation of the reticular dermis. *The Journal of Investigative Dermatology, 136*, 1130–1142.

Mesa, K. R., Rompolas, P., Zito, G., Myung, P., Sun, T. Y., Brown, S., et al. (2015). Niche-induced cell death and epithelial phagocytosis regulate hair follicle stem cell pool. *Nature, 522*, 94–97.

Michiels, C., Toussaint, O., & Remacle, J. (1990). Comparative study of oxygen toxicity in human fibroblasts and endothelial cells. *Journal of Cellular Physiology, 144*, 295–302.

Mills, K. M., Szczerkowski, J. L. A., & Habib, S. J. (2017). Wnt ligand presentation and reception: From the stem cell niche to tissue engineering. *Open Biology, 7*, 170140.

Mine, S., Fortunel, N. O., Pageon, H., & Asselineau, D. (2008). Aging alters functionally human dermal papillary fibroblasts but not reticular fibroblasts: A new view of skin morphogenesis and aging. *PLoS One, 3*, e4066.

Mok, K. W., Saxena, N., Heitman, N., Grisanti, L., Srivastava, D., Muraro, M. J., et al. (2019). Dermal condensate niche fate specification occurs prior to formation and is placode progenitor dependent. *Developmental Cell, 48*, 32-48.e35.

Muller-Rover, S., Handjiski, B., van der Veen, C., Eichmuller, S., Foitzik, K., McKay, I. A., et al. (2001). A comprehensive guide for the accurate classification of murine hair follicles in distinct hair cycle stages. *The Journal of Investigative Dermatology, 117*, 3–15.

Nauroy, P., Barruche, V., Marchand, L., Nindorera-Badara, S., Bordes, S., Closs, B., et al. (2017). Human dermal fibroblast subpopulations display distinct gene signatures related to cell behaviors and matrisome. *The Journal of Investigative Dermatology, 137*, 1787–1789.

Niemann, C., & Horsley, V. (2012). Development and homeostasis of the sebaceous gland. *Seminars in Cell & Developmental Biology, 23*, 928–936.

Oro, A. E., & Higgins, K. (2003). Hair cycle regulation of Hedgehog signal reception. *Developmental Biology, 255*, 238–248.

Page, M. E., Lombard, P., Ng, F., Gottgens, B., & Jensen, K. B. (2013). The epidermis comprises autonomous compartments maintained by distinct stem cell populations. *Cell Stem Cell, 13*, 471–482.

Petersson, M., & Niemann, C. (2012). Stem cell dynamics and heterogeneity: Implications for epidermal regeneration and skin cancer. *Current Medicinal Chemistry, 19*, 5984–5992.

Philippeos, C., Telerman, S. B., Oules, B., Pisco, A. O., Shaw, T. J., Elgueta, R., et al. (2018). Spatial and single-cell transcriptional profiling identifies functionally distinct human dermal fibroblast subpopulations. *The Journal of Investigative Dermatology, 138*, 811–825.

Pierce, G. F., Yanagihara, D., Klopchin, K., Danilenko, D. M., Hsu, E., Kenney, W. C., et al. (1994). Stimulation of all epithelial elements during skin regeneration by keratinocyte growth factor. *The Journal of Experimental Medicine, 179*, 831–840.

Plikus, M. V., Guerrero-Juarez, C. F., Ito, M., Li, Y. R., Dedhia, P. H., Zheng, Y., et al. (2017). Regeneration of fat cells from myofibroblasts during wound healing. *Science (New York, N.Y.), 355*, 748–752.

Plikus, M. V., Mayer, J. A., de la Cruz, D., Baker, R. E., Maini, P. K., Maxson, R., et al. (2008). Cyclic dermal BMP signalling regulates stem cell activation during hair regeneration. *Nature, 451*, 340–344.

Plisko, A., & Gilchrest, B. A. (1983). Growth factor responsiveness of cultured human fibroblasts declines with age. *Journal of Gerontology*, *38*, 513–518.

Puram, S. V., Tirosh, I., Parikh, A. S., Patel, A. P., Yizhak, K., Gillespie, S., et al. (2017). Single-cell transcriptomic analysis of primary and metastatic tumor ecosystems in head and neck cancer. *Cell*, *171*. 1611-1624.e1624.

Quan, T., He, T., Kang, S., Voorhees, J. J., & Fisher, G. J. (2004). Solar ultraviolet irradiation reduces collagen in photoaged human skin by blocking transforming growth factor-beta type II receptor/Smad signaling. *The American Journal of Pathology*, *165*, 741–751.

Quan, T., He, T., Shao, Y., Lin, L., Kang, S., Voorhees, J. J., et al. (2006). Elevated cysteine-rich 61 mediates aberrant collagen homeostasis in chronologically aged and photoaged human skin. *The American Journal of Pathology*, *169*, 482–490.

Quan, T., Qin, Z., Robichaud, P., Voorhees, J. J., & Fisher, G. J. (2011). CCN1 contributes to skin connective tissue aging by inducing age-associated secretory phenotype in human skin dermal fibroblasts. *Journal of Cell Communication and Signaling*, *5*, 201–207.

Rahmani, W., Abbasi, S., Hagner, A., Raharjo, E., Kumar, R., Hotta, A., et al. (2014). Hair follicle dermal stem cells regenerate the dermal sheath, repopulate the dermal papilla, and modulate hair type. *Developmental Cell*, *31*, 543–558.

Rendl, M., Lewis, L., & Fuchs, E. (2005). Molecular dissection of mesenchymal-epithelial interactions in the hair follicle. *PLoS Biology*, *3*, e331.

Ressler, S., Bartkova, J., Niederegger, H., Bartek, J., Scharffetter-Kochanek, K., Jansen-Dürr, P., et al. (2006). p16INK4A is a robust in vivo biomarker of cellular aging in human skin. *Aging Cell*, *5*, 379–389.

Rezza, A., Sennett, R., Tanguy, M., Clavel, C., & Rendl, M. (2015). PDGF signalling in the dermis and in dermal condensates is dispensable for hair follicle induction and formation. *Experimental Dermatology*, *24*, 468–470.

Rezza, A., Wang, Z., Sennett, R., Qiao, W., Wang, D., Heitman, N., et al. (2016). Signaling networks among stem cell precursors, transit-amplifying progenitors, and their niche in developing hair follicles. *Cell Reports*, *14*, 3001–3018.

Rhie, G.-e., Shin, M. H., Seo, J. Y., Choi, W. W., Cho, K. H., Kim, K. H., et al. (2001). Aging- and photoaging-dependent changes of enzymic and nonenzymic antioxidants in the epidermis and dermis of human skin in vivo. *Journal of Investigative Dermatology*, *117*, 1212–1217.

Rinkevich, Y., Walmsley, G. G., Hu, M. S., Maan, Z. N., Newman, A. M., Drukker, M., et al. (2015). Skin fibrosis. Identification and isolation of a dermal lineage with intrinsic fibrogenic potential. *Science (New York, N.Y.)*, *348*, aaa2151.

Rinn, J. L., Bondre, C., Gladstone, H. B., Brown, P. O., & Chang, H. Y. (2006). Anatomic demarcation by positional variation in fibroblast gene expression programs. *PLoS Genetics*, *2*, e119.

Rivera-Gonzalez, G., Shook, B., & Horsley, V. (2014). Adipocytes in skin health and disease. *Cold Spring Harbor Perspectives in Medicine*, *4*. pii: a015271.

Rognoni, E., Gomez, C., Pisco, A. O., Rawlins, E. L., Simons, B. D., Watt, F. M., et al. (2016). Inhibition of beta-catenin signalling in dermal fibroblasts enhances hair follicle regeneration during wound healing. *Development (Cambridge, England)*, *143*, 2522–2535.

Rognoni, E., Pisco, A. O., Hiratsuka, T., Sipilä, K. H., Belmonte, J. M., Mobasseri, S. A., et al. (2018). Fibroblast state switching orchestrates dermal maturation and wound healing. *Molecular Systems Biology*, *14*, e8174.

Rompolas, P., Deschene, E. R., Zito, G., Gonzalez, D. G., Saotome, I., Haberman, A. M., et al. (2012). Live imaging of stem cell and progeny behaviour in physiological hair-follicle regeneration. *Nature*, *487*, 496–499.

Rowlatt, U. (1979). Intrauterine wound healing in a 20 week human fetus. *Virchows Archiv. A, Pathological Anatomy and Histology*, *381*, 353–361.

Ruchti, C., Haller, D., Nuber, M., & Cottier, H. (1983). Regional differences in renewal rates of fibroblasts in young adult female mice. *Cell and Tissue Research*, *232*, 625–636.

Salzer, M. C., Lafzi, A., Berenguer-Llergo, A., Youssif, C., Castellanos, A., Solanas, G., et al. (2018). Identity noise and adipogenic traits characterize dermal fibroblast aging. *Cell*, *175*, 1575–1590.

Schipper, H. S., Prakken, B., Kalkhoven, E., & Boes, M. (2012). Adipose tissue-resident immune cells: Key players in immunometabolism. *Trends in Endocrinology and Metabolism*, *23*, 407–415.

Schlake, T. (2007). Determination of hair structure and shape. *Seminars in Cell & Developmental Biology*, *18*, 267–273.

Schmidt, B. A., & Horsley, V. (2013). Intradermal adipocytes mediate fibroblast recruitment during skin wound healing. *Development (Cambridge, England)*, *140*, 1517–1527.

Schor, S. L., Schor, A. M., Rushton, G., & Smith, L. (1985). Adult, foetal and transformed fibroblasts display different migratory phenotypes on collagen gels: Evidence for an isoformic transition during foetal development. *Journal of Cell Science*, *73*, 221–234.

Sennett, R., & Rendl, M. (2012). Mesenchymal-epithelial interactions during hair follicle morphogenesis and cycling. *Seminars in Cell & Developmental Biology*, *23*, 917–927.

Sennett, R., Wang, Z., Rezza, A., Grisanti, L., Roitershtein, N., Sicchio, C., et al. (2015). An integrated transcriptome atlas of embryonic hair follicle progenitors, their niche, and the developing skin. *Developmental Cell*, *34*, 577–591.

Serrano, M., Lin, A. W., McCurrach, M. E., Beach, D., & Lowe, S. W. (1997). Oncogenic ras provokes premature cell senescence associated with accumulation of p53 and p16INK4a. *Cell*, *88*, 593–602.

Sharov, A. A., Sharova, T. Y., Mardaryev, A. N., Tommasi di Vignano, A., Atoyan, R., Weiner, L., et al. (2006). Bone morphogenetic protein signaling regulates the size of hair follicles and modulates the expression of cell cycle-associated genes. *Proceedings of the National Academy of Sciences of the United States of America*, *103*, 18166–18171.

Shook, B. A., Wasko, R. R., Rivera-Gonzalez, G. C., Salazar-Gatzimas, E., Lopez-Giraldez, F., Dash, B. C., et al. (2018). Myofibroblast proliferation and heterogeneity are supported by macrophages during skin repair. *Science*, *362*, pii: eaar2971.

Shook, B., Xiao, E., Kumamoto, Y., Iwasaki, A., & Horsley, V. (2016). CD301b+ macrophages are essential for effective skin wound healing. *The Journal of Investigative Dermatology*, *136*, 1885–1891.

Sick, S., Reinker, S., Timmer, J., & Schlake, T. (2006). WNT and DKK determine hair follicle spacing through a reaction-diffusion mechanism. *Science (New York, N.Y.)*, *314*, 1447–1450.

Siersbaek, R., Nielsen, R., & Mandrup, S. (2010). PPARgamma in adipocyte differentiation and metabolism—Novel insights from genome-wide studies. *FEBS Letters*, *584*, 3242–3249.

Silva-Vargas, V., Lo Celso, C., Giangreco, A., Ofstad, T., Prowse, D. M., Braun, K. M., et al. (2005). Beta-catenin and hedgehog signal strength can specify number and location of hair follicles in adult epidermis without recruitment of bulge stem cells. *Developmental Cell*, *9*, 121–131.

Simpson, C. L., Patel, D. M., & Green, K. J. (2011). Deconstructing the skin: Cytoarchitectural determinants of epidermal morphogenesis. *Nature Reviews. Molecular Cell Biology*, *12*, 565–580.

Singer, A. J., & Clark, R. A. F. (1999). Cutaneous wound healing. *New England Journal of Medicine*, *341*, 738–746.

Snippert, H. J., Haegebarth, A., Kasper, M., Jaks, V., van Es, J. H., Barker, N., et al. (2010). Lgr6 marks stem cells in the hair follicle that generate all cell lineages of the skin. *Science (New York, N.Y.)*, *327*, 1385–1389.

Song, Y., Boncompagni, A. C., Kim, S. S., Gochnauer, H. R., Zhang, Y., Loots, G. G., et al. (2018). Regional control of hairless versus hair-bearing skin by Dkk2. *Cell Reports, 25*, 2981-2991.e2983.

Sorrell, J. M., & Caplan, A. I. (2004). Fibroblast heterogeneity: More than skin deep. *Journal of Cell Science, 117*, 667–675.

Sorrell, J. M., & Caplan, A. I. (2009). Fibroblasts-a diverse population at the center of it all. *International Review of Cell and Molecular Biology, 276*, 161–214.

Sriram, G., Bigliardi, P. L., & Bigliardi-Qi, M. (2015). Fibroblast heterogeneity and its implications for engineering organotypic skin models in vitro. *European Journal of Cell Biology, 94*, 483–512.

Stanulis-Praeger, B., & Gilchrest, B. A. (1986). Growth factor responsiveness declines during adulthood for human skin-derived cells. *Mechanisms of Ageing and Development, 35*, 185–198.

Stone, S. J., Myers, H. M., Watkins, S. M., Brown, B. E., Feingold, K. R., Elias, P. M., et al. (2004). Lipopenia and skin barrier abnormalities in DGAT2-deficient mice. *The Journal of Biological Chemistry, 279*, 11767–11776.

Sundberg, J. P., Peters, E. M., & Paus, R. (2005). Analysis of hair follicles in mutant laboratory mice. *The Journal of Investigative Dermatology. Symposium Proceedings, 10*, 264–270.

Tabib, T., Morse, C., Wang, T., Chen, W., & Lafyatis, R. (2018). SFRP2/DPP4 and FMO1/LSP1 define major fibroblast populations in human skin. *The Journal of Investigative Dermatology, 138*, 802–810.

Tan, D. W., Jensen, K. B., Trotter, M. W., Connelly, J. T., Broad, S., & Watt, F. M. (2013). Single-cell gene expression profiling reveals functional heterogeneity of undifferentiated human epidermal cells. *Development (Cambridge, England), 140*, 1433–1444.

Tang, A., & Gilchrest, B. A. (1996). Regulation of keratinocyte growth factor gene expression in human skin fibroblasts. *Journal of Dermatological Science, 11*, 41–50.

Telerman, S. B., Rognoni, E., Sequeira, I., Pisco, A. O., Lichtenberger, B. M., Culley, C. J., et al. (2017). Dermal Blimp1 acts downstream of epidermal TGFbeta and Wnt/beta-catenin to regulate hair follicle formation and growth. *The Journal of Investigative Dermatology, 137*, 2270–2281.

Thangapazham, R. L., Klover, P., Wang, J. A., Zheng, Y., Devine, A., Li, S., et al. (2014). Dissociated human dermal papilla cells induce hair follicle neogenesis in grafted dermal-epidermal composites. *The Journal of Investigative Dermatology, 134*, 538–540.

Tigges, J., Krutmann, J., Fritsche, E., Haendeler, J., Schaal, H., Fischer, J. W., et al. (2014). The hallmarks of fibroblast ageing. *Mechanisms of Ageing and Development, 138*, 26–44.

Tobin, D. J., Gunin, A., Magerl, M., Handijski, B., & Paus, R. (2003). Plasticity and cytokinetic dynamics of the hair follicle mesenchyme: Implications for hair growth control. *The Journal of Investigative Dermatology, 120*, 895–904.

Vaculik, C., Schuster, C., Bauer, W., Iram, N., Pfisterer, K., Kramer, G., et al. (2012). Human dermis harbors distinct mesenchymal stromal cell subsets. *The Journal of Investigative Dermatology, 132*, 563–574.

Varani, J., Dame, M. K., Rittie, L., Fligiel, S. E., Kang, S., Fisher, G. J., et al. (2006). Decreased collagen production in chronologically aged skin: Roles of age-dependent alteration in fibroblast function and defective mechanical stimulation. *The American Journal of Pathology, 168*, 1861–1868.

Varani, J., Spearman, D., Perone, P., Fligiel, S. E., Datta, S. C., Wang, Z. Q., et al. (2001). Inhibition of type I procollagen synthesis by damaged collagen in photoaged skin and by collagenase-degraded collagen in vitro. *The American Journal of Pathology, 158*, 931–942.

Viswanathan, P., Guvendiren, M., Chua, W., Telerman, S. B., Liakath-Ali, K., Burdick, J. A., et al. (2016). Mimicking the topography of the epidermal-dermal interface with elastomer substrates. *Integrative Biology: Quantitative Biosciences From Nano to Macro, 8*, 21–29.

Walmsley, G. G., Hu, M. S., Maan, Z. N., Rinkevich, Y., Weissman, I. L., & Longaker, M. T. (2015). En1 fibroblasts and melanoma. *Melanoma Management, 2*, 191–192.

Wang, Z., Wang, Y., Farhangfar, F., Zimmer, M., & Zhang, Y. (2012). Enhanced keratinocyte proliferation and migration in co-culture with fibroblasts. *PLoS One, 7*, e40951.

Watt, F. M. (2014). Mammalian skin cell biology: At the interface between laboratory and clinic. *Science (New York, N.Y.), 346*, 937–940.

Watt, F. M., & Fujiwara, H. (2011). Cell-extracellular matrix interactions in normal and diseased skin. *Cold Spring Harbor Perspectives in Biology, 3*, a00512.

Weger, N., & Schlake, T. (2005). Igf-I signalling controls the hair growth cycle and the differentiation of hair shafts. *The Journal of Investigative Dermatology, 125*, 873–882.

Werner, S., Krieg, T., & Smola, H. (2007). Keratinocyte-fibroblast interactions in wound healing. *The Journal of Investigative Dermatology, 127*, 998–1008.

Werner, S., Peters, K. G., Longaker, M. T., Fuller-Pace, F., Banda, M. J., & Williams, L. T. (1992). Large induction of keratinocyte growth factor expression in the dermis during wound healing. *Proceedings of the National Academy of Sciences of the United States of America, 89*, 6896–6900.

Werner, S., Smola, H., Liao, X., Longaker, M. T., Krieg, T., Hofschneider, P. H., et al. (1994). The function of KGF in morphogenesis of epithelium and reepithelialization of wounds. *Science (New York, N.Y.), 266*, 819–822.

Zhang, Y., Tomann, P., Andl, T., Gallant, N. M., Huelsken, J., Jerchow, B., et al. (2009). Reciprocal requirements for EDA/EDAR/NF-kappaB and Wnt/beta-catenin signaling pathways in hair follicle induction. *Developmental Cell, 17*, 49–61.

Zhang, B., Tsai, P. C., Gonzalez-Celeiro, M., Chung, O., Boumard, B., Perdigoto, C. N., et al. (2016). Hair follicles' transit-amplifying cells govern concurrent dermal adipocyte production through Sonic Hedgehog. *Genes & Development, 30*, 2325–2338.

Zhou, L., Xu, M., Yang, Y., Yang, K., Wickett, R. R., Andl, T., et al. (2016). Activation of beta-catenin signaling in CD133-positive dermal papilla cells drives postnatal hair growth. *PLoS One, 11*, e0160425.

Zhou, L., Yang, K., Xu, M., Andl, T., Millar, S. E., Boyce, S., et al. (2016). Activating beta-catenin signaling in CD133-positive dermal papilla cells increases hair inductivity. *The FEBS Journal, 283*, 2823–2835.

Zhu, J., Woods, D., McMahon, M., & Bishop, J. M. (1998). Senescence of human fibroblasts induced by oncogenic Raf. *Genes & Development, 12*, 2997–3007.

Zouboulis, C. C., & Makrantonaki, E. (2011). Clinical aspects and molecular diagnostics of skin aging. *Clinics in Dermatology, 29*, 3–14.

CHAPTER SEVEN

Immune cells and the epidermal stem cell niche

Etienne C.E. Wang[a,b,*]
[a]National Skin Centre, Singapore
[b]Skin Research Institute of Singapore, Singapore
*Corresponding author: e-mail address: etienne@nsc.com.sg

Contents

Advances in Stem Cells and their Niches, Volume 3
ISSN 2468-5097
https://doi.org/10.1016/bs.asn.2019.06.001

1. Introduction

The specific and conducive niche microenvironment is essential to the function and survival of stem cells in vivo. The niche was defined by Schofield in 1978 to describe the "association with other cells which determines [a stem cell's] behavior" (Schofield, 1978). The anatomical milieu of stem cells positions them to receive signals from paracrine factors, adhesion molecules and other cell-cell interactions that support their longevity, quiescence, self-renewal, and pluripotency (Fuchs, Tumbar, & Guasch, 2004; Goldstein & Horsley, 2012; Li & Xie, 2005). The niche also provides protection for stem cells from exogenous and endogenous insults such as pathogens, chemotherapy, and autoimmunity (Niederkorn & Stein-Streilein, 2010). Removing stem cells from their niche (e.g., by physiological competition, ex vivo or in vitro culture, or genetic methods to ablate portions of the niche) frequently leads to activation of cellular programs that push the cells toward proliferation and differentiation, and results in loss of stemness (Hsu, Pasolli, & Fuchs, 2011; Rompolas, Mesa, & Greco, 2013). The cells and structures that play a role in creating the stem cell niche are usually surrounding cells (epithelial or stromal), extracellular matrix (ECM), endothelial and neural cells, in a manner that is conserved across many organs like the bone marrow, gut, and brain (Jones & Wagers, 2008). Comprehensive deconstruction and understanding of stem cell-niche interactions are essential for the advancement of regenerative medicine, and the ability to produce viable tissue replacements ex vivo (Lane, Williams, & Watt, 2014). This chapter will review the role of immune cells in the maintenance of the epidermal stem cell niche, and the mechanisms through which they help to support stem cell activity in the skin.

2. Stem cell niches of the skin

The skin is the first line of defense from environmental insults, and a vital organ involved in thermoregulation and preventing dehydration. Different epithelial compartments of the skin have different rates of turnover, and different responses to wounding; this is reflected in the diverse stem cell compartments found in the skin and its appendages, which include hair, nails, and sweat/sebaceous/apocrine glands (Fig. 1) (Braun & Prowse, 2006).

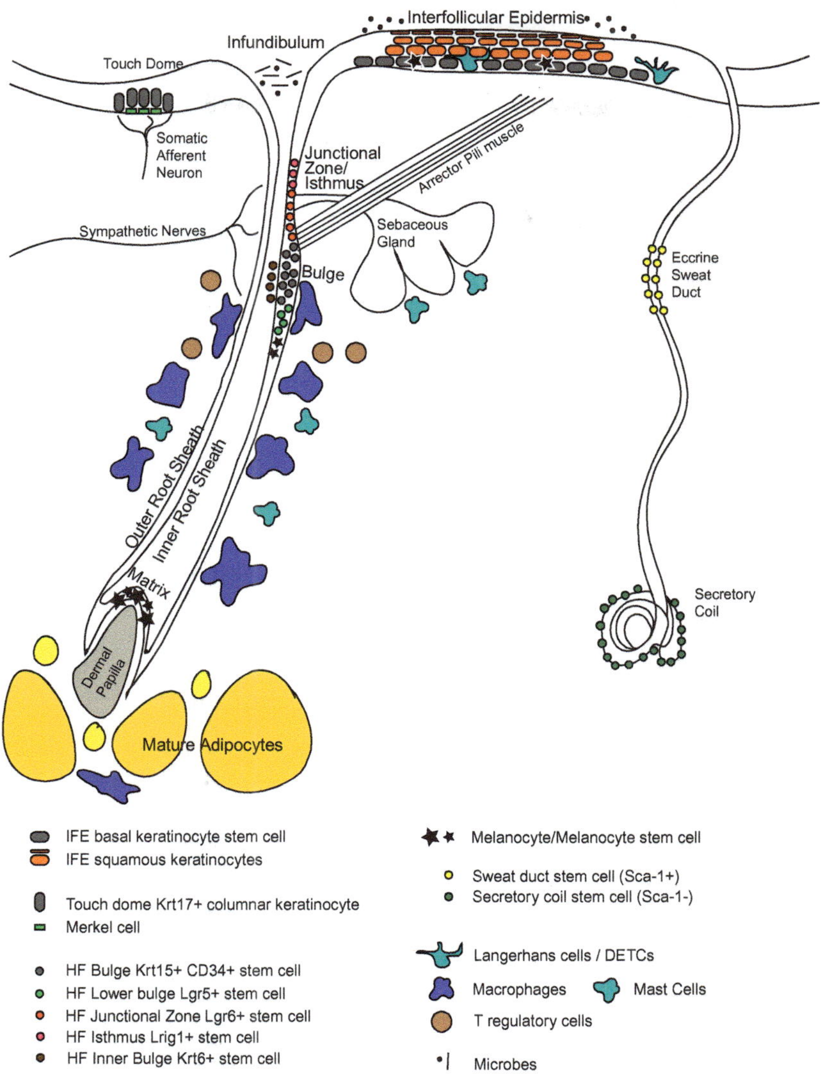

Fig. 1 Schematic of epidermal stem cells and their niche. Epidermal stem cells are found in the basal layer of the interfollicular epidermis (IFE), touch dome, sweat glands, and the hair follicles (HF). IFE stem cells give rise to keratinocytes that undergo a cornification programme that produces the stratified epithelium, and the basal niche is comprised of the basement membrane, basal layer melanocytes/melanocyte stem cells, dendritic epidermal T cells (DETCs) and Langerhans cells. Specialized Krt17+ epidermal stem cells serve as progenitor cells for Merkel cells in touch domes, which confer mechanosensitivity to soft touch in the epidermis. Epidermal stem cells that are recruited into skin appendages take on more specialized roles. The eccrine sweat gland is formed from direct invagination of the IFE during embryogenesis, with Sca-1+ stem cells in the

(Continued)

2.1 Interfollicular epidermis

The interfollicular epidermis (IFE) is a stratified squamous epithelium that regenerates from the stochastic competitive stem cell pool in the basal layer (Blanpain & Simons, 2013). The progeny of this basal pool of Krt5+ Krt14+ keratinocyte stem cells are maintained by autocrine Wnt signals (Lim et al., 2013), and differentiate upward and outward when they leave the basal niche, and turns over approximately every 28 days (Halprin, 1972). Basement membrane ECM is a vital component of the IFE stem cell niche, and includes α6, β1, β4 integrins, and laminin-5 (Hsu, Li, & Fuchs, 2014). Collagen 17A in the basement membrane, which is associated with junctional epidermolysis bullosa and bullous pemphigoid, also regulates the proliferation of basal keratinocyte stem cells (Watanabe et al., 2017). Defects in the IFE are resolved by a wound healing process that involves keratinocytes, stromal mesenchymal cells and immune cells. During the steady state, antigen-presenting cells like Langerhans cells are present within the epidermis where they function as immune sentinels (Merad, Ginhoux, & Collin, 2008).

2.2 Hair follicle

The hair follicle (HF) has several stem cell compartments that are tightly regulated in order to orchestrate the hair cycle, whereby each follicle undergoes phases of growth (anagen), regression (catagen) and rest (telogen) throughout the life of the mammal. The hierarchical nature of the stem cell niche in the HF reflects this function, with the HF Krt15+ bulge (CD34+ in the mouse) containing the most long-lived self-renewing population of stem cells. A precise balance of Wnt and BMP signaling in the niche is essential

Fig. 1—Cont'd duct and Sca-1- stem cells in the secretory coil. The ductal Sca-1+ cells are capable to contributing to wound healing in the IFE. Both ductal and glandular stem cells are further arranged radially as myoepithelial (outer) or luminal (inner) unipotent progenitors to give rise to cells that replenish their respective layers with almost no transdifferentiation. HF stem cells are also established during embryogenesis, with the most long-lived, slow-cycling stem cells in the bulge (CD34+). Other stem cells compartments in the HF include the inner bulge (Krt6+), lower bulge (Lgr5+), junctional zone/isthmus (Lrig1+/Lgr6+). The lower bulge cells give rise to outer root sheath keratinocytes that proliferate during anagen to give rise to the matrix of the HF bulb, where it undergoes epithelial-mesenchymal interactions with the dermal papilla (DP) to differentiate into specialized layers that form the growing hair shaft. The DP itself is replenished from stem cells likely arising from the dermal sheath (out of the scope of this chapter). The HF stem cell niche is complex and has been shown to include mesenchymal cells like the DP, inner bulge, sympathetic nerve endings, smooth muscle attachments, adipocytes, and immune cells like macrophages, DETCs, and T regs.

for the control of HF stem cell activity (Greco et al., 2009; Kobielak et al., 2007; Lien et al., 2014; Lim et al., 2016; Plikus et al., 2008). As stem cells leave the bulge, their stemness and pluripotency are reduced, and establish other stem cell populations in the lower bulge (Lgr5 +), isthmus and infundibulum (Lrig1 +), sebaceous duct (Lrig1 Lgr6 +), and inner bulge (Krt6 +) of the HF (Hsu et al., 2011; Kretzschmar & Watt, 2014). These stem cells are more readily mobilized to differentiate into keratinocytes for the outer root sheath and sebaceous ducts during physiological anagen (Hsu et al., 2011), and the IFE during wound healing (Ito et al., 2005). The HF stem cell niche in the bulge is made up of neighboring keratinocytes (i.e., the inner bulge), the basement membrane ECM, as well as other cell types in the local microenvironment such as the arrector pili muscle attachment and sympathetic nerves. Collagen 17A, which is important for IFE stem cell homeostasis, also plays a role in HF stem cell homeostasis (Matsumura et al., 2016). Stem cells in the lower bulge are juxtaposed to the dermal papilla (DP) during telogen and give rise to keratinocytes of the outer root sheath during anagen, during which they extend into the adipose layer (Festa et al., 2011).

The HF stem cell compartment is a site of immune privilege, with reduced expression of MHC molecules (Meyer et al., 2008). It is postulated that it avoids the autoimmune destruction of stem cell compartments (keratinocyte and melanocyte) at a site of increased microbial traffic (i.e., the HF infundibulum). This immune privilege also extends to the bulb region of the HF and protects the HF matrix and DP as well. Immunohistochemistry of anagen HF in humans shows a paucity of natural killer (NK) cells and γδ T cells around the HF, but prominence of Langerhans cells, macrophages, mast cells, and mature T and B cells (Christoph et al., 2000). The relevance of most of these immune cell types concerning the HF stem cell niche and immune privilege has only recently been interrogated and will be discussed in this chapter.

Epidermal stem cells are capable of regenerating HFs in large experimental wounds (Ito et al., 2007), likely an evolutionary adaptation for replacing tissue injury in the wild and the maintenance of protective fur. This de novo regeneration of HFs in a wound bed involves TNF signals from macrophages (Wang et al., 2017), and FGF9 produced by infiltrating γδ T cells (Gay et al., 2013).

2.3 Melanocytes

The skin also houses melanocyte stem cells in both the IFE and HF. These stem cells are derived from the neural crest, and migrate into the epidermis during embryogenesis (Dupin & Le Douarin, 2003). The IFE melanocyte stem cells give rise to melanocytes that transfer melanin to IFE keratinocytes to control skin pigmentation, while the melanocyte stem cells in the HF reside in the lower bulge, and give rise to melanocytes in the HF matrix that ensure hair shaft pigmentation. The keratinocytes of the lower bulge also provide a nice for the melanocyte stem cells (Tanimura et al., 2011). The cyclical activation of the HF keratinocytes via paracrine signals from bulge and DP are coupled with melanocyte stem cell activation to ensure constant hair pigmentation (Rabbani et al., 2011). Age-related loss of the melanocyte stem cell niche results in pigment loss in the skin and hair reflected clinically by idiopathic guttate hypomelanosis ("white spots" in aged skin) and hair graying.

2.4 Sweat glands

Mammalian skin contains eccrine sweat glands that open onto the interfollicular epidermis and produce sweat that aids in thermoregulation. Stem cells from the sweat gland arise from multipotent epidermal progenitors, and are unipotent, giving rise to the myoepithelial (Krt14+ basal layer) and luminal cells ($Krt14^{lo}$ Krt18+ suprabasal layer) that make up the eccrine sweat duct and gland (Lu et al., 2012), and also contribute to IFE wound healing under certain conditions (Rittie et al., 2013). The eccrine sweat gland niche is likely composed of specialized ECM (Huang et al., 2016), with possible contributions from bone marrow-derived mesenchymal stem cells (Sheng et al., 2009). Sebaceous glands (SGs) that are associated with most HFs are derived from the Lrig1+ epidermal progenitors in the HF junctional zone (Frances & Niemann, 2012; Jensen et al., 2009), and express Blimp1 as they become committed to an SG fate (Sellheyer & Krahl, 2010). Apocrine glands are holocrine glands that are only associated with HFs in areas of secondary sexual maturation (i.e., axillae, groin, nipples), and produce a lipid-rich compound that attains a characteristic odor when processed by skin microbial commensals. The stem cells of the apocrine gland remain unexplored. All sweat glands are richly innervated by sympathetic autonomic nerves and have distinct ECM and stromal interactions for their niche.

2.5 Merkel cells

The epidermis contains specialized structures known as touch domes that contain touch-sensitive neuroendocrine Merkel cells. Touch domes are richly innervated by somatic afferents from Aβ nerve fibers (Maricich et al., 2009).

Merkel cells are derived from epidermal Krt17+ stem cells, which are non-keratinized and long-lived in mammalian skin, and are the main cell type regulating the Merkel cell niche (Doucet et al., 2013; Van Keymeulen et al., 2009). Merkel cell carcinoma, though extremely rare, is encountered more frequently in patients with systemic immunosuppression (Clarke et al., 2015; Engels et al., 2002), suggesting that immune surveillance of the touch domes is essential for Merkel cell homeostasis.

3. Immune cells and their interactions with the niche

Cells of the immune system have been recognized as crucial players in the function and homeostasis of many tissues and organs. Their ability to respond to various tissue-specific chemokines allow them to home specifically to certain target tissues, and their capacity for migration within and between tissues allows them to mediate long-range cell-cell communication. These distinct properties of the immune system have allowed them to take on non-immune, non-canonical functions in the skin and other organs to support tissue homeostasis.

The skin, like the gut, provides an interface with the microbe-rich environment and thus has an important immunological role (Abdallah, Mijouin, & Pichon, 2017). Immune cells in the skin are also poised to participate in wound healing and epidermal regeneration. Many immune cells are resident in the epidermis and dermis, but there is also a constant flux of cells to and from the bloodstream. Many cell types are specifically skin-homing and respond to skin-specific chemokines (Butcher & Picker, 1996; Schaerli et al., 2004).

The immune system is arbitrarily divided into innate and adaptive arms. The innate immune system consists of cells that are generally defined by the lack of specificity of their pattern recognition receptors (PRR) which recognize conserved microbial motifs and respond to non-specific signals from tissue injury and pathogenic insults. The innate immune system is usually recruited early to sites of injury and includes neutrophils, macrophages, dendritic cells, and natural killer T cells. Conversely, the adaptive immune system consists of cells that have undergone germline modifications to increase the specificity of their pathogen receptors, and include mature T and B cells (including helper, cytotoxic, regulatory and memory subtypes). The activity and cytokines produced by the innate immune system during the acute phase of inflammation have been shown to influence the fate of adaptive immune cells (Iwasaki & Medzhitov, 2015). During tissue homeostasis, tissue-resident immune cells include many cells of the innate immune system, as well as memory cells of the adaptive immune system. These cells have recently been shown to take part in tissue homeostasis in many unexpected ways.

3.1 Innate immune system

As the main sentinels of the immune system, many innate immune cells are positioned in peripheral tissues as the first line of defense against physiological and environmental insults. They play an essential role in wound healing by producing cytokines that support epithelial and mesenchymal cells in repairing tissue injury by activating stem cell populations, as well as fending off pathogenic intruders to maintain a relatively sterile wound bed (MacLeod & Mansbridge, 2016; Portou et al., 2015). The interactions between innate immune cells and various tissue cell types (particularly stem cells) are also present during the steady-state as part of routine tissue homeostasis.

3.2 Langerhans cells

Langerhans cells are abundant in the epidermis and express Langerin (CD207). The primary function of Langerhans cells is to sample the epidermis for potential exogenous infiltration and present antigens to T cells in the skin-associated lymphoid tissue (SALT) or draining lymph nodes. Immunohistochemistry has shown them to be closely apposed to the basal keratinocyte stem cells of the IFE (Parkinson, 1992), as well as Merkel cells (Taira et al., 2002). Their specific functions in these stem cell niches are currently unknown.

3.3 Dendritic epidermal T cells

In murine skin, dendritic epidermal γδ T cells (DETCs) play a role in immune surveillance via their restricted T cell receptor (TCR) devoid of functional diversity (Tigelaar & Lewis, 1995) and are present in large numbers. DETCs support the IFE niche via insulin growth factor (IGF)-1 (Sharp et al., 2005), and also contribute to wound healing by producing fibroblast growth factor (FGF)-7 (Jameson et al., 2004). In aging mouse skin, while the epidermal stem cell compartment is relatively stable, the rate of proliferation decreases and is associated with a reduction in DETC numbers in the epidermis (Giangreco et al., 2008). An increase in HF-associated DETCs is also observed with anagen progression in mice (Hashizume, Tokura, & Takigawa, 1994; Paus et al., 1994), and may play a role in HF stem cell activation.

Similar γδ T cells may be present at much lower levels in the human epidermis (with the majority expressing the αβ TCR), and have been implicated in contact hypersensitivity and wound healing (Toulon et al., 2009). Whether epidermal T cells modulate stem cell niches in human skin remains unexplored.

3.4 Neutrophils

Neutrophils are the most abundant immune cell with a very short half-life and are recruited to sites of acute injury to provide the first line of defense against extrinsic pathogens (Mocsai, 2013). Neutrophils also support the peripheral function of B cells, NK cells, and T cells (Jaeger et al., 2012; Pillay et al., 2012; Puga et al., 2011). Degranulation of neutrophils results in the release of acute-phase cytokines, chemokines, proteases and reactive oxygen species into the tissue microenvironment—all of which contribute to the early inflammatory response but also sets up the target tissue for wound healing and resolution, and may in turn influence the stem cell niche to regulate stem cell activity during times of stress (Borregaard, Sorensen, & Theilgaard-Monch, 2007; Hager, Cowland, & Borregaard, 2010). In the bone marrow, aged neutrophils migrating back to the stem cell niche is coupled to the circadian release of hematopoietic progenitors into the blood stream (Casanova-Acebes et al., 2013; Mendez-Ferrer et al., 2008). This process is dependent on CXCR4 expression on the neutrophils (Martin et al., 2003).

3.5 Macrophages

Myeloid cells like macrophages and other antigen-presenting cells have phagocytic functions central to their immune function as tissue sentinels and surveillance of exogenous antigen but are also involved in the clearance of cellular debris during wound healing (DiPietro, 1995).

Tissue macrophages are diverse and highly heterogeneous (Qian et al., 2019), and may maintain homeostasis via multiple mechanisms. Due to this heterogeneity, categorization of tissue macrophages is difficult, and many distinct subsets have been identified through recent advances of genetic profiling at a single-cell level (Guilliams, Mildner, & Yona, 2018; Zhao et al., 2018). Tissue macrophages can be broadly dividing into the long-lived tissue-resident macrophages which originate from the embryonic yolk sac or fetal liver and seeded into tissues in utero, and the bone-marrow derived macrophages which are derived from circulating monocytes in adulthood via tissue chemotactic factors (Ginhoux & Guilliams, 2016). Most macrophages in the skin are bone marrow-derived, unlike other organs such as the brain where microglia are mostly long-lived residents derived from the embryonic yolk sac (Epelman, Lavine, & Randolph, 2014).

Macrophages have also been classified according to their function, namely the "classical" or "inflammatory" macrophages (previously called M1) that produce interferon, TNF-α and nitric oxide synthase (iNOS) in response to pathogen, and the "alternate" or "anti-inflammatory" (M2) macrophages that promote tissue repair by producing IL-4, IL-13, IL-10 and ornithine/polyamines (Wang, Liang, & Zen, 2014). Detailed analyses of macrophage subsets have shown that this classification is overly simplistic, and macrophages are tremendously diverse in phenotype and function (Knudsen & Lee, 2016; Martinez & Gordon, 2014). While it is preferred to discuss macrophages according to their precise function, the terms "M1-like" and "M2-like" have been used as shorthand to orientate the reader.

Macrophages are associated with the HF at almost every stage of the hair cycle, with different subsets carrying out specific functions to influence the stem cell niche (Muneeb, Hardman, & Paus, 2019). For example, macrophages increase in numbers in telogen (rest) skin in the mouse, and their depletion is associated with anagen (hair growth) (Castellana, Paus, & Perez-Moreno, 2014). Recent single-cell analyses coupled with functional analysis uncovered a TREM2+ subset of "trichophages" that maintain HF stem cell quiescence via oncostatin M (OSM) during telogen (Wang et al., 2019). Intriguingly, macrophages (likely other subsets) are also implicated in HF stem cell activation and thus anagen in certain conditions such as mechanical stretching of the skin (resulting in microtrauma) (Chu et al., 2019). Other studies have identified subsets of macrophages that may induce anagen via FGF-5 (Suzuki et al., 1998) or release of Wnt signals during the telogen-to-anagen transition (Castellana et al., 2014). Further, macrophages also clear cellular debris from HF regression during catagen (Eichmuller et al., 1998; Parakkal, 1969). It is very likely that different subsets and activation states of macrophages are responsible for the control of the HF stem cell niches at various stages of the hair cycle (Amberg et al., 2016). TNF-producing "M1-like" macrophages are implicated in de novo hair regeneration in wounds (Wang et al., 2017), as well as anagen initiation after plucking (Chen et al., 2015), while "M2-like" anti-inflammatory tissue-resident macrophages are involved in stem cell quiescence during telogen (Wang et al., 2019). Future studies which reveal more specific markers for these subsets will aid in understanding their roles more clearly.

3.6 Mast cells

Mast cells typically mediate IgE allergic inflammatory responses, and have been shown to be essential in some models of wound healing (Komi, Khomtchouk, & Santa Maria, 2019). Mast cells are also found surrounding HFs and sebaceous glands. IL-6 and TNF-α produced by mast cells have been postulated to influence the innervation of sebocytes, which in turn affects the local substance P concentration of the SG thus modulating the proliferation and differentiation of SG stem cells (Toyoda, Nakamura, & Morohashi, 2002).

3.7 Other innate immune cells

Eosinophils, basophils, innate lymphoid cells (ILCs) and NK cells are other members of the innate immune system. Innate lymphoid cells (ILCs) are derived from the common lymphoid progenitor but do not express mature T/B/NKT markers or antigen receptors, and are involved in wound healing and tissue homeostasis in various epithelial tissues (Rak et al., 2016). Specific interactions between the other cell types and stem cells have not been closely interrogated.

3.8 Adaptive immune system

In contrast to the innate immune system, the hallmarks of the adaptive immune system reflect its mutability and potential for changing in response to environmental or self-antigens. Prime examples are mature T and B cells which produce polyclonal receptors and antibodies that are edited down to match antigen presented to the organism by way of infection, vaccination or mutation (in the case of cancer). The adaptive immune system is also capable of memory and retains T and B cells with high-affinity receptors, which are able to respond in a more efficient and specific manner to re-exposure to the same antigen. Memory cells reside in peripheral lymphoid and somatic tissue, and have recently been shown to participate in tissue homeostasis. Skin-homing lymphocytes express CLA (cutaneous lymphocyte antigen) (Fuhlbrigge et al., 1997), and are also involved in establishing skin-associated lymphoid tissue (SALT), which also involve mast cells and cutaneous nerves (Streilein, Alard, & Niizeki, 1999).

3.9 Mature T and B cells

Depletion of CD4+ and CD8+ T cells with monoclonal antibodies have been shown to impair wound healing (Barbul et al., 1989), as does thymectomy in mice (Davis et al., 2001). However, these experiments were performed before the discovery of innate lymphoid cells, which bear the same CD4 and CD8 markers as mature T cells. It was an open question whether antigen-specific mature T cells contributed to skin homeostasis.

The skin microbiome plays a significant role in "training" the immune system and thus helping to dictate a specific immune vigor and tone. The presence of microbiota in the skin leads to antigen presentation via MHC molecules, which leads to maturation of T and B cells. Mature CD8+ T cells (Tc17 and Tc1 subsets) have been shown to accelerate wound healing in the skin in an MHC- and commensal-dependent manner (Linehan et al., 2018). This suggests that the experience of the immune system plays a role in tissue homeostasis. Similarly, in other systems, immunological age (i.e., the balance between naïve/mature lymphocytes) influences the degree of proliferation and differentiation of osteogenic precursors after fractures, and had a significant impact on the rate of bone healing and remodeling (Bucher et al., 2019).

3.10 Regulatory T cells

CD4+ CD25+ FoxP3+ T regulatory (T reg) cells patrol the peripheral immune system to prevent autoimmunity and maintain self-tolerance (Fehervari & Sakaguchi, 2004). These processes are essential for stem cell niches, which are typically sites of immune privilege (Niederkorn & Stein-Streilein, 2010). T regs express cell surface and secretory factors that modulate the activity of other cells in its vicinity and are thus poised to be important players in the stem cell niche of many tissues (Lane et al., 2014; Taams et al., 2005). T regs have been found to be in close association with the hematopoietic stem cell (HSC) niche in the bone marrow (Fujisaki et al., 2011), and are known to help maintain the donor stem cells pool after bone marrow transplantation (Edinger et al., 2003). In the skin, resident T regs (which represent about 20% of resident T cells) migrate to the skin during HF development and the acquisition of the skin microbiome in a CCL20-CCR6 dependent manner (Scharschmidt et al., 2017). T reg impairment in Scurfy mice results in profound autoimmunity and skin inflammation and alopecia (Yang et al., 2015). After hair plucking, resident T regs may mediate the subsequently induced anagen via Jagged1-Notch interactions with the HF stem cell niche (Ali et al., 2017).

4. Immune dysregulation and stem cells

Dysfunction of the stem cell niche results in the impairment of essential stem cell properties. For example, loss of attachment to vital basement membrane ECM proteins affects the proliferative state of stem cells.

Dysregulation or exhaustion of stem cells in a compromised niche ultimately leads to impairment in wound healing and tissue homeostasis, senescence, and loss of skin and hair integrity. The collapse of immune privilege of stem cell niches also leads to premature depletion of crucial stem cell populations (Harries et al., 2013; Paus & Bertolini, 2013). Epidermal stem cells are sensitive to changes in their microenvironment, and cytokines produced by immune cells can influence the balance between proliferative or quiescence signals in the stem cell niche.

4.1 Psoriasis

Psoriasis provides an intriguing model to study immune cell-niche interactions in the IFE. IL-17 and IL-22 produced by Th17 cells are believed to drive epidermal hyperproliferation in psoriatic lesions (van der Fits et al., 2009). The same cytokines have also been shown to increase the expression of stemness markers in the basal layer of keratinocytes, which is expanded to include the suprabasal Krt10 layer in psoriasis (Ekman et al., 2019).

4.2 Cancer stem cells

T regs, antigen presenting cells and NK-T cells which contribute to the relative immune privilege of stem cells during the steady state are also involved in cancer immunosurveillance, whereby they remove cells that might have accumulated DNA damage (Moodycliffe et al., 2000). Disruption in this process, either by hijacking and weaponizing cell-autonomous immune privilege tools (i.e., downregulating MHC molecules) or by defects in the immune surveillance mechanism (Browning & Bodmer, 1992; Hahne et al., 1996), allows for "immune escape" and proliferation of skin tumors (de Visser, Eichten, & Coussens, 2006).

Established malignant tumors have their own cancer stem cell niches, whose microenvironment is also regulated by immune cells, in particular, tumor-associated macrophages (TAMs) (Fujimura et al., 2018; Noy & Pollard, 2014). Squamous cell carcinoma arising from oncogenic mutations in the IFE basal stem cells contains TAMs that support the cancer stem cell niche by secreting vascular endothelial growth factor (VEGF) (Moussai et al., 2011). The inflammatory/anti-inflammatory balance (i.e., "M1-like" vs. "M2-like" skew) of TAMs in melanoma is also associated with survival and prognosis, with "M2-like" anti-inflammatory TAMs having a poorer prognosis—possibly by enabling immune evasion (Falleni et al., 2017).

In addition to TAMs, mast cells are also implicated in stimulating angiogenesis in a wide variety of cancers (Ribatti & Crivellato, 2012). Mast cells are commonly found within a variety of skin tumors (Cawley & Hoch-Ligeti, 1961), and a higher number of mast cells is associated with a poorer prognosis in Merkel cell carcinoma (Beer, Ng, & Murray, 2008).

4.3 Autoimmunity

Conversely, the collapse of immune privilege results in autoimmune attack of stem cells in their niche by cytotoxic NKT cells (Ito et al., 2008). Autoimmune destruction of IFE melanocyte stem cells results in vitiligo (Rork, Rashighi, & Harris, 2016). In the HF, depending on which stem cell population is affected, this can results in non-scarring alopecia areata if the attack occurs around the bulb or matrix keratinocytes (Paus & Bertolini, 2013), or scarring alopecia (i.e., lichen planopilaris or discoid lupus) is the attack is directed to the HF bulge region (Harries et al., 2013). The mechanism of immune privilege is believed to be the relative expression of MHC molecules on the stem cells, which is in turn modulated by niche signals. Whether anti-inflammatory "M2-like" macrophages or T reg cells in the skin are involved in the establishment of physiological immune privilege, as they do in pathological tumors (Zhao et al., 2012), remains an open question.

4.4 Aging and senescence

Age-related physiological changes in the skin and hair are a result of intrinsic aging of stem cells (e.g., waning DNA repair and replication infidelity), coupled with extrinsic aging mediated by environmental insults such an accumulation of UV mutations and oxidative damage. This results in a weakened skin barrier, increased transepidermal water loss, inadequate wound healing, wrinkles, pigmentary dyschromia of skin and hair, hair thinning, and increased risk of malignancies (Mimeault & Batra, 2010). Almost every part of the immune system has been implicated in interacting with senescent cells in most tissues (Prata et al., 2019). While much research has focused on the senescence of epidermal and mesenchymal cells in the skin, the role of the senescent immune system has not been explored in the maintenance of epidermal stem cell niches. Aging is associated with a reduction in Langerhans cells and DETCs (Giangreco et al., 2008; Sunderkotter, Kalden, & Luger, 1997), which provide immune surveillance of the epidermal stem cells and participate in the clearance of potentially pathogenic or deleterious mutations. Thus, impairment of immune-niche interactions with age should gain more scientific attention as the global population ages and lifespans continue to lengthen.

5. Studying immune-stem cell interactions

The interactions between immune cells and the epidermal stem cell niche have been alluded to for decades, with exhaustive descriptive studies associating the distribution of various immune cells in the epidermis and associated appendages (Christoph et al., 2000). However, the functional relationship between immune cells and the niche remains elusive, as sophisticated tools required to dissect their interaction have only been developed in the last few years. As immune cells are migratory and transient in tissues, and usually at much lower proportions compared with epithelial and stromal cells in unperturbed tissue, their representation on bulk transcriptomic analyses (i.e., RNA seq) may be masked.

5.1 Single-cell transcriptomics

The ability to study RNA expression at the single-cell level has allowed researchers to circumvent the dilution of immune cell signals in bulk RNA-sequencing experiments (Chattopadhyay et al., 2014). Briefly, tissue is enzymatically dissociated into a single-cell suspension, which is then parsed into single cell units labeled with a nucleotide barcode tag (usually with microfluidics or emulsification technology), and the reverse transcriptase PCR reaction is performed on each cell individually (Fig. 2A). The data from each cell is pooled with all the other cells from the tissue and analyzed with algorithms that compare their similarities. This allows these differences to be visualized as clusters. Single-cell RNA-sequencing has also been used to analyze the epithelial and stromal compartments of the skin and HF (Jensen & Watt, 2006; Joost et al., 2016; Yang et al., 2017). By performing single-cell RNA-sequencing on just the immune cells (i.e., by sorting for CD45+ cells in the niche), potentially rare subsets of immune cells can be identified (Nguyen et al., 2018). For example, a rare subset of TREM2+ macrophages associated with HFs was discovered by this method (Wang et al., 2019). Further validation showed that these macrophages produce OSM that inhibits HF stem cell activation, and plays a role in maintaining quiescence in the HF niche. Additionally, transition states of immune cell subsets can be identified with this technology, and "pseudotime" or "pseudodifferentiation" analyses can be carried out to generate hypotheses for further validation.

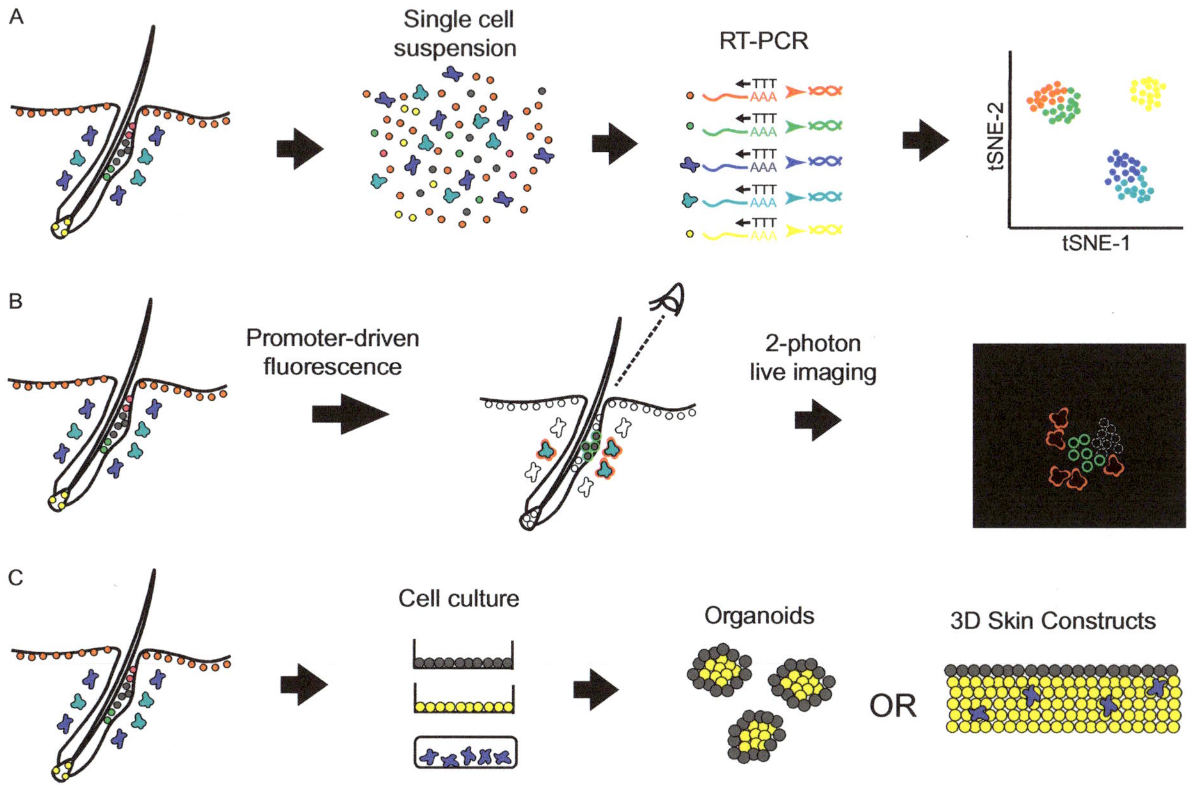

Fig. 2 See figure legend on opposite page.

For the bone marrow, the opposite strategy was employed (i.e., analyzing the CD45^{low}) cells to studying the hematopoietic niche *sans* the immune cells (Tikhonova et al., 2019). Single-cell RNA-sequencing of entire tumors has also identified the immune composition specific to certain cancers, like melanoma (Tirosh et al., 2016), but other experiments are still required to determine their significance in vivo.

5.2 In-vivo live imaging

Two-photon live imaging has revolutionized our perspective on cellular behavior in vivo and in vitro. Fluorescent labelling of live cells (usually via H2 histone promoter—green fluorescent protein (GFP) expression) coupled with sophisticated rigs has allowed researchers to visualize cellular

Fig. 2 Technologies for studying niche interactions. Representative cartoons of methods that are currently utilized in studying the interaction between immune cells and the stem cell niche. (A) Single-cell RNA sequencing. Tissues are dissociated mechanically and enzymatically into a single-cell suspension. Single cells are then individually analyzed with an RT-PCR reaction to create barcoded cDNA libraries, which are sequenced in a high-throughput manner. The gene expression profile of each cell is compared with all other cells; these differences are represented as Euclidean distances on a tSNE (t-distributed stochastic neighbor embedding) plot. In this example, epidermal stem cells (orange and green circles) resemble each other more than they do the immune cells (dark and light blue), which cluster separately. Dermal cells (yellow) are represented by yet another cluster. Increased resolution of immune cells can be achieved by selecting immune cells by FACS, so as to remove the noise from epidermal and stromal cells. This method has also been used for single-cell genomic, proteomic and epigenomic studies. (B) Intravital fluorescence imaging. Stem cells and immune cells can be labeled with fluorescent reporter proteins using a promoter driven approach. These cells will fluorescence under a two-photon microscopy, and their behavior can be observed in situ. In this figure, stem cells are labeled with a green fluorescence reporter, while immune cells have a red fluorescence reporter. However, this technique is limited to experimental model animals, and these animals need to be anesthetized and immobilized for up to 16–20 h in specialized rigs for the intravital images to be acquired. Results from these experiments allow us to observe the interactions, motility, and dynamics of cells in their native niche, and are mesmerizing and fascinating when presented at scientific meetings. (C) Organoid culture and 3D skin constructs. Sophisticated culture techniques and intricate engineering platforms have pushed the limitations of 2-dimensional monolayer culture of stem cells. Self-assembly of several cell types into organoids, or arranging cells into 3-dimensional constructs have allowed researchers to try to recreate the stem cell niche ex vivo. These platforms allow us to manipulate niche dynamics, test drugs on the system, and hopefully, at some point, replace animal testing. Induced pluripotent stem cells have also been used to generate different tissue types for organoids/skin constructs, and may lead to a more non-invasive way of utilizing patient cells. Addition of immune cells into these systems will be the next step to dissecting their relationships and interactions with stem cell niches.

interactions and motility in culture, as well as in situ in a living organism (Fig. 2B). The skin has the advantage of being easily accessible, and requires an immobilized, anesthetized animal with a window chamber. Using this technology, epidermal stem cell competition and behavior within their niche have been observed for the very first time (Rompolas et al., 2013; Rompolas & Greco, 2014). Other cells of the stem cell niche can be similarly labeled with more complex combinations of fluorescent reporter proteins coupled to the appropriate promoters, and this has been used to visualize immune cells (Pittet et al., 2018), endothelial cells and stromal cells during the steady-state, and during tissue injury. Combining this technology with multiphoton or multispectral imaging will allow visualization of more than two cell types at once.

Differential labelling of macrophages and neutrophils revealed that "M2-like" resident macrophages prevent neutrophil-led inflammation during microtrauma caused by single-cell laser ablation (Uderhardt et al., 2019). This is a form of immune privilege that is conferred to tissues that may undergo frequent remodeling so as to prevent unnecessary immune attention. Using this technology to study macrophage dynamics during the hair cycle will yield fascinating insight into the immune-niche interactions in this system. However, this technology is currently limited to experimental animal models, which are amenable to genetic fluorescence manipulation.

5.3 Organoid culture and 3D constructs

By placing a stem cell type in 2-dimensional culture, it is removed from its niche and surrounding interactions (Fig. 2C). Moreover, the nature of its attachment is changed from a complex ECM, which promotes stemness (Trappmann et al., 2012), to at best-coated polystyrene with one or two ECM components (Reichelt, 2007). This artificial in vitro system, while convenient, changes the epigenetics and phenotype of the stem cell so that it does not truly represent the in vivo version of itself. Both keratinocytes (Vollmers et al., 2012) and DP cells (Higgins et al., 2013) undergo drastic transcriptomic shifts when placed in 2-dimensional culture, losing their stemness and hair inductive abilities, and this is partially recovered by 3-dimensional spheroid culture. Co-cultures of keratinocytes and fibroblasts provide a better model to study wound healing (Wojtowicz et al., 2014), but are currently still do not incorporate immune cells. While ex vivo techniques like whole skin explant culture (Frade et al., 2015) and HF organ culture (i.e., the Philpott assay (Philpott, 2018) may shed some light on tissue interactions, these tissues are hard to obtain and may have patient- and site-specific variations that are hard to control.

Attempts to re-create the stem cell niche in organoid culture in a controlled manner, whereby multiple cell types from the native tissue are allowed to aggregate and replicate with in vivo relationships, have been pioneered in the study of intestinal stem cells (Date & Sato, 2015). Skin and hair organoids have also been utilized to study the skin niche (Lee et al., 2018), but the hierarchical and complex nature of stem cell compartments in the skin and hair and other appendages may require more complex and bio-engineered culture systems to truly study them ex vivo. Efforts to incorporate multiple cell types into a skin construct have recently been achieved (Abaci et al., 2015; Abaci et al., 2016; Abaci et al., 2018; Gledhill et al., 2015). The current limitation in this field is the inability to recreate and model the hair cycle ex vivo, possibly due to the loss of specific niche signals from immune cells. More advanced platforms in the near future that include immune cells will allow dissection and close study of the epidermal stem niche in unprecedented ways.

6. Conclusion

The contributions of immune cells to the regulation and maintenance of stem cell niches are beginning to be appreciated. Not just in the skin, but also in the bone marrow, gut, skeletal muscle, and lungs. It is likely that both innate immune cells and resident memory cells of the adaptive immune system, are intimately involved in most stem cell niches. Understanding these interactions will allow us to predict adverse effects of immunosuppressants or small molecule drugs on tissue homeostasis (e.g., for adverse effects like ulceration, carcinogenesis), harness the immune system in the management of stem cells disorders like cancer and aging, as well as reverse engineer the stem cell niche ex vivo to accelerate achievements in regenerative medicine.

References

Abaci, H. E., et al. (2015). Pumpless microfluidic platform for drug testing on human skin equivalents. *Lab on a Chip*, *15*(3), 882–888.

Abaci, H. E., et al. (2016). Human skin constructs with spatially controlled vasculature using primary and iPSC-derived endothelial cells. *Advanced Healthcare Materials*, *5*(14), 1800–1807.

Abaci, H. E., et al. (2018). Tissue engineering of human hair follicles using a biomimetic developmental approach. *Nature Communications*, *9*(1), 5301.

Abdallah, F., Mijouin, L., & Pichon, C. (2017). Skin immune landscape: Inside and outside the organism. *Mediators of Inflammation*, *2017*, 5095293.

Ali, N., et al. (2017). Regulatory T cells in skin facilitate epithelial stem cell differentiation. *Cell*, *169*(6), 1119–1129 e11.
Amberg, N., et al. (2016). Effects of Imiquimod on hair follicle stem cells and hair cycle progression. *The Journal of Investigative Dermatology*, *136*(11), 2140–2149.
Barbul, A., et al. (1989). The effect of in vivo T helper and T suppressor lymphocyte depletion on wound healing. *Annals of Surgery*, *209*(4), 479–483.
Beer, T. W., Ng, L. B., & Murray, K. (2008). Mast cells have prognostic value in Merkel cell carcinoma. *The American Journal of Dermatopathology*, *30*(1), 27–30.
Blanpain, C., & Simons, B. D. (2013). Unravelling stem cell dynamics by lineage tracing. *Nature Reviews. Molecular Cell Biology*, *14*(8), 489–502.
Borregaard, N., Sorensen, O. E., & Theilgaard-Monch, K. (2007). Neutrophil granules: A library of innate immunity proteins. *Trends in Immunology*, *28*(8), 340–345.
Braun, K. M., & Prowse, D. M. (2006). Distinct epidermal stem cell compartments are maintained by independent niche microenvironments. *Stem Cell Reviews*, *2*(3), 221–231.
Browning, M. J., & Bodmer, W. F. (1992). MHC antigens and cancer: Implications for T-cell surveillance. *Current Opinion in Immunology*, *4*(5), 613–618.
Bucher, C. H., et al. (2019). Experience in the adaptive immunity impacts bone homeostasis, remodeling, and healing. *Frontiers in Immunology*, *10*, 797.
Butcher, E. C., & Picker, L. J. (1996). Lymphocyte homing and homeostasis. *Science*, *272*(5258), 60–66.
Casanova-Acebes, M., et al. (2013). Rhythmic modulation of the hematopoietic niche through neutrophil clearance. *Cell*, *153*(5), 1025–1035.
Castellana, D., Paus, R., & Perez-Moreno, M. (2014). Macrophages contribute to the cyclic activation of adult hair follicle stem cells. *PLoS Biology*, *12*(12), e1002002.
Cawley, E. P., & Hoch-Ligeti, C. (1961). Association of tissue mast cells and skin tumors. *Archives of Dermatology*, *83*, 92–96.
Chattopadhyay, P. K., et al. (2014). Single-cell technologies for monitoring immune systems. *Nature Immunology*, *15*(2), 128–135.
Chen, C. C., et al. (2015). Organ-level quorum sensing directs regeneration in hair stem cell populations. *Cell*, *161*(2), 277–290.
Christoph, T., et al. (2000). The human hair follicle immune system: Cellular composition and immune privilege. *The British Journal of Dermatology*, *142*(5), 862–873.
Chu, S. Y., et al. (2019). Mechanical stretch induces hair regeneration through the alternative activation of macrophages. *Nature Communications*, *10*(1), 1524.
Clarke, C. A., et al. (2015). Risk of merkel cell carcinoma after solid organ transplantation. *Journal of the National Cancer Institute*, *107*(2), 1–9.
Date, S., & Sato, T. (2015). Mini-gut organoids: Reconstitution of the stem cell niche. *Annual Review of Cell and Developmental Biology*, *31*, 269–289.
Davis, P. A., et al. (2001). Effect of CD4(+) and CD8(+) cell depletion on wound healing. *The British Journal of Surgery*, *88*(2), 298–304.
de Visser, K. E., Eichten, A., & Coussens, L. M. (2006). Paradoxical roles of the immune system during cancer development. *Nature Reviews. Cancer*, *6*(1), 24–37.
DiPietro, L. A. (1995). Wound healing: The role of the macrophage and other immune cells. *Shock*, *4*(4), 233–240.
Doucet, Y. S., et al. (2013). The touch dome defines an epidermal niche specialized for mechanosensory signaling. *Cell Reports*, *3*(6), 1759–1765.
Dupin, E., & Le Douarin, N. M. (2003). Development of melanocyte precursors from the vertebrate neural crest. *Oncogene*, *22*(20), 3016–3023.
Edinger, M., et al. (2003). CD4+CD25+ regulatory T cells preserve graft-versus-tumor activity while inhibiting graft-versus-host disease after bone marrow transplantation. *Nature Medicine*, *9*(9), 1144–1150.

Eichmuller, S., et al. (1998). Clusters of perifollicular macrophages in normal murine skin: Physiological degeneration of selected hair follicles by programmed organ deletion. *The Journal of Histochemistry and Cytochemistry*, *46*(3), 361–370.
Ekman, A. K., et al. (2019). IL-17 and IL-22 promote keratinocyte stemness in the Germinative compartment in psoriasis. *The Journal of Investigative Dermatology* pii: S0022-202X(19)30034-X. https://doi.org/10.1016/j.jid.2019.01.014.
Engels, E. A., et al. (2002). Merkel cell carcinoma and HIV infection. *Lancet*, *359*(9305), 497–498.
Epelman, S., Lavine, K. J., & Randolph, G. J. (2014). Origin and functions of tissue macrophages. *Immunity*, *41*(1), 21–35.
Falleni, M., et al. (2017). M1 and M2 macrophages' clinicopathological significance in cutaneous melanoma. *Melanoma Research*, *27*(3), 200–210.
Fehervari, Z., & Sakaguchi, S. (2004). CD4+ Tregs and immune control. *The Journal of Clinical Investigation*, *114*(9), 1209–1217.
Festa, E., et al. (2011). Adipocyte lineage cells contribute to the skin stem cell niche to drive hair cycling. *Cell*, *146*(5), 761–771.
Frade, M. A., et al. (2015). Prolonged viability of human organotypic skin explant in culture method (hOSEC). *Anais Brasileiros de Dermatologia*, *90*(3), 347–350.
Frances, D., & Niemann, C. (2012). Stem cell dynamics in sebaceous gland morphogenesis in mouse skin. *Developmental Biology*, *363*(1), 138–146.
Fuchs, E., Tumbar, T., & Guasch, G. (2004). Socializing with the neighbors: stem cells and their niche. *Cell*, *116*(6), 769–778.
Fuhlbrigge, R. C., et al. (1997). Cutaneous lymphocyte antigen is a specialized form of PSGL-1 expressed on skin-homing T cells. *Nature*, *389*(6654), 978–981.
Fujimura, T., et al. (2018). Tumor-associated macrophages: Therapeutic targets for skin Cancer. *Frontiers in Oncology*, *8*, 3.
Fujisaki, J., et al. (2011). In vivo imaging of Treg cells providing immune privilege to the haematopoietic stem-cell niche. *Nature*, *474*(7350), 216–219.
Gay, D., et al. (2013). Fgf9 from dermal gammadelta T cells induces hair follicle neogenesis after wounding. *Nature Medicine*, *19*(7), 916–923.
Giangreco, A., et al. (2008). Epidermal stem cells are retained in vivo throughout skin aging. *Aging Cell*, 7(2), 250–259.
Ginhoux, F., & Guilliams, M. (2016). Tissue-resident macrophage ontogeny and homeostasis. *Immunity*, *44*(3), 439–449.
Gledhill, K., et al. (2015). Melanin transfer in human 3D skin equivalents generated exclusively from induced pluripotent stem cells. *PLoS One*, *10*(8), e0136713.
Goldstein, J., & Horsley, V. (2012). Home sweet home: Skin stem cell niches. *Cellular and Molecular Life Sciences*, *69*(15), 2573–2582.
Greco, V., et al. (2009). A two-step mechanism for stem cell activation during hair regeneration. *Cell Stem Cell*, *4*(2), 155–169.
Guilliams, M., Mildner, A., & Yona, S. (2018). Developmental and functional heterogeneity of monocytes. *Immunity*, *49*(4), 595–613.
Hager, M., Cowland, J. B., & Borregaard, N. (2010). Neutrophil granules in health and disease. *Journal of Internal Medicine*, *268*(1), 25–34.
Hahne, M., et al. (1996). Melanoma cell expression of Fas(Apo-1/CD95) ligand: Implications for tumor immune escape. *Science*, *274*(5291), 1363–1366.
Halprin, K. M. (1972). Epidermal "turnover time"—a re-examination. *The British Journal of Dermatology*, *86*(1), 14–19.
Harries, M. J., et al. (2013). Lichen planopilaris is characterized by immune privilege collapse of the hair follicle's epithelial stem cell niche. *The Journal of Pathology*, *231*(2), 236–247.

Hashizume, H., Tokura, Y., & Takigawa, M. (1994). Increased number of dendritic epidermal T cells associated with induced anagen phase of hair cycles. *Journal of Dermatological Science*, *8*(2), 119–124.

Higgins, C. A., et al. (2013). *Proceedings of the National Academy of Sciences of the United States of America, 110*(49), 19679–19688.

Hsu, Y. C., Li, L., & Fuchs, E. (2014). Emerging interactions between skin stem cells and their niches. *Nature Medicine*, *20*(8), 847–856.

Hsu, Y. C., Pasolli, H. A., & Fuchs, E. (2011). Dynamics between stem cells, niche, and progeny in the hair follicle. *Cell*, *144*(1), 92–105.

Huang, S., et al. (2016). 3D bioprinted extracellular matrix mimics facilitate directed differentiation of epithelial progenitors for sweat gland regeneration. *Acta Biomaterialia*, *32*, 170–177.

Ito, M., et al. (2005). Stem cells in the hair follicle bulge contribute to wound repair but not to homeostasis of the epidermis. *Nature Medicine*, *11*(12), 1351–1354.

Ito, M., et al. (2007). Wnt-dependent de novo hair follicle regeneration in adult mouse skin after wounding. *Nature*, *447*(7142), 316–320.

Ito, T., et al. (2008). Immune privilege and the skin. *Current Directions in Autoimmunity*, *10*, 27–52.

Iwasaki, A., & Medzhitov, R. (2015). Control of adaptive immunity by the innate immune system. *Nature Immunology*, *16*(4), 343–353.

Jaeger, B. N., et al. (2012). Neutrophil depletion impairs natural killer cell maturation, function, and homeostasis. *The Journal of Experimental Medicine*, *209*(3), 565–580.

Jameson, J. M., et al. (2004). Regulation of skin cell homeostasis by gamma delta T cells. *Frontiers in Bioscience*, *9*, 2640–2651.

Jensen, K. B., & Watt, F. M. (2006). Single-cell expression profiling of human epidermal stem and transit-amplifying cells: Lrig1 is a regulator of stem cell quiescence. *Proceedings of the National Academy of Sciences of the United States of America*, *103*(32), 11958–11963.

Jensen, K. B., et al. (2009). Lrig1 expression defines a distinct multipotent stem cell population in mammalian epidermis. *Cell Stem Cell*, *4*(5), 427–439.

Jones, D. L., & Wagers, A. J. (2008). No place like home: Anatomy and function of the stem cell niche. *Nature Reviews. Molecular Cell Biology*, *9*(1), 11–21.

Joost, S., et al. (2016). Single-cell transcriptomics reveals that differentiation and spatial signatures shape epidermal and hair follicle heterogeneity. *Cell Systems*, *3*(3), 221–237 e9.

Knudsen, N. H., & Lee, C. H. (2016). Identity crisis: CD301b(+) mononuclear phagocytes blur the M1-M2 macrophage line. *Immunity*, *45*(3), 461–463.

Kobielak, K., et al. (2007). Loss of a quiescent niche but not follicle stem cells in the absence of bone morphogenetic protein signaling. *Proceedings of the National Academy of Sciences of the United States of America*, *104*(24), 10063–10068.

Komi, D. E. A., Khomtchouk, K., & Santa Maria, P. L. (2019). A review of the contribution of mast cells in wound healing: Involved molecular and cellular mechanisms. *Clinical Reviews in Allergy and Immunology*.

Kretzschmar, K., & Watt, F. M. (2014). Markers of epidermal stem cell subpopulations in adult mammalian skin. *Cold Spring Harbor Perspectives in Medicine*, *4*(10).

Lane, S. W., Williams, D. A., & Watt, F. M. (2014). Modulating the stem cell niche for tissue regeneration. *Nature Biotechnology*, *32*(8), 795–803.

Lee, J., et al. (2018). Hair follicle development in mouse pluripotent stem cell-derived skin organoids. *Cell Reports*, *22*(1), 242–254.

Li, L., & Xie, T. (2005). Stem cell niche: Structure and function. *Annual Review of Cell and Developmental Biology*, *21*, 605–631.

Lien, W. H., et al. (2014). In vivo transcriptional governance of hair follicle stem cells by canonical Wnt regulators. *Nature Cell Biology*, *16*(2), 179–190.

Lim, X., et al. (2013). Interfollicular epidermal stem cells self-renew via autocrine Wnt signaling. *Science*, *342*(6163), 1226–1230.

Lim, X., et al. (2016). Axin2 marks quiescent hair follicle bulge stem cells that are maintained by autocrine Wnt/beta-catenin signaling. *Proceedings of the National Academy of Sciences of the United States of America*, *113*(11), E1498–E1505.

Linehan, J. L., et al. (2018). Non-classical immunity controls microbiota impact on skin immunity and tissue repair. *Cell*, *172*(4), 784–796 e18.

Lu, C. P., et al. (2012). Identification of stem cell populations in sweat glands and ducts reveals roles in homeostasis and wound repair. *Cell*, *150*(1), 136–150.

MacLeod, A. S., & Mansbridge, J. N. (2016). The innate immune system in acute and chronic wounds. *Advances in Wound Care (New Rochelle)*, *5*(2), 65–78.

Maricich, S. M., et al. (2009). Merkel cells are essential for light-touch responses. *Science*, *324*(5934), 1580–1582.

Martin, C., et al. (2003). Chemokines acting via CXCR2 and CXCR4 control the release of neutrophils from the bone marrow and their return following senescence. *Immunity*, *19*(4), 583–593.

Martinez, F. O., & Gordon, S. (2014). The M1 and M2 paradigm of macrophage activation: Time for reassessment. *F1000Prime Reports*, *6*, 13.

Matsumura, H., et al. (2016). Hair follicle aging is driven by transepidermal elimination of stem cells via COL17A1 proteolysis. *Science*, *351*(6273), aad4395.

Mendez-Ferrer, S., et al. (2008). Haematopoietic stem cell release is regulated by circadian oscillations. *Nature*, *452*(7186), 442–447.

Merad, M., Ginhoux, F., & Collin, M. (2008). Origin, homeostasis and function of Langerhans cells and other langerin-expressing dendritic cells. *Nature Reviews. Immunology*, *8*(12), 935–947.

Meyer, K. C., et al. (2008). Evidence that the bulge region is a site of relative immune privilege in human hair follicles. *The British Journal of Dermatology*, *159*(5), 1077–1085.

Mimeault, M., & Batra, S. K. (2010). Recent advances on skin-resident stem/progenitor cell functions in skin regeneration, aging and cancers and novel anti-aging and cancer therapies. *Journal of Cellular and Molecular Medicine*, *14*(1–2), 116–134.

Mocsai, A. (2013). Diverse novel functions of neutrophils in immunity, inflammation, and beyond. *The Journal of Experimental Medicine*, *210*(7), 1283–1299.

Moodycliffe, A. M., et al. (2000). Immune suppression and skin cancer development: Regulation by NKT cells. *Nature Immunology*, *1*(6), 521–525.

Moussai, D., et al. (2011). The human cutaneous squamous cell carcinoma microenvironment is characterized by increased lymphatic density and enhanced expression of macrophage-derived VEGF-C. *The Journal of Investigative Dermatology*, *131*(1), 229–236.

Muneeb, F., Hardman, J. A., & Paus, R. (2019). Hair growth control by innate immunocytes: Perifollicular macrophages revisited. *Experimental Dermatology*, *28*(4), 425–431.

Nguyen, A., et al. (2018). Single cell RNA sequencing of rare immune cell populations. *Frontiers in Immunology*, *9*, 1553.

Niederkorn, J. Y., & Stein-Streilein, J. (2010). History and physiology of immune privilege. *Ocular Immunology and Inflammation*, *18*(1), 19–23.

Noy, R., & Pollard, J. W. (2014). Tumor-associated macrophages: From mechanisms to therapy. *Immunity*, *41*(1), 49–61.

Parakkal, P. F. (1969). Role of macrophages in collagen resorption during hair growth cycle. *Journal of Ultrastructure Research*, *29*(3), 210–217.

Parkinson, E. K. (1992). Epidermal keratinocyte stem cells: Their maintenance and regulation. *Seminars in Cell Biology*, *3*(6), 435–444.

Paus, R., & Bertolini, M. (2013). The role of hair follicle immune privilege collapse in alopecia areata: Status and perspectives. *The Journal of Investigative Dermatology. Symposium Proceedings*, *16*(1), S25–S27.
Paus, R., et al. (1994). Distribution and changing density of gamma-delta T cells in murine skin during the induced hair cycle. *The British Journal of Dermatology*, *130*(3), 281–289.
Philpott, M. P. (2018). Culture of the human pilosebaceous unit, hair follicle and sebaceous gland. *Experimental Dermatology*, *27*(5), 571–577.
Pillay, J., et al. (2012). A subset of neutrophils in human systemic inflammation inhibits T cell responses through mac-1. *The Journal of Clinical Investigation*, *122*(1), 327–336.
Pittet, M. J., et al. (2018). Recording the wild lives of immune cells. *Science immunology*, (3), 27.
Plikus, M. V., et al. (2008). Cyclic dermal BMP signalling regulates stem cell activation during hair regeneration. *Nature*, *451*(7176), 340–344.
Portou, M. J., et al. (2015). The innate immune system, toll-like receptors and dermal wound healing: A review. *Vascular Pharmacology*, *71*, 31–36.
Prata, L., et al. (2019). Senescent cell clearance by the immune system: Emerging therapeutic opportunities. *Seminars in Immunology* pii: S1044-5323(18)30065-4. https://doi.org/10.1016/j.smim.2019.04.003.
Puga, I., et al. (2011). B cell-helper neutrophils stimulate the diversification and production of immunoglobulin in the marginal zone of the spleen. *Nature Immunology*, *13*(2), 170–180.
Qian, C., et al. (2019). Heterogeneous macrophages: Supersensors of exogenous inducing factors. *Scandinavian Journal of Immunology*, e12768.
Rabbani, P., et al. (2011). Coordinated activation of Wnt in epithelial and melanocyte stem cells initiates pigmented hair regeneration. *Cell*, *145*(6), 941–955.
Rak, G. D., et al. (2016). IL-33-dependent group 2 innate lymphoid cells promote cutaneous wound healing. *The Journal of Investigative Dermatology*, *136*(2), 487–496.
Reichelt, J. (2007). Mechanotransduction of keratinocytes in culture and in the epidermis. *European Journal of Cell Biology*, *86*(11 – 12), 807–816.
Ribatti, D., & Crivellato, E. (2012). Mast cells, angiogenesis, and tumour growth. *Biochimica et Biophysica Acta*, *1822*(1), 2–8.
Rittie, L., et al. (2013). Eccrine sweat glands are major contributors to reepithelialization of human wounds. *The American Journal of Pathology*, *182*(1), 163–171.
Rompolas, P., & Greco, V. (2014). Stem cell dynamics in the hair follicle niche. *Seminars in Cell & Developmental Biology*, *25-26*, 34–42.
Rompolas, P., Mesa, K. R., & Greco, V. (2013). Spatial organization within a niche as a determinant of stem-cell fate. *Nature*, *502*(7472), 513–518.
Rork, J. F., Rashighi, M., & Harris, J. E. (2016). Understanding autoimmunity of vitiligo and alopecia areata. *Current Opinion in Pediatrics*, *28*(4), 463–469.
Schaerli, P., et al. (2004). A skin-selective homing mechanism for human immune surveillance T cells. *The Journal of Experimental Medicine*, *199*(9), 1265–1275.
Scharschmidt, T. C., et al. (2017). Commensal microbes and hair follicle morphogenesis coordinately drive Treg migration into neonatal skin. *Cell Host & Microbe*, *21*(4), 467–477 e5.
Schofield, R. (1978). The relationship between the spleen colony-forming cell and the haemopoietic stem cell. *Blood Cells*, *4*(1–2), 7–25.
Sellheyer, K., & Krahl, D. (2010). Blimp-1: A marker of terminal differentiation but not of sebocytic progenitor cells. *Journal of Cutaneous Pathology*, *37*(3), 362–370.
Sharp, L. L., et al. (2005). Dendritic epidermal T cells regulate skin homeostasis through local production of insulin-like growth factor 1. *Nature Immunology*, *6*(1), 73–79.

Sheng, Z., et al. (2009). Regeneration of functional sweat gland-like structures by transplanted differentiated bone marrow mesenchymal stem cells. *Wound Repair and Regeneration*, *17*(3), 427–435.

Streilein, J. W., Alard, P., & Niizeki, H. (1999). A new concept of skin-associated lymphoid tissue (SALT): UVB light impaired cutaneous immunity reveals a prominent role for cutaneous nerves. *The Keio Journal of Medicine*, *48*(1), 22–27.

Sunderkotter, C., Kalden, H., & Luger, T. A. (1997). Aging and the skin immune system. *Archives of Dermatology*, *133*(10), 1256–1262.

Suzuki, S., et al. (1998). Localization of rat FGF-5 protein in skin macrophage-like cells and FGF-5S protein in hair follicle: Possible involvement of two Fgf-5 gene products in hair growth cycle regulation. *The Journal of Investigative Dermatology*, *111*(6), 963–972.

Taams, L. S., et al. (2005). Modulation of monocyte/macrophage function by human CD4+CD25+ regulatory T cells. *Human Immunology*, *66*(3), 222–230.

Taira, K., et al. (2002). Spatial relationship between Merkel cells and Langerhans cells in human hair follicles. *Journal of Dermatological Science*, *30*(3), 195–204.

Tanimura, S., et al. (2011). Hair follicle stem cells provide a functional niche for melanocyte stem cells. *Cell Stem Cell*, *8*(2), 177–187.

Tigelaar, R. E., & Lewis, J. M. (1995). Immunobiology of mouse dendritic epidermal T cells: A decade later, some answers, but still more questions. *The Journal of Investigative Dermatology*, *105*(1 Suppl), 43S–49S.

Tikhonova, A. N., et al. (2019). The bone marrow microenvironment at single-cell resolution. *Nature*, *569*(7755), 222–228.

Tirosh, I., et al. (2016). Dissecting the multicellular ecosystem of metastatic melanoma by single-cell RNA-seq. *Science*, *352*(6282), 189–196.

Toulon, A., et al. (2009). A role for human skin-resident T cells in wound healing. *The Journal of Experimental Medicine*, *206*(4), 743–750.

Toyoda, M., Nakamura, M., & Morohashi, M. (2002). Neuropeptides and sebaceous glands. *European Journal of Dermatology*, *12*(5), 422–427.

Trappmann, B., et al. (2012). Extracellular-matrix tethering regulates stem-cell fate. *Nature Materials*, *11*(7), 642–649.

Uderhardt, S., et al. (2019). Resident macrophages cloak tissue microlesions to prevent neutrophil-driven inflammatory damage. *Cell*, *177*(3), 541–555 e17.

van der Fits, L., et al. (2009). Imiquimod-induced psoriasis-like skin inflammation in mice is mediated via the IL-23/IL-17 axis. *Journal of Immunology*, *182*(9), 5836–5845.

Van Keymeulen, A., et al. (2009). Epidermal progenitors give rise to Merkel cells during embryonic development and adult homeostasis. *The Journal of Cell Biology*, *187*(1), 91–100.

Vollmers, A., et al. (2012). Two- and three-dimensional culture of keratinocyte stem and precursor cells derived from primary murine epidermal cultures. *Stem Cell Reviews*, *8*(2), 402–413.

Wang, N., Liang, H., & Zen, K. (2014). Molecular mechanisms that influence the macrophage m1-m2 polarization balance. *Frontiers in Immunology*, *5*, 614.

Wang, X., et al. (2017). Macrophages induce AKT/beta-catenin-dependent Lgr5(+) stem cell activation and hair follicle regeneration through TNF. *Nature Communications*, *8*, 14091.

Wang, E. C. E., et al. (2019). A subset of TREM2(+) dermal macrophages secretes Oncostatin M to maintain hair follicle stem cell quiescence and inhibit hair growth. *Cell Stem Cell*, *24*(4), 654–669 e6.

Watanabe, M., et al. (2017). Type XVII collagen coordinates proliferation in the interfollicular epidermis. *eLife*, *6*.

Wojtowicz, A. M., et al. (2014). The importance of both fibroblasts and keratinocytes in a bilayered living cellular construct used in wound healing. *Wound Repair and Regeneration*, *22*(2), 246–255.

Yang, S., et al. (2015). Immune tolerance. Regulatory T cells generated early in life play a distinct role in maintaining self-tolerance. *Science*, *348*(6234), 589–594.

Yang, H., et al. (2017). Epithelial-mesenchymal micro-niches govern stem cell lineage choices. *Cell*, *169*(3), 483–496 e13.

Zhao, Q., et al. (2012). Activated CD69+ T cells foster immune privilege by regulating IDO expression in tumor-associated macrophages. *Journal of Immunology*, *188*(3), 1117–1124.

Zhao, Y., et al. (2018). The origins and homeostasis of monocytes and tissue-resident macrophages in physiological situation. *Journal of Cellular Physiology*, *233*(10), 6425–6439.

Kolibri 360 GmbH
Ferdinand-Jühlke-Straße 7
99095 Erfurt, Deutschland
produktsicherheit@kolibri360.de